The NMR of Polymers

The NMR of Polymers

I.Ya. Slonim
Director, Radiospectroscopy Group
Plastics Research Institute
Moscow

and

A. N. Lyubimov
Head, Radiospectroscopy Design Section
Institute of Organic Chemistry
Academy of Sciences of the USSR
Moscow

Translated from Russian by
C. Nigel Turton and Tatiana I. Turton

ℚ PLENUM PRESS · NEW YORK · 1970

First Printing - December 1970
Second Printing - July 1972

The Russian text, originally published by Khimiya Press in
Moscow in 1966, has been corrected by the authors for this
edition. The present translation is published under an agreement with Mezhdunarodnaya Kniga, the Soviet book export
agency.

THE NMR OF POLYMERS
YADERNYI MAGNITNYI REZONANS V POLIMERAKH

ЯДЕРНЫЙ МАГНИТНЫЙ РЕЗОНАНС В ПОЛИМЕРАХ
И. Я. Слоним, А. Н. Любимов

Library of Congress Catalog Card Number 69-12543

SBN 306-30415-5

© 1970 Plenum Press, New York
A Division of Plenum Publishing Corporation
227 West 17th Street, New York, N. Y. 10011

Printed in the United States of America

Preface to the American Edition

In the time since this book was written, the application of NMR in the study of polymers has continued to develop rapidly. The main trends of the work have remained as before, namely, the study of the structure of polymers, molecular motion in them, and chemical conversions of high-molecular-weight substances. However, as a result of the refinement of experimental techniques and development of the theory, new progress has been achieved, particularly in the NMR spectroscopy of polymer solutions.

We therefore decided that it was worthwhile to provide an additional list of literature, covering papers published in 1965, 1966, and the beginning of 1967. By using the bibliographic directory appended to the list, the reader can readily find references to the latest literature for all sections of the book.

I. Ya. Slonim
A. N. Lyubimov

Preface to the Russian Edition

The aim of this book is mainly to answer the questions which inevitably arise for a chemist who wishes to use NMR in his work on polymers.

What is NMR? What is the physical nature of the phenomenon? What are the basic quantitative rules of the processes occurring in a sample which is placed in a steady magnetic field and subjected to a magnetic field of resonance frequency?

How do we determine the NMR spectrum of a polymer? How, in general terms, is an NMR spectrometer constructed, and how do we select suitable conditions for determining the spectrum in relation to the nature of the sample and the problem at hand? How is a spectrum analyzed so as to obtain the information required?

What problems in the chemistry and physical chemistry of polymers may be solved by means of NMR, and what are the main results of work in this field?

The application of NMR in polymer chemistry is increasing. The refinement of NMR spectrometers and in particular the development of high-resolution instruments which operate at elevated temperatures offer great new possibilities. It would obviously be risky to attempt to give at the present time a critical generalization of all available material and a complete picture of the state of work in this field. The authors undertook a more modest task, namely, to give a systematic account of published work and to clarify the main trends in the study of high polymers of NMR. We hope that the book will be useful in this form.

The book is divided into two parts.

Part One is theoretical and contains the physical bases of the NMR method and the appropriate mathematical procedures. In the opinion of the authors this material is quite necessary for understanding methods of calculating high-resolution spectra and for interpreting them. The presentation is quite simple and intelligible to a reader who has acquired the fundamentals of higher mathematics and physics at the levels of courses given to chemists in higher institutes of learning. Introductory information on the theory of matrices is of necessity brief. The reader requiring more information on this theory is referred to any book on higher algebra.

In addition, in Part One we describe the principles of construction of NMR spectrometers and methods of recording and interpreting spectra.

Part Two of the book consists of a review of papers on NMR of high polymers published through 1963 and the most important and interesting papers which have appeared in 1964-65.

The distribution of the material between Chapters III-VI is arbitrary to some extent. For example, in the examination of structural problems we also discuss molecular motion. Naturally, the selection of material is also partly arbitrary; work carried out by the authors or of interest to them is examined in greatest detail.

At the end of the book there is a bibliographic directory of books and reviews on NMR and original work on NMR of individual polymers.

The introduction to the book and Chapters I and II were written by A. N. Lyubimov, and Chapters III-VI by I. Ya. Slonim.

The authors consider it their pleasant duty to thank their co-workers in the Plastics Research Institute and the Central Automation Laboratory, Ya. G. Urman, A. G. Konovalov, A. F. Varenik, and V. I. Belitskaya for help in the planning of the book, and M. B. Neiman, N. M. Pomerantsev, D. Ya. Tsvankin, E. I. Fedin, and T. N. Khazanovich, who checked the manuscript.

Contents

Introduction

Nuclear magnetic resonance (NMR) is one of a group of phenomena which constitute a field of physics known as r a d i o s p e c - t r o s c o p y. The characteristic feature of these phenomena is that in them we observe induced transitions of microparticles between energy levels, which arise under definite conditions. These transitions are accompanied by electromagnetic emission or absorption in the r a d i o f r e q u e n c y r a n g e. In the case of NMR, we are concerned with the behavior of atomic nuclei in e x t e r n a l magnetic fields.

Experimental data accumulated in physics show unequivocally that many atomic nuclei have their own angular momentum and an associated magnetic moment which characterize the magnetic field of the nucleus.*

The vectors of the magnetic, $\overleftarrow{\mu}$, and mechanical, $\overleftarrow{J}$, moments are collinear, and are related by the equation

$$\overleftarrow{\mu} = \gamma \overleftarrow{J}$$

where γ is a coefficient known as the gyromagnetic ratio and is a nuclear characteristic.

Initial attempts to explain the existence of nuclear moments by rotation of a nucleus with mass and charge were unsuccessful. The observed properties of nuclear moments were contradictory to the theory based on this assumption. Thus, according to this theory, the magnetic moment of a proton should be exactly equal to the nuclear magneton:

$$m_n = \frac{e\hbar}{2cm_p}$$

---

*Some atomic nuclei also have an electric quadrupole moment, which is a measure of the deviation of the charge distribution inside the nucleus from a spherically symmetrical distribution.

1

where $\hbar$ is the reduced Planck constant $(\hbar = h/2\pi)$; e and m_p are the charge and mass of a proton, respectively; c is the velocity of light.*

The magnetic moment of a proton actually measured corresponded to the nuclear magneton only in order of magnitude. Moreover, the gyromagnetic ratios of some nuclei were even found to be negative, which is contradictory to the theory mentioned.

According to contemporary ideas, the nature of nuclear moments is more complex. In view of the fact that the source of nuclear moments plays no part in the NMR phenomena interesting us, here we will examine only the observed properties of nuclear moments.

It has been established experimentally that for any atomic nucleus the length of the vector of the intrinsic angular momentum is always expressed by the relation

$$J = \hbar \sqrt{I(I+1)}$$

The number I, which is called the spin number, or simply the spin of the nucleus, may be an integer, half-integer, or zero, depending on the type of the nucleus. The spin of the elementary particles which form a nucleus (protons and neutrons) equals $\frac{1}{2}$.† Experimental data also make it possible to assume that the spin of an atomic nucleus is in a certain sense a combination of the spins of the elementary particles forming the nucleus. Thus, if the mass number of the nucleus (i.e., the total number of protons and neutrons in the nucleus) is odd, the spin is a half-integer. If the mass number is even, while the atomic number of the nucleus (i.e., the number of protons present in the nucleus) is odd, then the spin is an integer. Finally, if the mass number and the atomic number of the nucleus are even, the spin equals zero.

Even more interesting properties of nuclear moments were observed by studying the behavior of atomic nuclei in strong steady

---

* In analogy with the Lande g-factor for electrons, the relation between the nuclear moments $\overleftarrow{\mu}$ and $\overleftarrow{J}$ is sometimes expressed by the equation

$$\overleftarrow{\mu} = g_n m_n \cdot \frac{\overleftarrow{J}}{\hbar}, \quad \text{so that} \quad g_n = \gamma \cdot \frac{\hbar}{m_n}$$

† Nuclei with a spin of $\frac{1}{2}$ have no electric quadrupole moment.

magnetic fields. Even in the historic experiment of Stern and Ger-
lach (1924) it was established that the projections of the magnetic
moments of atoms on the direction of the external magnetic field
do not form a continuous set. Somewhat later (Frisch and Stern,
1933) analogous properties were found for the magnetic moments
of atomic nuclei. It was found that the projection μ_H of the vector
of the magnetic moment $\overleftarrow{\mu}$ of an atomic nucleus with a spin I on
the direction of the external magnetic field $\overleftarrow{H}$ may assume only
one of the values:

$$\mu_H = \gamma \hbar I; \quad \gamma \hbar (I-1); \quad \ldots; \quad -\gamma \hbar I$$

and, in connection with this, the energy of an atomic nucleus in this
field, which equals

$$E = -\overleftarrow{\mu}\overleftarrow{H} = -\mu_H H$$

may have only discrete values, belonging to the following finite
series:

$$E = -\gamma H \hbar I; \quad -\gamma H \hbar (I-1); \quad \ldots; \quad +\gamma H \hbar I$$

The possible values of the energy of an atomic nucleus in an
external magnetic field are called the e n e r g y   l e v e l s.   The
energy levels are equidistant and the energy difference of two adjacent levels equals $\gamma H \hbar$.

In 1939, Rabi showed that it is possible to induce transitions
of atomic nuclei between adjacent energy levels if the nucleus is
also subjected to the action of a weak a l t e r n a t i n g magnetic
field of definite frequency and polarization. Transitions arise if
the frequency ω of this exciting field coincides with the frequency
of the quantum corresponding to the distance between adjacent energy levels, i.e., if $\omega = \gamma H$. This phenomenon of induced transitions of atomic nuclei between energy levels under the action of an
alternating field of resonance frequency $(\omega = \gamma H)$ is called
n u c l e a r   m a g n e t i c   r e s o n a n c e.

The transition of a nucleus from one level to the next is accompanied by the absorption or emission of energy by the nucleus
with the result that there is exchange of energy between resonating
atomic nuclei and the external alternating field. The resonance
frequency is determined solely by the gyromagnetic ratio of the

nucleus and the magnitude of the external steady magnetic field. For example, in a field with a strength of 15,000 G, the resonance frequency of protons, for which $\gamma = 2.67 \cdot 10^4$, is about 63 MHz.

Rabi's experiments were the first in which nuclear magnetic resonance was observed. Despite their great value in experimental physics, the practical importance of these experiments remained limited because of the specific state of the material containing the nuclei investigated (atomic beams).

In 1945, two groups of scientists under the direction of Purcell and Bloch simultaneously carried out the first successful experiments to observe NMR in macroscopic matter in a condensed state. It is interesting to note that the possibility of observing NMR in macroscopic matter was predicted (Gorter, 1936) long before the first successful experiments. Apparently, only the inadequacy of the experimental technique and the unsuccessful choice of the experimental material delayed this discovery by a decade.

The possibility of observing NMR in macroscopic matter is explained by the fact that in such matter there are always conditions for the exchange of energy between a system of atomic nuclei in different energy levels (spin system) and a molecular system containing the nuclei and involved in thermal motion with it (lattice *).

The process of energy exchange between the spin system and the molecular system which is in thermal motion is called s p i n — l a t t i c e   r e l a x a t i o n, and the mechanism of this will be explained below. Moreover, it will be shown that under the conditions where NMR is observed, the presence of spin—lattice relaxation leads to direct transfer of energy from the alternating external field which excites resonance to the spin system and from the spin system to particles of matter in thermal motion. This absorption of energy of the exciting field may be measured with special instruments, and by means of the same instruments it is possible to investigate the relation of the energy absorbed in unit time to the so-called "field detuning," i.e., to the difference $\omega - \gamma H$ at different values of the steady field H. This relation is called the r e s o - n a n c e   l i n e.

---

*In NMR theory the lattice refers to the combination of degrees of freedom of all the microparticles in the matter with which the given spin interacts.

From the above it is clear that the study of spin—lattice relaxation processes is one of the main problems of nuclear magnetic resonance, since these processes are responsible for the very possibility of observing NMR.

The investigation of the fine forms of resonance lines soon led to two discoveries which made NMR the most powerful instrument for analyzing the structure of matter. It was established that in solids, because of the low mobility of the particles, there may act on the atomic nuclei constant internal magnetic fields which depend on the disposition of adjacent magnetic moments in the crystal lattice. These internal fields, which are superimposed on the external steady field, affect the resonance frequencies of the various atomic nuclei of the substance investigated to different extents.

Moreover, as a result of the direct magnetic interaction of the nuclear spins, there arise energy exchange processes between spins of the spin system itself (spin—spin relaxation). Then the energy equilibrium inside the spin system is established more rapidly than the equilibrium between the spin system and the lattice. Therefore, such a system of interacting spins behaves not simply as a system of independent spins in a homogeneous external field, but as a single system with a broad spectrum of resonance frequencies.

In solids the resonance lines are normally very broad and their form is determined by the structure of the crystal lattice. It is possible to establish quantitative relations between the lattice parameters and the parameters (moments) of the resonance line.

Because of the intensive thermal motion of the molecules in liquids, the magnetic interaction of this type (direct spin—spin interaction) is practically averaged out to zero. As a result of this, it is possible to observe the small difference in the resonance frequencies of nuclei belonging to chemically nonequivalent atoms of the molecule which is due to the effect of the different electronic environment, and is called the c h e m i c a l s h i f t o f t h e r e s o n a n c e (or simply the chemical shift). Moreover, thermal motion does not average out the bonds between nuclei through paired electrons of the valence shells (indirect spin—spin interaction). Therefore, the NMR spectra of pure liquids usually consist of a series of narrow lines, whose relative distribution and intensities are determined by the structure of the molecules. The lines in such spectra are usually very close and recording the spectra of liquids requires the use of a so-called high-resolution instrument.

In accordance with what has been stated, in nuclear resonance spectroscopy there are three main methods of investigating a substance: the broad-line method, the high-resolution method, and the relaxation method (spin echo).

The present book is devoted to the application of NMR to the investigation of polymers. It should be emphasized first of all that a very important characteristic of polymers is the complex character of the spin—lattice relaxation processes. The study of the dielectric, dynamic, and other properties of polymers shows that the spectrum of frequencies of molecular motions in them may be very wide. In many cases the molecular motions have a cooperative character. Therefore, the very simple theory of spin—lattice relaxation which applies to low-molecular substances may be applied to polymers only with reservations, and sometimes it does not even give qualitative agreement with experiment. Despite this, all the experimental methods mentioned have been used more or less successfully in the study of polymers. However, as a rule, the interpretation of the data obtained is difficult. Different conclusions are sometimes drawn from the same experimental results.

The broad-line method is used for studying polymers in bulk. The form, width, and second-order moment (or simply second moment) of an NMR line depend on the structure of the chain, i.e., the presence of branching and stereoregularity, and the degree of crystallinity of the polymer. By comparing the experimental value of the second moment of a line with the theoretical value calculated for a definite structure, it is possible to obtain information on the molecular structure of the polymer. In oriented polymers (fibers and films), the NMR spectrum depends on the angle of the sample in the magnetic field, and by means of NMR it is possible to obtain information on the character of the orientation of the macromolecules or crystallites in the polymer. By observing the change in the width of a line with temperature it is possible to obtain data on molecular motion in a polymer. The width and form of an NMR line also change when chemical and physical processes such as polymerization, cross linking of chains, degradation, etc., occur in a polymer. The NMR method makes it possible to study the kinetics and mechanism of these processes.

High-resolution spectra may be obtained by observing NMR in polymer solutions and melts. From the chemical shifts and spin—spin splitting it is possible to assess the structure of the macromolecules of the polymer. Particularly great success has been achieved in recent years in the study of the stereoregularity of polymers. For a series of polymers and copolymers it has been possible to determine completely the order of attachment of units in the chain. The change in the high-resolution NMR spectrum of a polymer solution with temperature gives information on the character of the molecular motions in solution. Chemical reactions of the functional groups of a polymer, ion-exchange reactions, the formation of hydrogen bonds, and other processes in solution may also be studied by the high-resolution NMR method.

The spin-echo method makes it possible to determine the relaxation times in polymers with great accuracy. This makes it possible to study molecular motions in more detail than by using only broad lines. In some cases, such as in the study of the spin echo in polyethylene melts, it is possible to determine the molecular weight and estimate the width of the molecular weight distribution. Moreover, by this method it is possible to measure the self-diffusion coefficients in polymer systems.

Naturally, the NMR method should not be used in isolation, but in combination with other physical and chemical methods.

PART ONE

Chapter I

Physical Bases of
Nuclear Magnetic Resonance

1. Nuclear Spin and
Magnetic Moment

Nuclear moments have a series of special properties which are observed in different physical experiments, but cannot be explained by the laws of classical physics. These properties are mainly as follows.

1. The length of the vector of the spin angular momentum of any atomic nucleus is always expressed by the relation

$$J = \hbar \sqrt{I(I+1)} \tag{I-1}$$

where the spin number I is constant for the given nucleus and may be an integer, half-integer, or zero, depending on the type of the nucleus.

2. If we measure the projection of a spin with a spin number I on any chosen direction, then it is possible to obtain only some number from the following finite series:

$$I\hbar; \quad (I-1)\hbar; \quad \ldots; \quad -I\hbar \tag{I-2}$$

The numbers of this series are called the s p i n e i g e n v a l u e s.

Under certain measurement conditions the same number from this series is obtained each time, and under other conditions, any of them may be obtained with certain probabilities. Intermediate values of the measured projection are never obtained. The physical meaning of the number I is thus the maximum length of the projection of the spin on any direction which may be obtained by measurement.

3. An attempt at the successive measurement of two projections of the spin on two coordinate axes, i.e., an attempt at the exact determination of the direction in space of a spin vector of known magnitude, leads to indefinite results. After the exact measurement of the first projection, any of the intrinsic values is obtained for the second within certain probabilities.

Due to the proportionality of the spin and magnetic moment vectors, the quantum properties of the latter are analogous to the quantum properties of spin.

To explain the observed mechanical and magnetic properties of nuclei, it is necessary to use quantum mechanics.

The observed behavior of spin is explained in the following way by quantum mechanics. It is quite impossible to assign to the intrinsic angular momentum simultaneously a definite direction in space and a definite magnitude in the same way that it is impossible to assign to an electron a definite coordinate and a definite momentum simultaneously. Therefore, there is no point in making measurements aimed at determining this orientation (for example, measuring two projections of the spin on coordinate axes).

Together with the concept of orientation, quantum mechanics introduces the concept of the s p i n   s t a t e .   If the spin is in a state characterized by the fact that its projection on a selected direction equals one of the eigenvalues, then the process of measuring this projection does not perturb the spin state and precisely this eigenvalue will be obtained with confidence in a measurement. These states are called the s p i n   e i g e n s t a t e s .

In other states the projection of the spin on a selected direction is indeterminate. In this case the measurement process converts the spin randomly into one of the eigenstates and the **mea**sured projection is found to equal the corresponding eigenvalue. In this case there is a change of state, i.e., it is impossible to predict which of the eigenvalues will be obtained as a result of the measurement. It is only possible to calculate the probability of a particular result.

The mechanical and magnetic properties of nuclei may be described by means of special mathematical methods and a series of assumptions of a physical character, which form the postulates or axioms of quantum mechanics.

The peculiarity of the mathematical language of quantum mechanics is explained by the peculiarity of the laws of motion of microscopic systems. The main elements of this language are the concepts of a matrix and an operator.

A matrix is a rectangular or square table of real or complex numbers of the form

$$A = \begin{pmatrix} a_{11} & a_{12} & \cdots & a_{1n} \\ a_{21} & a_{22} & \cdots & a_{2n} \\ \cdot & \cdot & \cdots & \cdot \\ a_{m1} & a_{m2} & \cdots & a_{mn} \end{pmatrix} \tag{I-3}$$

The numbers a_{ij} forming the matrix are called the e l e - m e n t s o f t h e m a t r i x.

The sum of two matrices A and B is the matrix C, whose elements equal the sums of the corresponding elements of the matrices A and B:

$$c_{ij} = a_{ij} + b_{ij} \tag{I-4}$$

From this definition it follows that two matrices may be added if they have the same number of rows and the same number of columns.

For example, if the matrices A and B have the form

$$A = \begin{pmatrix} a_{11} & a_{12} & a_{13} \\ a_{21} & a_{22} & a_{23} \end{pmatrix} \qquad B = \begin{pmatrix} b_{11} & b_{12} & b_{13} \\ b_{21} & b_{22} & b_{23} \end{pmatrix}$$

then their sum equals

$$C = A + B = \begin{pmatrix} a_{11} + b_{11} & a_{12} + b_{12} & a_{13} + b_{13} \\ a_{21} + b_{21} & a_{22} + b_{22} & a_{23} + b_{23} \end{pmatrix}$$

The product of the matrix A and the matrix B is the matrix C in which the element in the i-th row and in the j-th column equals the sum of the products of the i-th row of matrix A and the corresponding elements of the j-th column of matrix B:

$$c_{ij} = \sum_{m=1}^{n} a_{im} b_{mj} \qquad\qquad \text{(I-5)}$$

From this definition it follows that it is necessary to distinguish between the left and right products of matrix A and matrix B, i.e., the products AB and BA. The matrix A may be multiplied by matrix B from the right if the number of columns of the former equals the number of rows of the latter.

For it to be possible to construct products AB and BA simultaneously, matrices A and B must be square and the number of rows and columns in them identical.

For example, if matrices A and B have the form

$$A = \begin{pmatrix} a_{11} & a_{12} \\ a_{21} & a_{22} \end{pmatrix} \qquad\qquad B = \begin{pmatrix} b_{11} & b_{12} \\ b_{21} & b_{22} \end{pmatrix}$$

then the products AB and BA are

$$\begin{pmatrix} a_{11}b_{11} + a_{12}b_{21} & a_{11}b_{12} + a_{12}b_{22} \\ a_{21}b_{11} + a_{22}b_{21} & a_{21}b_{12} + a_{22}b_{22} \end{pmatrix} \text{ and } \begin{pmatrix} b_{11}a_{11} + b_{12}a_{21} & b_{11}a_{12} + b_{12}a_{22} \\ b_{21}a_{11} + b_{22}a_{21} & b_{21}a_{12} + b_{22}a_{22} \end{pmatrix}$$

It is obvious that in the general case products AB and BA will not equal eath other, i.e., multiplication of the matrices is noncommutative. If these products are identical in special cases, then it is said that the matrices A and B commute.

The matrix A* is said to be Hermitian-conjugate with matrix A if it is obtained from matrix A by replacing the rows by the columns and all the elements by the complex-conjugate numbers. For example,

$$A = \begin{pmatrix} a_{11} & a_{12} \\ a_{21} & a_{22} \\ a_{31} & a_{32} \end{pmatrix}$$

then

$$A^* = \begin{pmatrix} a_{11}^* & a_{21}^* & a_{31}^* \\ a_{12}^* & a_{22}^* & a_{32}^* \end{pmatrix}$$

By an o p e r a t o r we mean a symbol which represents a combination of mathematical operations which are performed on a given function of certain independent variables. It is said that the operator **A** operates on the given function φ and this situation is represented by the expression **A** φ.

As regards the function φ, it is assumed that it is defined over a certain range of changes of continuous or discrete independent variables.

We will subsequently be concerned only with operators, the result of whose operation on the function φ will be again some function of the same independent variables, which is defined in the same region. Moreover, we will limit ourselves to considering only so-called linear operators, i.e., those for which the following relations are valid:

$$\mathbf{A}\,(\varphi_1 + \varphi_2) = \mathbf{A}\varphi_1 + \mathbf{A}\varphi_2$$
$$\mathbf{A}\,(c\varphi) = c \cdot \mathbf{A}\varphi \qquad\qquad \text{(I-6)}$$

where c is a constant.

If the result of the operation of the operator **A** on the function φ is the same function multiplied by a constant α,

$$\mathbf{A}\varphi = \alpha \cdot \varphi \qquad\qquad \text{(I-7)}$$

then this function is called an e i g e n f u n c t i o n of the operator **A** and the number α is its e i g e n v a l u e .

It may be demonstrated that a system of eigenfunctions of a linear operator has the property of o r t h o g o n a l i t y , which means that the s c a l a r   p r o d u c t of any pair of eigenfunctions equals zero:

$$(\varphi_i^* \varphi_j) = 0 \quad (i \neq j) \qquad\qquad \text{(I-8)}$$

The scalar product is the sum of the products of the complex-conjugate values of one function and the values of the other (for the same values of the independent variables), extended over the whole region of definition of the function.

By using the property of linearity of the operator, this system may always be normalized, i.e., it is possible to choose constant coefficients for the eigenvalues so that for each of them the following relation holds:

$$\varphi_i^* \varphi_i = 1 \quad (i = 1, 2, \ldots) \tag{I-9}$$

Moreover, the system of eigenfunctions of a linear operator is c o m p l e t e .   This means that any function v, which is defined in the same region of the independent variables, may be expanded into a series with respect to the eigenfunctions of the operator:

$$v = \sum_{i=1}^{n} c_i \varphi_i$$

The coefficients of this series are calculated on the basis of the property of orthogonality of the function φ_i . By scalar multiplication of both parts of the previous equation, for example, by the function φ_k^* and by using the property of orthogonality, we obtain

$$c_k = (\varphi_k^* v)$$

There is a very close relation between operators and matrices. For example, let $\varphi_1, \varphi_2, \ldots, \varphi_n$ be a normalized system of eigenfunctions of the operator $\mathbf{A}$ and let the numbers $\alpha_1, \alpha_2, \ldots, \alpha_n$ be the corresponding eigenvalues of this operator, so that

$$\mathbf{A}\varphi_i = a_i \varphi_i$$

Then let the system of functions $v_1, v_2, \ldots, v_n$ form some orthonormalized system. Then any eigenfunctions of the operator $\mathbf{A}$, for example, φ_i , may be represented in the form of the series

$$\varphi_i = \sum_{j=1}^{n} c_{ji} v_j$$

It is obvious that with known functions v_j, this series is completely defined by the system of numbers c_{ji}, i.e., by the column matrix

$$\varphi_i = \begin{pmatrix} c_{1i} \\ c_{2i} \\ \vdots \\ c_{ji} \\ \vdots \\ c_{ni} \end{pmatrix}$$

By substituting the series for φ_i in the expression for determining the eigenfunctions of the operator **A**

$$A \left(\sum_{j=1}^{n} c_{ji} v_j \right) = a_i \left(\sum_{j=1}^{n} c_{ji} v_j \right) \tag{I-10}$$

and scalar multiplication of the two parts of this equation by any of the functions v_k^*, we obtain

$$\sum_{j=1}^{n} (v_k^* A v_j) c_{ji} = a_i c_{ki} \quad (k=1, 2, \ldots n) \tag{I-11}$$

By writing an analogous expression for all values of k from 1 to n, we obtain a system of equations, which may be represented in the following form by means of the concept of the product of matrices:

$$\begin{pmatrix} v_1^* A v_1 & v_1^* A v_2 & \ldots & v_1^* A v_n \\ v_2^* A v_1 & v_2^* A v_2 & \ldots & v_2^* A v_n \\ \cdot & \cdot & & \cdot \\ \cdot & \cdot & & \cdot \\ v_n^* A v_1 & v_n^* A v_2 & \ldots & v_n^* A v_n \end{pmatrix} \begin{pmatrix} c_{1i} \\ c_{2i} \\ \vdots \\ c_{ni} \end{pmatrix} = a_i \begin{pmatrix} c_{1i} \\ c_{2i} \\ \vdots \\ c_{ni} \end{pmatrix} \tag{I-12}$$

This equation coincides with the operator equation

$$A\varphi_i = a_i \varphi_i$$

if in the latter the operator **A** is replaced by a matrix with the elements $a_{kj} = v_k^* \mathbf{A} v_j$, the function φ_i is replaced by a column matrix with the elements c_{ji}, and the operation of the operator on the function is replaced by matrix multiplication. It is therefore said that the operator **A** is represented by the matrix (a_{kj}) by means of the function v_j. The numbers $(v_k^* \mathbf{A} v_j)$ forming the matrix which represents the operator are called the matrix elements of the operator in v representation:

$$a_{kj} = v_k^* \mathbf{A} v_j \tag{I-13}$$

The system of equations (I-11) written for all values of k represent a system of linear homogeneous equations with respect to the values c_{ki} (k = 1, 2, . . . , n)

$$(a_{11} - a_i) c_{1i} + a_{12} c_{2i} + \ldots + a_{1n} c_{ni} = 0$$
$$a_{21} c_{1i} + (a_{22} - a_i) c_{2i} + \ldots + a_{2n} c_{ni} = 0$$

$$. \quad . \quad . \quad . \quad . \quad . \quad . \quad . \quad . \quad . \quad . \quad . \quad . \quad . \quad .$$

$$a_{n1} c_{1i} + a_{n2} c_{2i} + \ldots + (a_{nn} - a_i) c_{ni} = 0$$

This system has a solution if its determinant

$$\Delta = \begin{vmatrix} (a_{11} - a_i) \, a_{12} \, \ldots\ldots\ldots \, a_{1n} \\ a_{21} \, (a_{22} - a_i) \, \ldots\ldots\ldots \, a_{2n} \\ . \quad . \quad . \quad . \quad . \quad . \quad . \quad . \quad . \quad . \quad . \\ a_{n1} \, \ldots\ldots\ldots\ldots \, (a_{nn} - a_i) \end{vmatrix} \tag{I-14}$$

equals zero. The equation $\Delta = 0$ is an equation of the n-th degree with respect to α_i. By solving it it is possible to find all the eigenvalues of the operator **A**.

The physical system of quantum mechanics is based on the following postulates [3]:

1. With each dynamic variable (particularly the spin of the nucleus) there may be coordinated some operator **L**, operating on the function φ, which definitely describes the state of the dynamic variable.

2. Between the operators expressing the different dynamic variables there are established the same identity relations which exist between the variables themselves in classical physics.

3. The eigenvalues α_i of the dynamic variable represented by the operator $\mathbf{L}$ and the functions φ_i, which describe the eigenstates of this variable, are found from the equation

$$\mathbf{L}\varphi_i = \alpha_i\varphi_i$$

i.e., are the eigenvalues and eigenfunctions, respectively, of the operator $\mathbf{L}$.

4. Any improper state of the dynamic variable may be described by the function ψ, which is written in the form of a series in eigenfunctions of its operator

$$\psi = \sum_{i=1}^{n} c_i\varphi_i$$

in which the squares of the values of the coefficients c_i give the probabilities p_i of obtaining in a measurement of the dynamic variable the eigenvalue α_i:

$$p_i = |c_i|^2 \quad \text{with} \quad \sum_{i=1}^{n} |c_i|^2 = 1$$

5. The relation between the energy of the system and the function of its state ψ is described by the differential equation

$$\mathscr{H}\psi = j\hbar \frac{\partial\psi}{\partial t} \tag{I-15}$$

where $\mathscr{H}$ is the energy operator of the system. This equation is called the Schrödinger time equation.

We will try, as far as possible, to explain the meaning of these postulates and also the mathematical concepts introduced by using them step by step for deriving the mathematical system, which describes the behavior of an atomic nucleus with a spin of $\frac{1}{2}$.

Two eigenvalues are possible for a spin with a spin number of $\frac{1}{2}$. This means that two possible spin eigenstates correspond to each direction in space.

Let us examine the two spin states which are eigenstates with respect to the coordinate axis z. In one of these states measurement of the projection of the spin on the z axis gives the number $+\hbar/2$ with confidence, while in the other $-\hbar/2$ is obtained with confidence.

These states may be described by the two column matrices,

$$\varphi_1 = \begin{pmatrix} 1 \\ 0 \end{pmatrix} \quad \text{"spin upward"} \tag{I-16}$$

$$\varphi_2 = \begin{pmatrix} 0 \\ 1 \end{pmatrix} \quad \text{"spin downward"} \tag{I-17}$$

which are called the functions of state. By using the rules of multiplication of matrices, it is readily demonstrated that the system of functions φ_1 and φ_2 is orthogonal and normalized, i.e., it satisfies relations (I-8) and (I-9).

In all other spin states measurement of the spin projection on the z axis gives one of the eigenvalues with definite probabilities p_1 and p_2, and in the measurement process the spin is converted into the corresponding eigenstate. Therefore, in accordance with the fourth postulate, any other state may be described by a function of the form

$$\psi = c_1 \begin{pmatrix} 1 \\ 0 \end{pmatrix} + c_2 \begin{pmatrix} 0 \\ 1 \end{pmatrix} = \begin{pmatrix} c_1 \\ c_2 \end{pmatrix} \tag{I-18}$$

where

$$p_1 = |c_1|^2 \quad p_2 = |c_2|^2$$

and

$$|c_1|^2 + |c_2|^2 = 1$$

If the operators $\mathbf{I}_x$, $\mathbf{I}_y$, and $\mathbf{I}_z$, which represent the three corresponding spin projections, are represented by matrices and their operation on the function of state by matrix multiplication, then these matrices will have two columns each [otherwise it would be impossible to multiply them by the function $\binom{c_1}{c_2}$, as is required by the third postulate] and two rows each [otherwise this multiplication would not give a function of the type $\binom{c_1}{c_2}$ consisting of two rows].

Thus, these matrices must have the form

$$\mathbf{I}_x = \begin{pmatrix} x_{11} & x_{12} \\ x_{21} & x_{22} \end{pmatrix} \quad \mathbf{I}_y = \begin{pmatrix} y_{11} & y_{12} \\ y_{21} & y_{22} \end{pmatrix} \quad \mathbf{I}_z = \begin{pmatrix} z_{11} & z_{12} \\ z_{21} & z_{22} \end{pmatrix} \tag{I-19}$$

The numbers forming the matrix $\mathbf{I}_z$ are readily found. On the basis of the third postulate, the matrix $\mathbf{I}_z$ and the functions ψ_1 and ψ_2 must satisfy the relations

$$\begin{pmatrix} z_{11} & z_{12} \\ z_{21} & z_{22} \end{pmatrix} \begin{pmatrix} 1 \\ 0 \end{pmatrix} = +\frac{1}{2}\hbar \begin{pmatrix} 1 \\ 0 \end{pmatrix}$$

$$\begin{pmatrix} z_{11} & z_{12} \\ z_{21} & z_{22} \end{pmatrix} \begin{pmatrix} 0 \\ 1 \end{pmatrix} = -\frac{1}{2}\hbar \begin{pmatrix} 0 \\ 1 \end{pmatrix}$$

By multiplying out the matrix products and equating the corresponding elements of the matrices in the right- and left-hand parts of the equations, we find

$$z_{12} = z_{21} = 0$$

$$z_{11} = \frac{1}{2}\hbar$$

$$z_{22} = -\frac{1}{2}\hbar$$

Thus, the operator $\mathbf{I}_z$ is a diagonal matrix on whose diagonal lie the eigenvalues of the z projection of spin:

$$\mathbf{I}_z = \frac{1}{2}\hbar \begin{pmatrix} 1 & 0 \\ 0 & -1 \end{pmatrix} \tag{I-20}$$

To find the numbers forming the matrices $\mathbf{I}_x$ and $\mathbf{I}_y$, let us examine the projection J_n of the spin on any direction in space, defined by the angles α, β, and γ with respect to the axes x, y, and z.

In accordance with the classical relation between this projection and the spin projections on the coordinate axes

$$J_n = J_x \cos \alpha + J_y \cos \beta + J_z \cos \gamma$$

and on the basis of the second postulate, it must be assumed that an analogous relation exists between the operators of the projections:

$$\mathbf{I}_n = \mathbf{I}_x \cos \alpha + \mathbf{I}_y \cos \beta + \mathbf{I}_z \cos \gamma$$

The possible values of the projection J_n in spin states which are eigenstates with respect to the selected direction should first equal $\pm \hbar/2$, while the eigenfunctions will have the form $\psi =$ $\begin{pmatrix} c_1 \\ c_2 \end{pmatrix}$ and must satisfy the equation

$$\mathbf{I}_n \begin{pmatrix} c_1 \\ c_2 \end{pmatrix} = \pm \frac{1}{2} \hbar \begin{pmatrix} c_1 \\ c_2 \end{pmatrix} = \pm a \begin{pmatrix} c_1 \\ c_2 \end{pmatrix} \qquad \text{(I-21)}$$

By substituting in the expression for $\mathbf{I}_n$ the matrices $\mathbf{I}_x$ and $\mathbf{I}_y$ from (I-19) and the matrix $\mathbf{I}_z$ from (I-20), multiplying out the matrix product in the left-hand part of (I-21), and equating the corresponding elements in the right- and left-hand parts, we obtain a system of homogeneous linear equations for finding the numbers c_1 and c_2:

$$\left. \begin{aligned} (x_{11} \cos \alpha + y_{11} \cos \beta + a \cos \gamma \mp a) c_1 + (x_{12} \cos \alpha + y_{12} \cos \beta) c_2 = 0 \\ (x_{21} \cos \alpha + y_{21} \cos \beta) c_1 + (x_{22} \cos \alpha + y_{22} \cos \beta - a \cos \gamma \mp a) c_2 = 0 \end{aligned} \right\} \qquad \text{(I-22)}$$

This system has a solution if its determinant equals zero. For a solution to exist for any chosen direction of the projection of J_n, it is necessary for the determinant to become zero at any values of the angles α, β, and γ, which satisfy the relation $\cos^2 \alpha + \cos^2 \beta + \cos^2 \gamma = 1$.

This system is analyzed conveniently by making the angles α, β, and γ equal to $\pi/2$ in turn and finding for each case the solutions of the system (I-22). For example, when $\beta = \pi/2$ ($\cos \beta = 0$, $\cos \alpha = \sin \gamma$) the solution will have the form

$$(x_{11}x_{22} - x_{12}x_{21} + a^2) \sin^2 \gamma + a (x_{22} - x_{11}) \frac{1}{2} \sin 2\gamma \mp a (x_{11} + x_{22}) \sin \gamma = 0$$

Since this equation must hold for any value of the angle γ, it must be that

$$x_{11}x_{22} - x_{12}x_{21} + a^2 = 0$$

$$x_{22} - x_{11} = 0$$

$$x_{22} + x_{11} = 0$$

Analogous relations are obtained for the elements of the matrix I_y by examining the solutions of the system (I-22) for $\alpha = \pi/2$.

From these relations we find

$$\left.\begin{aligned} x_{11} = x_{22} = y_{11} = y_{22} = 0 \\ x_{12}x_{21} = y_{12}y_{21} = a^2 \end{aligned}\right\} \tag{I-23}$$

For $\gamma = \pi/2$, taking into account (I-23), we obtain from relations (I-22) the following system of equations for the numbers c_1 and c_2:

$$\left.\begin{aligned} (x_{12}\cos\alpha + y_{12}\sin\alpha)\,c_2 = \pm\,ac_1 \\ (x_{21}\cos\alpha + y_{21}\sin\alpha)\,c_1 = \pm\,ac_2 \end{aligned}\right\} \tag{I-24}$$

In this case the spin eigenstates correspond to directions perpendicular to the z axis.

Therefore it may be assumed that in measurements of the spin projection on the z axis in these states we should obtain the numbers $\pm\hbar/2$ with equal probabilities, i.e., $|c_1|^2 = |c_2|^2$.

Therefore we will have

$$\left.\begin{aligned} (x_{12}\cos\alpha + y_{12}\sin\alpha)(x_{12}^*\cos\alpha + y_{12}^*\sin\alpha) = a^2 \\ (x_{21}\cos\alpha + y_{21}\sin\alpha)(x_{21}^*\cos\alpha + y_{21}^*\sin\alpha) = a^2 \end{aligned}\right\} \tag{I-25}$$

These equations must hold for any values of the angle α and, consequently,

$$\left.\begin{aligned} |x_{12}|^2 = |x_{21}|^2 = |y_{12}|^2 = |y_{21}|^2 = a^2 \\ x_{12}y_{12}^* + x_{12}^*y_{12} = 0 \\ y_{21}x_{21}^* + x_{21}y_{21}^* = 0 \end{aligned}\right\} \tag{I-26}$$

From relations (I-23) and (I-26) it is obvious that, generally speaking, elements of the matrices I_x and I_y differing from zero must be complex numbers with a modulus $a = \hbar/2$. Moreover, the numbers x_{12} and x_{21} and also y_{12} and y_{21} must be complex-conjugate in pairs. Therefore we may write

$$x_{12} = ae^{j\varphi}$$
$$y_{12} = ae^{j\psi}$$
$$x_{21} = ae^{-j\varphi}$$
$$y_{21} = ae^{-j\psi}$$

In addition, from the system of equations (I-26) we obtain

$$\cos(\varphi - \psi) = 0 \qquad \varphi - \psi = \pm\frac{\pi}{2}$$

Thus, the matrices I_x and I_y will have the form

$$I_x = \frac{1}{2}\hbar \begin{pmatrix} 0 & e^{j\varphi} \\ e^{-j\varphi} & 0 \end{pmatrix} \qquad I_y = \pm\frac{1}{2}\hbar \begin{pmatrix} 0 & -je^{j\varphi} \\ je^{-j\varphi} & 0 \end{pmatrix}$$

The phase angle φ in these relations is arbitrary. Physically this means that the selection of the direction of the x axis in a plane perpendicular to the z axis makes no difference. Therefore we may write $\varphi = 0$.

As regards the signs in the expression for I_y, it may be shown that the plus sign should be chosen for the right system of coordinates, i.e., for the system in which movement from the x axis to the y axis is counterclockwise, looking from the positive direction of the z axis. The minus sign should be used for the left system of coordinates.

On the basis of the calculations presented above we may write

$$I_x = \frac{1}{2}\hbar \begin{pmatrix} 0 & 1 \\ 1 & 0 \end{pmatrix} \qquad I_y = \frac{1}{2}\hbar \begin{pmatrix} 0 & -j \\ j & 0 \end{pmatrix} \tag{I-27}$$

By using the matrices I_x, I_y, and I_z we may determine the matrix I^2 of the square of the spin length:

$$I^2 = I_x^2 + I_y^2 + I_z^2 = \frac{3}{4}\hbar^2\begin{pmatrix} 1 & 0 \\ 0 & 1 \end{pmatrix} \qquad (I\text{-}28)$$

It is readily verified that both functions φ_1 and φ_2 are eigenfunctions of the operator I^2 and correspond to its only eigenvalue, which equals $3\hbar^2/4$.

The fact that the operators I_z and I^2 have common eigenfunctions means that the z projection and the square of the total length of the spin may be measured simultaneously, since definite results of measurements are obtained only for eigenstates of the dynamic variable.

In the general case the two operators **A** and **B** have common eigenfunctions (and consequently the dynamic variables corresponding to them may be measured simultaneously), when the product of the operators is commutative, i.e., if **AB** = **BA** .

It is readily seen that the operator I^2 commutes with any operator of the projections, but no pair of projections commute with each other, forming the relations

$$\left.\begin{aligned} I_x I_y - I_y I_x &= j\hbar I_z \\ I_y I_z - I_z I_y &= j\hbar I_x \\ I_z I_x - I_x I_z &= j\hbar I_y \end{aligned}\right\} \qquad (I\text{-}29)$$

which are called the commutation rules.

The mathematical system introduced, together with the physical postulates, makes it possible to describe the observed behavior of spin completely, i.e., the discreteness of the values of its projection on a selected direction, the value of the spin modulus, and the impossibility of measuring simultaneously the length and direction of the spin follow automatically from the relations presented.

This mathematical system may be generalized for spin with any spin number I. In this case, the eigenfunctions of the operator I_z may be represented in the form of matrix columns with the number of lines equal to 2I + 1, in which all the elements equal zero with the exception of one which equals unity:

$$\cdot \varphi_1 = \begin{pmatrix} 1 \\ 0 \\ 0 \\ \vdots \\ 0 \end{pmatrix}; \quad \varphi_2 = \begin{pmatrix} 0 \\ 1 \\ 0 \\ \vdots \\ 0 \end{pmatrix}; \quad \ldots; \quad \varphi_{2I+1} = \begin{pmatrix} 0 \\ 0 \\ \vdots \\ 0 \\ 1 \end{pmatrix} \qquad (I-30)$$

The operator I_z is a diagonal square matrix, on whose diagonal lie the eigenvalues of the z projection of the spin: $I\hbar$; $(I - 1)\hbar$; $\ldots$; $-I\hbar$. The operators of the other projections are also square matrices. The commutation rules (I-29) also hold for any value of I.

2. Behavior of an Isolated

Atomic Nucleus in Magnetic Fields

If an atomic nucleus with a magnetic moment that is not zero is placed in an external magnetic field H_0, directed, for example, along the z axis, then, according to classical electrodynamics, the potential energy of the nucleus changes by a value

$$E = -\mu H = -\gamma J_z H_0 \qquad (I-31)$$

We are assuming that the atomic nucleus is isolated, i.e., its interaction with other nuclei may be neglected.

In accordance with the second postulate, the spin energy operator will have the form

$$\mathcal{H} = -\gamma H_0 I_z \qquad (I-32)$$

To find the eigenvalues E_m of the energy operator it is necessary (third postulate) to solve the equation

$$\mathcal{H} \varphi_m = E_m \varphi_m$$

By operating with the operator $\mathcal{H}$ on the eigenfunctions (I-30) of the operator I_z, we find that these functions are also eigenfunctions of the energy operator and that its eigenvalues equal

$$E_m = m\hbar\gamma H_0 \qquad (I-33)$$

where

$$m = I; \quad I-1; \quad \ldots; \quad -I$$

The energy eigenvalues are called the energy levels. The energy levels are equidistant; the distance between a pair of adjacent levels depends only on the magnitude of the external field and the gyromagnetic ratio and is given by

$$\Delta E = \hbar \gamma H_0$$

The number of levels equals 2I + 1.

Measurement of the energy of the nucleus should give one of its eigenvalues. However, a problem arises: does a nucleus remain at some definite energy level or does its state change?

According to the postulate of quantum mechanics mentioned above, a change in the state of a microsystem with time is described by the Schrödinger time equation:

$$\mathcal{H}\psi = j\hbar \frac{\partial \psi}{\partial t} \tag{I-34}$$

where $\mathcal{H}$ is the energy operator and ψ is the function of state of the system.

For the problem considered, the operator $\mathcal{H}$ is given by the expression (I-32), while the function ψ is to be determined.

For the sake of simplicity we will assume that the spin of the nucleus equals $\frac{1}{2}$. In this case,

$$I_z = \frac{1}{2}\hbar \begin{pmatrix} 1 & 0 \\ 0 & -1 \end{pmatrix}$$

We assume that the state of the nucleus may vary and, therefore, the function of state should have the form

$$\psi = \begin{pmatrix} c_1 \\ c_2 \end{pmatrix} \tag{I-35}$$

where c_1 and c_2 are functions of time.

Thus we have

$$-\frac{1}{2}\hbar\gamma H_0\begin{pmatrix}1 & 0\\ 0 & -1\end{pmatrix}\begin{pmatrix}c_1\\ c_2\end{pmatrix}=j\hbar\begin{pmatrix}\dot{c}_1\\ \dot{c}_2\end{pmatrix}\qquad\text{(I-36)}$$

Here the dots above c_1 and c_2 denote differentiation with respect to time.

By mutliplying out the matrix product and equating the elements of the matrices in the right- and left-hand parts of equation (I-36), we obtain two differential equations,

$$\gamma H_0 c_1 = -2j\dot{c}_1$$
$$\gamma H_0 c_2 = 2j\dot{c}_2$$

whose solutions have the form

$$c_1 = ae^{j\gamma H_0 t/2}$$
$$c_2 = be^{-j\gamma H_0 t/2}\qquad\text{(I-37)}$$

For the solution of the problem to be physically defined, it must be assumed that the energy of the nucleus is known at some moment of time, for example, the initial moment. If at the initial moment (t = 0) the nucleus was at the lower level, then in (I-37) we should put b = 0, and in connection with this the probability of finding it at the upper level similarly equals zero ($|c_2|^2 = 0$) at any subsequent moment of time.

Thus, depending on the initial conditions, the behavior of a free atomic nucleus in a magnetic field $H_z = H_0$ is described by one of the functions

$$\psi_1 = \begin{pmatrix}1\\ 0\end{pmatrix}e^{j\gamma H_0 t/2}\quad\text{or}\quad\psi_2 = \begin{pmatrix}0\\ 1\end{pmatrix}e^{-j\gamma H_0 t/2}$$

and although these functions are time-dependent, the energy of the nucleus is constant and equal to one of the eigenvalues.

What forces can change the energy of the nucleus and induce its transition from one energy level to another?

Since the question arises as to the effect of these external forces on the magnetic moment of the nucleus, it is to be expected that these forces must be external magnetic fields.

Then if there is a transition of the nucleus from one level to the next, then the nucleus either emits or absorbs a quantum of energy equal to the difference in the energies of the adjacent levels:

$$\Delta E = \hbar \omega = \gamma \hbar H_0$$

Therefore it must be assumed that the magnetic fields exciting transitions must be alternating and that their frequency must equal or be close to the frequency

$$\omega_0 = \gamma H_0 \tag{I-38}$$

Finally, in accordance with certain physical considerations, the vector of the field exciting transitions must have a component perpendicular to the direction of H_0.

In the simplest form the field satisfying these conditions is a field whose vector rotates with an angular frequency ω in the plane xy so that its projection on the coordinate axis changes in accordance with the law

$$H_x = H_1 \cos \omega t$$
$$H_y = - H_1 \sin \omega t \tag{I-39}$$

The minus sign in the expression for H_y denotes that the vector H_1 rotates clockwise if we look at the xy plane from above.

Now let this rotating field act on the atomic nucleus in addition to the constant field $H = H_0$. The potential energy of the nucleus in the total field equals

$$E = -\overleftarrow{\mu}\overleftarrow{H} = -\gamma (J_x H_x + J_y H_y + J_z H_z) \tag{I-40}$$

and, consequently, the energy operator is expressed in terms of the operators of the spin projections in the form

$$\mathcal{H} = -\gamma (I_x H_x + I_y H_y + I_z H_0) \tag{I-41}$$

By substituting in this relation the expression for the matrices of the spin projections, summing the matrices, and taking into account the fact that

$$H_x - jH_y = H_1 e^{j\omega t}$$
$$H_x + jH_y = H_1 e^{-j\omega t}$$

we obtain the matrix representation of the energy operator for a nucleus with a spin number $\frac{1}{2}$:

$$\mathscr{H} = -\frac{1}{2} \gamma\hbar \begin{pmatrix} H_0 & H_1 e^{j\omega t} \\ H_1 e^{-j\omega t} & -H_0 \end{pmatrix} \qquad (\text{I-42})$$

The eigenvalues α_i and eigenfunctions ψ_i of this operator should be determined as previously from the equation

$$\mathscr{H}\psi_i = \alpha_i \psi_i$$

and by solving this we obtain

$$\alpha_{1,2} = \mp \frac{1}{2} \hbar\gamma \sqrt{H_0^2 + H_1^2} \qquad (\text{I-43})$$

$$\psi_{1,2} = \frac{1}{\sqrt{2}} \cdot \frac{1}{\sqrt[4]{1+x^2}} \begin{pmatrix} \sqrt{\sqrt{1+x^2} \pm 1} & e^{j\omega t/2} \\ \sqrt{\sqrt{1+x^2} \mp 1} & e^{-j\omega t/2} \end{pmatrix}$$

where

$$x = \frac{H_1}{H_0}$$

The solution of this equation only shows that as a result of measuring the spin energy at a moment t we must obtain either the number α_1 or the number α_2 and that the spin state directly at the moment of measurement and depending on the result of it will be described by the function ψ_1 or the function ψ_2, respectively. The behavior of the spin with time, i.e., the probabilities of the possible results of the measurement and their time dependence, are defined in accordance with the fourth and fifth postulates by the coefficients of the expansion with respect to the functions ψ_1 and ψ_2 of the function ψ, which satisfies the Schrödinger equation

$$\mathscr{H}\psi = j\hbar \frac{\partial\psi}{\partial t} \qquad (\text{I-44})$$

with the energy operator (I-42).

The solution of equation (I-44) may be simplified considerably if we limit our examination to weak perturbation, i.e., if it is assumed that $H_1 \ll H_0$. It is precisely this case which is of greatest practical interest.

It is readily seen that with the assumption made, the exact eigenfunctions (I-43) of the energy operator may be replaced approximately by the functions

$$\psi_1 = \begin{pmatrix} 1 \\ 0 \end{pmatrix} e^{j\omega t/2} \qquad \psi_2 = \begin{pmatrix} 0 \\ 1 \end{pmatrix} e^{-j\omega t/2}$$

and the function ψ sought, which describes the behavior of the spin, will have the form

$$\psi = c_1(t)\,\psi_1 + c_2(t)\,\psi_2 \qquad (I\text{-}45)$$

By substituting (I-45) and (I-42) in equation (I-44), we obtain two differential equations with respect to the variables c_1 and c_2:

$$(\omega_0 - \omega)\,c_1 + \omega_1 c_2 = -2j\dot{c_1}$$
$$\omega_1 c_1 - (\omega_0 - \omega)\,c_2 = -2j\dot{c_2}$$

the solutions of which have the form

$$c_1(t) = a\cos\alpha t + b\sin\alpha t$$
$$c_2(t) = \frac{2\alpha}{\omega_1}\,[(a\delta - jb)\cos\alpha t + (b\delta + ja)\sin\alpha t] \qquad (I\text{-}46)$$

where

$$\omega_0 = \gamma H_0 \qquad \omega_1 = \gamma H_1 \qquad \alpha = \frac{1}{2}\sqrt{\omega_1^2 + (\omega - \omega_0)^2}$$
$$\delta = \frac{\omega - \omega_0}{\sqrt{\omega_1^2 + (\omega - \omega_0)^2}}$$

and the integration constants a and b must be determined from the initial conditions.

Let us assume that at the initial moment ($t = 0$) the nucleus was in the upper level, i.e.,

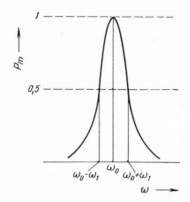

$$|c_1(0)|^2 = |a|^2 = 0$$

$$|c_2(0)|^2 = \frac{4a}{\omega_1^2}(a\delta - jb)(a^*\delta + jb^*) = 1$$

Then the probability of finding the nucleus in the lower level at a moment of time t is given by the relation

$$P(t) = |c_1(t)|^2 = \frac{\omega_1^2}{\omega_1^2 + (\omega - \omega_0)^2} \times$$

$$\times \sin^2 \frac{t}{2} \sqrt{(\omega - \omega_0)^2 + \omega_1^2} \qquad (I\text{-}47)$$

Fig. 1. Relation of the amplitude of the probability P_m to the frequency difference of the field $\Delta \omega$.

It is obvious that the probability of finding the nucleus in the upper level, as before at this moment, equals $1 - P(t)$. Exactly the same expression as (I-47) is obtained for the probability of finding the nucleus in the upper level at the moment t if it was in the lower level at the initial moment. We will call the probability $P(t)$ the probability of transition.

Formula (I-47) is valid for any moment of time if no measurement of the energy of the nucleus was made from the initial moment to this moment and the nucleus was subjected to the action of only the fields H_0 and H_1. The application of the perturbing field H_1 should not be regarded as a method of measuring the energy of the nucleus as we assumed that the amplitude and frequency of this field were constant and independent of what occurred with the nucleus.

If a measurement is made of the energy of the nucleus at any moment and it is found in a particular level, then the subsequent behavior of the probability of transition depends on the result of the measurement and is determined by the same relation (I-47) with the condition that the moment of measurement is taken as the initial moment.

The results obtained make it possible to describe the character of the change in the energy of a nucleus under the simultaneous action of the fields H_0 and H_1 in the following way.

At the moments of time

$$t_k = \frac{2k\pi}{\sqrt{(\omega - \omega_0)^2 + \omega_1^2}} \qquad (k = 0, 1, 2, \ldots)$$

the nucleus is with certainty in the initial energy level. In the intermediate moments the state of the nucleus varies continuously, but this does not at all imply a continuous variation in its energy. On the contrary, at these moments the energy of the nucleus equals either $+\frac{1}{2}\gamma H_0 h$ or $-\frac{1}{2}\gamma H_0 h$, i.e., the nucleus abruptly and randomly changes from one energy level to the other. On the basis of equation (I-47) it may be stated only that between the moments t_k the nucleus undergoes the same number of transitions up and down with the result that at the moments t_k it is always at the same level.

The transition of the nucleus from the bottom to the top level is accompanied by the absorption by the nucleus of a quantum of energy $\gamma H_0 h$, which is supplied by the exciting field H_1. In the reverse transition the nucleus emits this quantum.

At the moments of time

$$t_m = \frac{(2m + 1)\,\pi}{\sqrt{(\omega - \omega_0)^2 + \omega_1^2}} \qquad (m = 0, 1, 2, \ldots)$$

the probability of a transition is maximal and is closer to unity, the closer the frequency of the exciting field is to the value $\omega_0 = \gamma H_0$. The relation of the amplitude of the probability of transition to the difference $(\omega - \omega_0)$, which is called the "detuning of the field," is shown in Fig. 1.

When $\omega = \gamma H_0$, the maximum value of the probability of transition exactly equals unity at the moments

$$t_m = (2m + 1)\pi/\omega_1$$

i.e., the nucleus oscillates with certainty between two energy levels with a mean frequency, which depends only on the amplitude of the exciting field. This phenomenon is called nuclear magnetic resonance. No matter how small the amplitude of the exciting field, it produces definite transitions of the nucleus between the energy levels if only its frequency equals the resonance frequency $(\omega = \omega_0)$.

In conclusion, we should note that with low values of the probability of transition $P(t)$ the formula (I-47) is no longer valid because of the assumptions made above in the derivation of this formula. In particular, when $\omega = 0$, i.e., when the vector H_1 is stationary, formula (I-47) gives probabilities of transition which are not zero and are time-dependent, which is naturally contradictory to the true behavior of a nucleus in a steady field. However, as is readily seen, the error will be of the order of $(\omega_1/\omega_0)^2$, and it may be neglected in all regions where the probability of transition differs substantially from zero.

3. Nuclear Magnetic Resonance
in a Real Substance

The theory of nuclear resonance described in the previous section is idealized; in actual fact, atomic nuclei are never isolated. By participating in thermal motion together with atoms and molecules, they interact through their magnetic moments with each other and with other bearers of magnetic moments inside the substance, e.g., atomic nuclei of a different type, the electronic shells of atoms, paramagnetic ions, etc.

As a result of this, internal constant and alternating magnetic fields act on any atomic nucleus. The physical picture of the behavior of real nuclei under resonance conditions is also complicated by the fact that as a result of the interaction of a group of such nuclei complex magnetic systems may be formed with resonance frequencies which differ from the resonance frequencies of the separate nuclei. Moreover, as a result of the interaction, the energy of the nuclear moments may be transferred to thermal motion of the particles of the substance and redistributed between the nuclei.

In the first approximation, the interaction of atomic nuclei under NMR conditions may be replaced by the action on them of some additional constant and alternating magnetic fields with the same character as the internal fields. Interaction leading to the formation of complex magnetic systems and the characteristics of energy exchange between a system of nuclei and thermal motion will be examined and calculated in the appropriate sections below.

As has already been stated, the sources of internal magnetic fields may be any bearers of magnetic moments. Some of the magnetic moments are stationary relative to a given nucleus and produce at its point in space a constant magnetic field H_{loc} which is superimposed on the external field H_0 and changes the resonance frequency of the given nucleus. The internal magnetic fields are different for different nuclei of a spin system. The distribution of the internal steady magnetic field among the nuclei of the system is defined conveniently by means of some function $g(\omega_x)$ so that the value $g(\omega_x)d\omega_x$ represents the fraction of nuclei for which the frequency ω_x will be the resonance frequency with an external magnetic field H_0. It is obvious that, in this case,

$$\int_{-\infty}^{\infty} g(\omega_x)\, d\omega_x = 1 \qquad (I-48)$$

Other bearers of magnetic moments, which participate in thermal motion together with atoms and molecules, create at the point in space of the nucleus examined alternating magnetic fields with a very broad spectrum of frequencies. At certain moments in time these random alternating fields may have components of the required frequency and polarization and, consequently, they may change the state of the nuclei of the spin system.

The thermal motion of the particles has a random character. Therefore, the vector of the internal alternating field acting on the nucleus examined will be a random function of time. In examining its projection on the x axis, for example, it may be represented as consisting of the sum of periodic components with any similar frequencies, but the amplitudes and phases of these frequencies will be random functions of time.

For many random processes, which are called steady processes, it is found that the mean values of the squares of the amplitudes of these components are constant and depend only on the frequency of the component. A steady-state random process may be described, therefore, by means of a function $I(\omega)$ such that the value $I(\omega)d\omega$ expresses the mean value of the square of the amplitude of the periodic component with a frequency ω. This function is called the energy spectrum of the process.

The energy spectrum of a random process is related to another characteristic, called the correlation function of the process, which consists of the mean value of the product of pairs of values of the random variable, describing the process, shifted in time by the value τ:

$$\varphi(\tau) = \overline{f^*(t) f(t+\tau)} \tag{I-49}$$

where the line denotes averaging with respect to time t.

The energy spectrum $I(\omega)$ is defined mathematically in terms of the correlation function $\varphi(\tau)$ by means of the relation

$$I(\omega) = \int_{-\infty}^{\infty} \varphi(\tau) e^{-j\omega\tau} d\tau \tag{I-50}$$

which follows from the Fourier theory. By means of an inverse Fourier transformation the correlation function may be expressed in terms of the energy spectrum:

$$\varphi(\tau) = \frac{1}{2\pi} \int_{-\infty}^{\infty} I(\omega) e^{j\omega\tau} d\omega$$

The distribution functions of the internal steady and alternating magnetic fields $g(\omega_x)$ and $I(\omega)$ constitute valuable information, which may be obtained by observing NMR phenomena, since these functions are related to the intramolecular structure of the substance and internal motion in it.

According to statistical theory, the behavior of a system of spins with a spin number $\frac{1}{2}$ may be characterized by the probabilities P and Q of population of the lower and upper energy levels, which are functions of time in the general case. Since the state of each spin of the system is described by the function

$$\psi_i = \begin{pmatrix} c_{1i} \\ c_{2i} \end{pmatrix}$$

the probabilities P and Q consist of the squares of the coefficient of the numbers forming this matrix, averaged over the whole assembly of spins:

$$P = \frac{1}{N} \sum_{i=1}^{N} |c_{1i}|^2 = <|c_{1i}|^2>$$

$$Q = \frac{1}{N} \sum_{i=1}^{N} |c_{2i}|^2 = <|c_{2i}|^2>$$

(I-51)

where $P + Q = 1$.

To describe transitions between energy levels it is advantageous to introduce the concept of conditional probabilities of transition in unit time from the lower level to the upper, w_{12}, and from the upper to the lower, w_{21}. These probabilities are defined in such a way that if N_1 and N_2 are the populations of the lower and upper levels, respectively, at the moment t, then the products

$$N_1 w_{12} \, dt$$

$$N_2 w_{21} \, dt$$

represent the number of spins transferred in time dt from the lower level to the upper and from the upper level to the lower. It is obvious that the difference in these products represents the increase in the number of spins in the lower level in time dt:

$$dN_1 = (N_2 w_{21} - N_1 w_{12}) \, dt$$

or, in terms of the probabilities P and Q, and taking into account the fact that $N_1 + N_2 = N$:

$$\frac{dP}{dt} = Q w_{21} - P w_{12}$$

(I-52)

The description of the behavior of the system by means of the conditional probabilities w_{12} and w_{21} is of a more precise character than its description only by means of the value P and Q. A knowledge of the probabilities w_{12} and w_{21} makes it possible to integrate equation (I-52) and find the time dependence of the probabilities of population of the levels. However, generally speaking, it is impossible to find w_{12} and w_{21} from data on P and Q.

There is considerable physical interest in mechanisms of transitions in which the conditional probabilities are constants, i.e., are independent of time. In this case equation (I-52) is readily integrated and gives an exponential time dependence of the probabilities of population of the levels

$$P(t) - Q(t) = \frac{w_{21} - w_{12}}{w_{21} + w_{12}} + \left[P(0) - Q(0) - \frac{w_{21} - w_{12}}{w_{21} + w_{12}} \right] e^{-(w_{21} + w_{12})t} \qquad \text{(I-53)}$$

Systems described by time-independent conditional probabilities of transitions tend to an equilibrium state, which is characterized by a preponderance of the population of one of the levels

$$P(\infty) - Q(\infty) = \frac{w_{21} - w_{12}}{w_{21} + w_{12}} \qquad \text{(I-54)}$$

The value $1/(w_{21} + w_{12})$, which has the dimensions of time and determines the rate of establishment of the equilibrium state, is called the relaxation time (T_1).

Thus, we will examine a system which consists of identical spins $(I = \frac{1}{2})$, which are in a steady field H_0 and a perturbing field H_1, rotating in the xy plane with an angular velocity ω. To take into account the interaction between spins, we assume that, in addition to these fields, there acts on the spin system an internal steady field, which is not the same for different spins, but is directed along the z axis and given by the distribution function $g(\omega_x)$, and also internal alternating fields $h_x(t)$, $h_y(t)$, and $h_z(t)$, which have a stochastic character. The components of the internal steady field which lie in the xy plane will not be considered, since these components have only an insignificant effect on the behavior of the spins provided that they are sufficiently small.

The energy operator written for the i-th spin of the system will have the form

$$\mathcal{H}_i = -\frac{1}{2} \gamma \hbar \begin{pmatrix} H_z & H_x - jH_y \\ H_x + jH_y & -H_z \end{pmatrix}$$

while the state of this spin is described by the function ψ_i, which satisfies the Schrödinger equation

$$\mathcal{H}_i \psi_i = j\hbar \frac{\partial \psi_i}{\partial t}$$

We require a function ψ_i in the form

$$\psi_i = \begin{pmatrix} a_i e^{j\omega_0 t/2} \\ b_i e^{-j\omega_0 t/2} \end{pmatrix}$$

Then for the values a_i and b_i we obtain two differential equations

$$\left.\begin{aligned} \dot{a}_i(t) &= \frac{j}{2}\gamma(H_z - H_0)a_i(t) + \frac{j}{2}\gamma H_n b_i(t)e^{-j\omega_0 t} \\ \dot{b}_i(t) &= -\frac{j}{2}\gamma(H_z - H_0)b_i(t) + \frac{j}{2}\gamma H_n^* a_i(t)e^{j\omega_0 t} \end{aligned}\right\} \qquad \text{(I-55)}$$

where

$$H_n = H_x - jH_y; \qquad \omega_0 = \gamma H_0$$

These differential equations are completely equivalent to the initial operator of the Schrödinger equation.

The squares of the coefficients of the numbers a_i and b_i, which are functions of t, give the probabilities of finding the i-th spin in the lower or upper energy level, respectively, and since it is precisely these probabilities which are of physical interest, it is desirable to find equations containing directly the values $a_i a_i^*$ and $b_i b_i^*$.

Noting the fact that

$$\frac{d}{dt}(aa^*) = \dot{a}a^* + \dot{a}^* a \qquad \frac{d}{dt}(ab^*) = \dot{a}b^* + a\dot{b}^*$$

and using equation (I-55), we obtain

$$\begin{aligned} \frac{d}{dt}(a_i a_i^* - b_i b_i^*) &= j\gamma H_n e^{-j\omega_0 t}a_i^* b_i - j\gamma H_n^* e^{j\omega_0 t}a_i b_i^* \\ \frac{d(a_i b_i^*)}{dt} &= j\gamma(H_z - H_0)a_i b_i^* - \frac{j}{2}\gamma H_n e^{-j\omega_0 t}(a_i a_i^* - b_i b_i^*) \end{aligned} \qquad \text{(I-56)}$$

The averaging of these equations over all the spins of the system examined is the object of our analysis. *

---

*The set of pair products in these equations may be represented in the form of a matrix

$$\sigma_i = \begin{pmatrix} a_i a_i^* & a_i b_i^* \\ a_i^* b_i & b_i b_i^* \end{pmatrix}$$

The mean value of this matrix over all the spins

$$\sigma = \langle \sigma_i \rangle$$

is called the density matrix of the system examined. The diagonal elements of this matrix are the probabilities of population of the energy levels. The meaning of the off-diagonal elements will be explained below. By using the density matrix it would be possible to write further equations in the matrix form. However, for physical clarity we are not doing this.

In averaging equations (I-56) we assume first of all that there are no internal steady fields, i.e., $H_z = H_0 + h_z(t)$. We will introduce the notation

$$a_i a_i^* - b_i b_i^* = z_i \qquad <z_i> = z$$

$$a_i b_i^* = \varrho_i \qquad <\varrho_i> = \varrho$$

$$\omega_0 - \omega = \Delta\omega \qquad \gamma H_1 = \omega_1$$

$$h_x(t) + j h_y(t) = h(t)$$

Using these symbols we rewrite equations (I-56) in the form

$$\frac{dz_i}{dt} - j\omega_1 (\varrho_i^* e^{-j\Delta\omega t} - \varrho_i e^{j\Delta\omega t}) = j\gamma [h^*(t) e^{-j\omega_0 t} \varrho_i^* - h(t) e^{j\omega_0 t} \varrho_i]$$

$$\frac{d\varrho_i}{dt} + \frac{j}{2} \omega_1 e^{-j\Delta\omega t} z_i = j\gamma h_z(t) \varrho_i - \frac{j}{2} \gamma \hbar^*(t) e^{-j\omega_0 t} z_i$$

(I-57)

The left-hand parts of these equations may be averaged directly, since the coefficients of ρ_i and z_i and their products are the same for all spins. The following method is used for approximate averaging of the right-hand parts. By integrating the second of the equations (I-57) we may write

$$\varrho_i(t) = \varrho_i(0) - \frac{j}{2} \omega_1 \int_0^t e^{-j\Delta\omega x} z_i(x) \, dx +$$

$$+ j\gamma \int_0^t h_z(x) \varrho_i(x) \, dx - \frac{j}{2} \gamma \int_0^t h^*(x) e^{-j\omega_0 x} z_i(x) \, dx \qquad (I-58)$$

In averaging the first of the equations (I-57), we replace in the right-hand part of it the coefficients of ρ_i and ρ_i^* by the expression (I-58) and its complex-conjugate expression. In the averaging we will assume that the mean values of the random fields $h_z(t)$ and $h(t)$ equal zero and that the correlation between them may be neglected, i.e., they may be averaged independently of each other in products. Moreover, we assume that it is possible to neglect the correlation between the random value $h(t)$ and the values of the variables $\rho_i(x)$ and $z_i(x)$ in previous moments of time, i.e., when $x < t$.

With these assumptions, by averaging the first of the equations (I-57) we obtain

$$\frac{dz}{dt} - j\omega_1 \left(\varrho^* e^{-j\Delta\omega t} - \varrho e^{j\Delta\omega t}\right) = -\frac{\gamma^2}{2} \int_{-t}^{t} <h^*(t)\,h(x)> z(x)\, e^{-j\omega_0(x-t)} dx \qquad (\text{I-59})$$

Similarly, by integrating the first of the equations (I-57) and replacing the corresponding coefficient in the right-hand part of the second equation by the expression for z_i obtained, and also by replacing in this part the coefficient ρ_i by the expression (I-58), we obtain after averaging

$$\frac{d\varrho}{dt} + \frac{1}{2}\,\omega_1 e^{-j\Delta\omega t} z = -\gamma^2 \int_{0}^{t} <h_z(t)\,h_z(x)> \varrho(x)\, dx -$$

$$-\frac{1}{2}\gamma^2 \int_{0}^{t} <h(t)\,h^*(x)> e^{-j\omega_0(x-t)} \varrho(x)\, dx +$$

$$+\frac{1}{2}\gamma^2 \int_{0}^{t} <h^*(t)\,h^*(x)\, e^{-j\omega_0(t+x)}> \varrho^*(x)\, dx \qquad (\text{I-60})$$

We observe that in this expression it is possible to discard the last term, since, in contrast to others, it changes rapidly.

We will now assume that the moment t examined is so close to the initial moment (but not infinitely close) that the mean values of $z(x)$ and of $\rho(x)$ change little over the interval from 0 to t, so that they may be replaced by the values $z(t)$ and $\rho(t)$, and taken out of the integral signs.

We will then assume that at the points in space of the different spins of the system all forms of the functions $h(t)$ and $h_z(t)$ are realized, i.e., the mean values of the products $h(t)h^*(x)$ and $h_z(t)h_z(x)$ are, respectively, equal to the correlation functions of the random values $h(t)$ and $h_z(t)$:

$$<h(t)\,h^*(x)> = \varphi(t-x)$$
$$<h_z(t)\,h_z^*(x)> = \varphi_z(t-x)$$

Finally, we assume that the correlation functions $\varphi(\tau)$ and $\varphi_z(\tau)$ diminish with a change in τ over the interval from 0 to t so much that their contribution to the integrals with a further increase in the limits is insignificant with the result that in the integrals obtained we may make t = ∞.

With these conditions, in accordance with the definition (I-50) the integral appearing in equation (I-59) is the energy spectrum $I(\omega)$ of the random field $h(t)$ when $\omega = \omega_0$:

$$\int_{-\infty}^{\infty} \varphi(\tau)\, e^{-j\omega_0\tau}\, d\tau = I(\omega_0)$$

The first integral of equation (I-60) is evidently half of the value of the energy spectrum $I_z(\omega)$ of the random field $h_z(t)$ with zero frequencies:

$$\int_{0}^{\infty} \varphi_z(\tau)\, d\tau = \frac{1}{2} I_z(0)$$

Finally, the second integral of this equation may be represented in the form

$$\int_{0}^{\infty} \varphi(\tau)\, e^{-j\omega_0\tau}\, d\tau = \frac{1}{2} I(\omega_0) - jK(\omega_0)$$

where

$$K(\omega_0) = \int_{0}^{\infty} \varphi(\tau) \sin \omega_0\tau\, d\tau$$

Thus, the averaged equations of motion of the spin system assume the form

$$\left.\begin{aligned}
\frac{dz}{dt} - j\omega_1 (\varrho^* e^{-j\Delta\omega t} - \varrho e^{j\Delta\omega t}) &= -\frac{\gamma^2}{2} I(\omega_0)\, z \\
\frac{d\varrho}{dt} + \frac{j}{2}\, \omega_1 e^{-j\Delta\omega t} z &= -\frac{\gamma^2}{2} I_z(0)\, \varrho - \frac{\gamma^2}{4} I(\omega_0)\, \varrho + \frac{j}{2}\, \gamma^2 K(\omega_0)\, \varrho
\end{aligned}\right\} \qquad (I\text{-}61)$$

The equations obtained have the drawback that together with the real variable z (t) they also include complex variables $\rho(t)$ and $\rho^*(t)$. Only real numbers have a physical interpretation.

Noting the fact that the values

$$
x = j \left(\varrho^* e^{-j\Delta\omega t} - \varrho e^{j\Delta\omega t} \right)
$$
$$
y = \varrho^* e^{-j\Delta\omega t} + \varrho e^{j\Delta\omega t}
$$
(I-62)

are certainly real variables, and using the notation

$$
\frac{1}{T_1} = \frac{\gamma^2}{2} I(\omega_0)
$$
$$
\frac{1}{T_2} = \frac{\gamma^2}{2} I_z(0) + \frac{\gamma^2}{4} I(\omega_0)
$$
(I-63)

we obtain the equations of motion of the system in the form

$$
\left.
\begin{aligned}
&\frac{dz}{dt} + \frac{z}{T_1} - \omega_1 x = 0 \\
&\frac{dx}{dt} + \frac{x}{T_2} - [\Delta\omega - \gamma^2 K(\omega_0)] y = -\omega_1 z \\
&\frac{dy}{dt} + \frac{y}{T_2} + [\Delta\omega - \gamma^2 K(\omega_0)] x = 0
\end{aligned}
\right\}
$$
(I-64)

We should remember that the variable z in these equations is the difference in the probabilities of population of the lower and upper levels $z = P(t) - Q(t)$. As regards the auxiliary values x and y, their physical interpretation will be given in the next section.

In analyzing the system of equations (I-64), first of all we should note that the terms containing $K(\omega_0)$ are small and, moreover, may always be included in the main field H_0 so that we will not write them in future.

Then we should note that in the absence of an exciting field ($\omega_1 = 0$), the change in the probabilities of population of the levels is described by the equation

$$\frac{dz}{dt} + \frac{z}{T_1} = 0$$

As is evident from equation (I-53) and (I-63), this means that the stochastic fields produce transitions characterized by the same probabilities:

$$w_{12} = w_{21} = \frac{\gamma^2}{4} I(\omega_0) \qquad (I-65)$$

and that, as a result of this, the same population of the levels should correspond to the equilibrium state of the system.

This result is inaccurate since, in accordance with Boltzmann's theorem, the equilibrium populations of a two-level system, interacting with a thermal reservoir which is at a finite temperature, must be related by the equation

$$\frac{N_1}{N_2} = e^{-\frac{E_1 - E_2}{kT}}$$

In our case, $E_1 = -\frac{1}{2}\gamma\hbar H_0$, $E_2 = +\frac{1}{2}\gamma\hbar H_0$. In connection with the fact that under normal conditions (T = 300°K, $H_0 \approx 10^4$) the value of $\gamma\hbar H_0$ is much less than kT, the ratio of the equilibrium values of the populations must be equal to

$$\frac{N_1}{N_2} = 1 + \frac{\gamma\hbar H_0}{kT} \qquad (I-66)$$

The discrepancy between the results obtained above and Boltzmann's theorem is explained by the fact that in the theory presented no allowance was made for the characteristics of energy exchange between the spin system and the thermal motion of the particles of a substance at a finite temperature. Naturally, nothing prevents the spins of the system examined, which is excited by random field h(t), from donating energy to the surrounding space.

However, if a group of spins of the system must absorb energy of the field, then the energy of the resonance frequencies of this field must at this moment be not less than the number of quanta absorbed. It is clear that in a system in which the energy may be redistributed randomly between the frequencies of the spectrum as a result of random motion, the probability of this event does not equal unity. Therefore, it must be that

$$w_{12} < w_{21}$$

We satisfy this relation and the conditions of Boltzmann's theorem if we write

$$w_{12} = \left(1 - \frac{\gamma \hbar H_0}{2kT}\right) \frac{\gamma^2}{4} I(\omega_0)$$
$$w_{21} = \left(1 + \frac{\gamma \hbar H_0}{2kT}\right) \frac{\gamma^2}{4} I(\omega_0) \qquad \text{(I-67)}$$

Since, when $\omega_1 = 0$ the first of the equations (I-64) must change into equation (I-52) with the values of the probabilities w_{12} and w_{21} given by relations (I-67), to allow for the characteristics of energy exchange between the spin system and thermal motion, to the right-hand part of the first equation (I-64) we should add a constant term equal to $w_{21} - w_{12}$. This term is conveniently written in the form

$$w_{21} - w_{12} = \frac{\gamma \hbar H_0}{2kT} \cdot \frac{\gamma^2}{2} I(\omega_0) = \frac{z_0}{T_1} \qquad \text{(I-68)}$$

where

$$z_0 = \frac{\gamma \hbar H_0}{2kT}$$

Finally, the equations of motion of the system of spins under the action of the stochastic fields and the exciting rotating field H_1 will have the form

$$\frac{dz}{dt} + \frac{z}{T_1} - \omega_1 x = \frac{z_0}{T_1}$$

$$\frac{dx}{dt} + \frac{x}{T_2} - \Delta\omega y = -\omega_1 z \qquad (I-69)$$

$$\frac{dy}{dt} + \frac{y}{T_2} + \Delta\omega x = 0$$

Let us now examine the equilibrium state of the system in the presence of the exciting field H_1. In the equilibrium state all the products in equations (I-69) should be equal to zero. By solving the algebraic equations obtained for the equilibrium values $z(\infty)$ of the difference in probabilities of population of the levels, we find

$$z(\infty) \left[\frac{1}{T_1} + \frac{\omega_1^2 T_2}{1 + (\Delta\omega T_2)^2} \right] = \frac{z_0}{T_1}$$

Taking into account the fact that

$$\frac{1}{T_1} = w_{12} + w_{21}; \qquad \frac{z_0}{T_1} = w_{21} - w_{12}$$

we rewrite this relation in the form

$$z(\infty) \left[w_{12} + w_{21} + \frac{\omega_1^2 T_2}{1 + (\Delta\omega T_2)^2} \right] = w_{21} - w_{12} \qquad (I-70)$$

On the other hand, as follows from the equation (I-52), the equilibrium values of the conditional probabilities and the probabilities of population with any mechanism for the transition must be constant and must satisfy the relation

$$Q(\infty) w_{21}(\infty) - P(\infty) w_{12}(\infty) = 0$$

or

$$[P(\infty) - Q(\infty)] [w_{21}(\infty) + w_{12}(\infty)] = w_{21}(\infty) - w_{12}(\infty)$$

By comparing this expression with (I-70) we may conclude that at the equilibrium of the system examined the perturbing field H_1 produces equally probable transitions of the spins up and down with the conditional probabilities

$$p_{12} = p_{21} = \frac{1}{2} \cdot \frac{\omega_1^2 T_2}{1 + (\Delta\omega T_2)^2} \tag{I-71}$$

We conclude the account of the approximate theory of interaction between spins by taking into account internal steady fields,
which have been assumed to be equal to zero up to now. Here we
will limit ourselves to an examination of only the equilibrium state
of the spin system. It is evident from relations (I-67) that the
presence of internal steady fields has only a very insignificant effect on the probabilities of transitions caused by internal variable
fields, since these steady fields are several orders less than the
external field H_0. As regards transitions under the influence of the
exciting field H_1, the probabilities (I-71) depend substantially on
the resonance frequency. By averaging the probabilities $p_{12} = p_{21}$
over the whole of the distribution $g(\omega_x)$, we obtain

$$p = \int_{-\infty}^{\infty} g(\omega_x) \, d\omega_x \cdot \frac{1}{2} \cdot \frac{\omega_1^2 T_2}{1 + (\omega_x - \omega)^2 \, T_2^2} \tag{I-72}$$

If the width of the distribution $g(\omega_x)$ is much greater than the
value $1/T_2$, then this integral is readily calculated approximately

$$p = \frac{\pi}{2} \omega_1^2 g(\omega) \tag{I-73}$$

It should be emphasized again that this relation is unsuitable
for describing transient processes in the spin system: it may only
be used when the populations of the levels do not change.

In conclusion, we should point out yet another circumstance.
We will again turn to the system of equations (I-55) and will solve
it approximately, starting from the assumption that at the initial
moment all spins are in one level, for example, the upper level

$$\left.\begin{array}{c} a_i(0) = 0 \\ |b_i(0)|^2 = 1 \end{array}\right\} \quad \text{for all } i$$

Then for the initial time interval, during which the population of
the upper level still changes little, we may write

$$| b_i(t) |^2 \approx 1$$

and because of this the integration of the first of the equations (I-55) with subsequent averaging over the spins of the system gives

$$P(t) = \frac{1}{4} \gamma^2 \int_0^t \int_0^t [H_1^2 e^{j\Delta\omega(x-y)} + \langle h^*(x) h(y) \rangle e^{-j\omega_0(x-y)}] \, dx \, dy$$

The derivative of this expression will have the form

$$\frac{dP}{dt} = \frac{1}{4} \omega_1^2 \frac{\sin \Delta\omega t}{\Delta\omega} + \frac{1}{4} \gamma^2 \int_{-t}^t \varphi(\tau) e^{-j\omega_0\tau} \, d\tau$$

so that in analogy with what has been stated above, i.e., assuming a sufficiently rapid fall in $\varphi(\tau)$, we obtain for not too small values of t

$$\frac{dP}{dt} = \frac{1}{4} \omega_1^2 \frac{\sin \Delta\omega t}{\Delta\omega} + \frac{1}{4} \gamma^2 I(\omega_0)$$

By averaging this expression over an internal nonhomogeneous field, we will have

$$\frac{dP}{dt} = \frac{\pi}{4} \omega_1^2 g(\omega) + \frac{1}{4} \gamma^2 I(\omega_0)$$

Since at the initial moment from the condition $Q = 1$, $P = 0$, then in accordance with equation (I-52), this expression gives essentially the conditional probability of transition w_{21}.

Naturally it cannot be assumed that this probability remains constant with a further change in the state of the system. However, the fact that the expressions for the probabilities of transition at the initial moment coincide with the expressions (I-65) and (I-73) for the equilibrium values of the conditional probabilities of transition makes it possible to use the method described, which is called the first-order perturbation theory, when an exact solution of the problem involves difficulties.

We may now examine in detail the phenomenon of nuclear magnetic resonance in a real substance.

The substance containing the nuclei interesting us is in an external steady field H_0 and is excited by a field H_1 rotating with an angular velocity ω. In addition to these fields, the atomic nuclei are subject to the action of internal steady and alternating magnetic fields which are characterized by the distribution functions $g(\omega_x)$ and $I(\omega)$.

As has been shown, the presence of internal steady magnetic fields with a nonhomogeneous distribution in space leads to the appearance, under the action of the exciting field, of identical probabilities of transition in unit time from the lower level to the upper, and from the upper level to the lower. The internal alternating fields also produce transitions, but for the reasons examined above, the probability of a transition upward in unit time is less than the probability of a transition downward. Therefore, in the equilibrium state the lower level will have a higher population than the upper, though the preponderance of its population will be somewhat lower than in the absence of the exciting field. Due to the inequality of the populations of the levels and the equal probability of transitions produced by the exciting field H_1, the number of transitions in unit time from the lower level to the upper under the action of this field will be greater than the number of transitions from the upper level to the lower. On the whole, the exciting field donates energy to the spin system on an average and shifts spins to the upper level more rapidly, the higher the strength of this field.

On the other hand, the internal alternating field, which tends to bring the system to equilibrium, absorbs energy from the system of spins which this system obtains from the external field.

Let us examine the energy balance mathematically. If the equilibrium populations of the lower and upper levels equal N_1 and N_2, respectively, then the number of transitions in unit time from the simultaneous action of the external and internal alternating fields will be $N_1(p + w_{12})$ upward and $N_2(p + w_{21})$ downward.

At equilibrium these numbers are identical:

$$N_1(p + w_{12}) = N_2(p + w_{21})$$

Hence, taking into account the fact that $N_1 + N_2 = N$, we obtain

$$N_1 = \frac{p+w_{21}}{2p+w_{12}+w_{21}} \cdot N$$

$$N_2 = \frac{p+w_{12}}{2p+w_{12}+w_{21}} \cdot N \qquad \text{(I-74)}$$

$$n = N_1 - N_2 = \frac{w_{21}-w_{12}}{2p+w_{12}+w_{21}} \cdot N$$

The excess number of transitions upward in unit time under the action of the exciting field will be

$$m = pn = \frac{p(w_{21}-w_{12})}{2p+w_{12}+w_{21}} \cdot N$$

Each excess transition upward means the donation by the exciting field of a quantum of energy $\gamma \hbar H_0$. Consequently, the rate of energy donation, i.e., the amount of energy donated to the spin system by the exciting field in unit time, is given by

$$A = m\gamma\hbar H_0 = \gamma\hbar H_0 \frac{p(w_{21}-w_{12})}{2p+w_{21}+w_{12}} \cdot N \qquad \text{(I-75)}$$

By substituting here the value of p from the equation (I-73), and taking into account the fact that

$$w_{21} + w_{12} = \frac{1}{T_1}, \qquad \frac{w_{21}-w_{12}}{w_{12}+w_{21}} = \frac{\gamma\hbar H_0}{2kT}$$

we obtain

$$A = \frac{N\omega_1^2\omega_0^2\hbar^2}{4kT} \cdot \frac{\pi g(\omega)}{1+\pi\omega_1^2 g(\omega) T_1} \qquad \text{(I-76)}$$

In addition to the universal constants k and $\hbar$, the power absorbed is determined by the experimental conditions, namely, the amplitude of the exciting field H_1, the magnitude of the external steady field H_0, the gyromagnetic ratio of the nuclei investigated γ, the temperature of the substance T, and the frequency of the exciting field ω.

Moreover, the absorption is related to the distribution function of the internal steady magnetic fields in the substance $g(\omega)$ and the relaxation time T_1, which is determined by the statistical properties of the internal alternating fields.

It is interesting to calculate the power absorbed for a system in a homogeneous field. In this case, in formula (I-75) we substitute the values of the conditional probabilities of transitions (I-71). Then we obtain

$$A = \frac{N\omega_0^2\hbar^2\omega_1^2}{4kT} \cdot \frac{T_2}{1+(\Delta\omega T_2)^2+\omega_1^2 T_1 T_2} \qquad \text{(I-77)}$$

It is readily demonstrated that the absorbed power is proportional to the steady-state value of x in equation (I-69), which equals

$$x(\infty) = -\frac{\hbar\omega_0\omega_1}{2kT} \cdot \frac{T_2}{1+(\Delta\omega T_2)^2+\omega_1^2 T_1 T_2}$$

Equations (I-76) and (I-77) show that by measuring the power absorbed (or some value related to it) it is possible to obtain information on the behavior of the internal magnetic fields in a substance.

In an overwhelming majority of cases the determination of the exact form of the function $g(\omega_x)$ is the main problem of nuclear magnetic resonance spectroscopy.

We will show below the information which may be connected with the exact form of this function.

The relation of the absorbed power to the function $g(\omega)$ is complicated by the fact that this function appears in both the numerator and denominator of the expression for A.

If the magnitude of the exciting field is chosen such that

$$\pi\omega_1^2 g_{max}(\omega) T_1 \ll 1 \qquad \text{(I-78)}$$

[where $g_{max}(\omega)$ is the maximum value of $g(\omega)$], then the relation is simplified, and the power absorbed at different frequencies becomes proportional to the corresponding values of the function $g(\omega)$:

$$A \equiv g(\omega)$$

This method is called the we ak f i e l d m e t h o d.

We defined the function $g(\omega_x)$ as a measure of the probability of a spin of the spin system examined having the resonance frequency ω_x. Had the internal fields been absent, all the nuclei observed would have had the same resonance frequency of γH_0. Therefore, the function $g(\omega_x)$ would equal zero for all frequencies with the exception of $\omega = \gamma H_0$. At this frequency value the function $g(\omega_x)$ would be infinite.

The internal steady magnetic fields superimposed on the external field broaden the region of resonance frequencies. Therefore, $g(\omega_x)$ is not zero over a certain region of frequencies about the frequency $\omega = \gamma H_0$, at which it has a finite maximum.

In connection with this, under the conditions of a weak field, the absorption has a resonance character, i.e., it is observed only in the region of the resonance frequency and is maximal at this frequency.

When the weak field condition (I-78) is upset, the observed relation of the absorption to the frequency deviates from the function $g(\omega)$, and with very strong exciting fields resonance disappears completely.

If

$$\pi \omega_1^2 g_{max}(\omega) \, T_1 \gg 1 \qquad (I\text{-}79)$$

then the absorbed power is practically independent of the frequency of excitation.

This phenomenon is called r e s o n a n c e s a t u r a t i o n. The saturation is explained by the fact that the excess number of transitions, corresponding to the absorbed energy, is proportional to both the probability of transitions and also the difference in the populations of the levels. With very high probabilities of transition (a high value of H_1) the levels are populated practically the same and, therefore, with a further increase in the amplitude of the exciting field, the excess number of transitions does not increase.

In exactly the same way, at high values of H_1 the probabilities of transition are sufficiently great for practical equalization of the populations over a wide band of excitation frequencies and the energy absorbed in unit time is no longer dependent on frequency.

At saturation there is distortion or complete loss of information on $g(\omega)$. From what has been stated it is clear that this distortion consists of broadening of the region of intensive absorption.

4. Nuclear Magnetization

of a Substance

The physical value which can be measured under the conditions of nuclear magnetic resonance is not the magnetic moment of a single atomic nucleus, but the so-called nuclear magnetization of the substance, which consists of the sum of the mean values of the magnetic moments of all the atomic nuclei of a given type in unit volume of this substance:

$$\overleftarrow{M} = \frac{1}{V} \sum_{i=1}^{N_V} \overleftarrow{\mu}_i \text{ av} \qquad\qquad (I-80)$$

where V is the volume of the substance; N_V is the number of nuclei in this volume.

The concept of the mean value here requires elucidation. The meaning of the vector $\overleftarrow{M}$ is such that its projections on the coordinate axes are proportional to the components of the magnetic field created by the magnetic moments of the atomic nuclei in the corresponding directions:

$$M_x = \frac{1}{4\pi} B_x; \quad M_y = \frac{1}{4\pi} B_y; \quad M_z = \frac{1}{4\pi} B_z$$

The methods of measuring these macroscopic values do not involve any action on the atomic nuclei and, therefore, the state of the nuclear spins does not change during their measurement. In this connection the question arises as to what is the maximum information on the value of some dynamic variable which may be obtained with the condition that the measurement process must not change the state of this variable.

Let $\alpha_1, \alpha_2, \ldots, \alpha_n$ be the eigenvalues of some variable, and let it be in a state ψ such that $p_1, p_2, \ldots, p_n$ are the probabilities of obtaining the corresponding eigenvalues. With the measurement of a large number of these variables in a state ψ, the mean result of accurate measurements is

$$A_{av} = \sum_{i=1}^{n} p_i a_i$$

It is obvious that this value is the maximum accessible information on the value A in the state ψ if measurements are made of A without upsetting the state of this variable. The number A_{av} is called the mean value of the quantum dynamic variable A.

If **A** is the operator of the variable examined, while ψ is its function of state, then by expanding this function into a series with respect to the eigenfunctions of the operator it is readily demonstrated that the mean value will equal

$$A_{av} = \psi^* A \psi \tag{I-81}$$

Taking into account the fact that the spin and the magnetic moment of the nucleus are related by the equation

$$\overleftarrow{\mu} = \gamma \overleftarrow{J}$$

and that the operators of projections of the spin are represented by the matrices (I-20) and (I-27), it is easy to calculate the mean values of the projections of the magnetic moment of the nucleus if its spin is in a state described by the function

$$\psi_i = \begin{pmatrix} c_{1i} \\ c_{2i} \end{pmatrix}$$

By using these relations from equation (I-81) we obtain

$$\mu_{xi\ av} = \frac{1}{2}\ \gamma h\ (c_{1i}^* c_{2i} + c_{1i} c_{2i}^*)$$

$$\mu_{yi\ av} = \frac{1}{2}\ \gamma h j\ (c_{1i}^* c_{2i} - c_{1i} c_{2i}^*)$$

$$\mu_{zi\ av} = \frac{1}{2}\ \gamma h\ (c_{1i} c_{1i}^* - c_{2i} c_{2i}^*)$$

By summing these expressions over all the nuclei of any spin system which are present in unit volume of the substance, and taking into account the fact that

$$< c_{1i}c_{2i}^* > = \frac{1}{N} \sum_{i=1}^{N} c_{1i}c_{2i}^*$$

we will have

$$\left.\begin{array}{l} M_x = \frac{1}{2}\,\gamma\hbar N\,[\,<c_{1i}^*c_{2i}> + <c_{1i}c_{2i}^*>\,] \\[2mm] M_y = \frac{1}{2}\,\gamma\hbar N j\,[\,<c_{1i}^*c_{2i}> - <c_{1i}c_{2i}^*>\,] \\[2mm] M_z = \frac{1}{2}\,\gamma\hbar N\,[\,<c_{1i}c_{1i}^*> - <c_{2i}c_{2i}^*>\,] \end{array}\right\} \qquad \text{(I-82)}$$

It is easy to perceive the connection of the projections of the vector of the nuclear magnetization with the elements of the density matrix mentioned above. Likewise, by using equations (I-62) and (I-69), it is not difficult to find that with the assumptions made previously (see page 41), the projections of the magnetization vector are connected by the relations

$$\left.\begin{array}{l} M_x = u\cos\omega t - v\sin\omega t \\[2mm] M_y = -\,[u\sin\omega t + v\cos\omega t] \end{array}\right\} \qquad \text{(I-83)}$$

while

$$\left.\begin{array}{l} \dfrac{du}{dt} + \dfrac{u}{T_2} + \Delta\omega v = 0 \\[3mm] \dfrac{dv}{dt} + \dfrac{v}{T_2} - \Delta\omega u = -\,\omega_1 M_z \\[3mm] \dfrac{dM_z}{dt} + \dfrac{M_z}{T_1} - \omega_1 v = \dfrac{M_0}{T_1} \end{array}\right\} \qquad \text{(I-84)}$$

where

$$M_0 = \frac{1}{2}\,\gamma\hbar N z_0 = \frac{N\gamma^2\hbar^2 H_0}{4kT}$$

$$u = \frac{1}{2}\,\gamma\hbar N y; \quad v = \frac{1}{2}\,\gamma\hbar N x; \quad M_z = \frac{1}{2}\,\gamma\hbar N z$$

Finally, by eliminating from the equations the auxiliary values u and v and vector addition of the components of the magnetization vector and their products, it is possible to obtain

$$\frac{d\overleftarrow{M}}{dt} = \gamma\,[\overleftarrow{M}\overleftarrow{H}] - \overleftarrow{i}\,\frac{M_x}{T_2} - \overleftarrow{j}\,\frac{M_y}{T_2} - \overleftarrow{k}\,\frac{M_z - M_0}{T_1} \qquad (I\text{-}85)$$

where $\overleftarrow{i}$, $\overleftarrow{j}$, and $\overleftarrow{k}$ are single vectors along the coordinate axes.

The equation, which is called Bloch's equation, was original-ly proposed by F. Bloch to describe a system of interacting spins purely phenomenologically.

The values u and v which appear in the equations (I-83) and (I-84) are the projections of the vector $\overleftarrow{M}_n = M_x\overleftarrow{i} + M_y\overleftarrow{j}$ on the direction of the rotating vector $\overleftarrow{H}_1$ and on the direction perpendicu-lar to it, respectively.

For a system of independent spins, equation (I-86) assumes the form

$$\frac{d\overleftarrow{M}}{dt} = \gamma\,[\overleftarrow{M}\overleftarrow{H}] \qquad (I\text{-}86)$$

If this system is in a steady and homogeneous field $H_z = H_0$, then the projections of the vector $\overleftarrow{M}$ on the coordinate axes de-pend on time in the following way:

$$M_z = \text{const}; \quad M_x = M_n \cos\omega_0 t; \quad M_y = -M_n \sin\omega_0 t$$

i.e., the vector $\overleftarrow{M}$ precesses about the direction of the field so that its longitudinal component remains constant, while the trans-verse component rotates in the xy plane with an angular velo-city ω_0.

The presence of the interaction of the spins leads to damping of the precession of the magnetization vector with its components damped according to the exponential laws

$$M_z = M_0 + M_{z0}e^{-t/T_1}; \quad M_x = M_{x0}e^{-t/T_2}; \quad M_y = M_{y0}e^{-t/T_2}$$

The longitudinal component M_z, which determines the mean energy of the system, tends to an equilibrium value M_0 as a result of the exchange of energy between the system of spins and thermal motion.

The transverse component of the vector $\overleftarrow{M}_n$, rotating in the xy plane perpendicular to H_0, is quenched as a result of the diver-

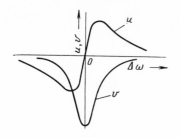

Fig. 2. Relation of the absorption v and dispersion u signals to the frequency difference of the field $\Delta\omega$.

gence of the phases of precession of the elementary components of the vector $\vec{M}$. Therefore, the equilibrium value of this component equals zero, while its rate of damping is greater than the rate of establishment of the equilibrium value of the longitudinal component, since this rate is additionally increased by the difference in the frequencies of precession of the components of the vector from the group of nuclei in nonidentical steady fields.

If a rotating exciting field H_1 is applied to a system of interacting spins, then the precession of the vector $\vec{M}$ is restored, but it will proceed with the frequency ω of the exciting field and not with the frequency ω_0. Then the transverse component rotates at the same rate as the vector H_1, but forms a constant angle with it, whose magnitude, like the values M_n and M_z, depends on the closeness of the frequency of the exciting field to the resonance frequency.

The projections M_x and M_y of the vector $\vec{M}$ on the coordinate axes each consists of two oscillating components with a phase difference of 90°.

For example,

$$M_x = u \cos \omega t - v \sin \omega t \tag{I-87}$$

As we saw previously (page 50), the amplitude of one of the components (v) is proportional to the power absorbed. With a given absorption, the amplitude of the other (u) is determined by the angle between M_n and H_1 and is called the dispersion signal.

In a system of interacting spins, excited by a rotating field H_1, there is the possibility of the establishment of a steady state if the magnitude of the field H_0 and the amplitude and frequency of the exciting field H_1 are constant. Then

$$M_z = M_0 \ \frac{1 + (\Delta\omega T_2)^2}{1 + (\Delta\omega T_2)^2 + \omega_1^2 T_1 T_2}$$

$$u = \omega_1 M_0 \frac{\Delta\omega T_2^2}{1 + (\Delta\omega T_2)^2 + \omega_1^2 T_1 T_2} \qquad (I-88)$$

$$v = -\omega_1 M_0 \frac{T_2}{1 + (\Delta\omega T_2)^2 + \omega_1^2 T_1 T_2}$$

Figure 2 shows the relation of the absorption signal v and the dispersion signal u to the frequency difference of the field $\Delta\omega = \omega - \gamma H_0$.

From the above it follows that the system of spins excited by a rotating field H_1 creates a macroscopic field, whose transverse component rotates with a frequency ω. Therefore, if the substance analyzed is placed in a coil so that its axis is perpendicular to the direction of H_0, then the alternating transverse components of the vector $\overleftarrow{M}$ will induce in this coil alternating electromotive forces, which may be measured by electronic means.

In particular, if the axis of the coil is directed along the x axis, then the magnitude of the induced emf will equal

$$e = -n \frac{d\Phi_x}{dt} \cdot 10^{-8} = -4\pi n S \frac{dM_x}{dt} \cdot 10^{-8} \text{ V}$$

where n is the number of turns; S is the cross section of the coil.

This is the electromagnetic principle of the observation of nuclear magnetic resonance signals.

5. Magnetic Interaction

of Atomic Nuclei

We saw above that in the first approximation the phenomenon of NMR is characterized by two parameters, namely, the longitudinal relaxation time T_1 and the distribution function of resonance frequencies $g(\omega_x)$.

The first of these parameters characterizes the mechanism of energy exchange between the spin system and the thermal motion of the magnetic moments inside the substance. It is related to the energy spectrum of the internal alternating fields by the first relation of equations (I-63) and, consequently, depends on the character of the molecular motion.

The second parameter, i.e., the distribution function of the resonance frequencies $g(\omega_x)$, depends on the internal steady fields.

By definition, the value $g(\omega_x)d\omega_x$ is the probability that with an external field H_0 the resonance frequency of the nucleus examined lies within the range $\omega_x \pm d\omega_x/2$. It is obvious that the same value is the probability that the internal steady field will equal

$$h = \frac{\omega_x}{\gamma} - H_0$$

If we introduce the distribution function of the internal fields $g(h)$, then this probability will be $g(h)dh$. Since $d\omega_x = \gamma dh$, there is the following relation between the two functions $g(h)$ and $g(\omega_x)$:

$$\gamma \cdot g(\omega_x) = g(h) \qquad \omega_x = \gamma(\boldsymbol{H_0 + h})$$

According to classical electrodynamics, the magnetic field produced by the magnetic moment $\overleftarrow{\mu_1}$ of one nucleus at the position of another equals

$$\overleftarrow{h} = \frac{3[\overleftarrow{\mu_1}\overleftarrow{r_0}] - \overleftarrow{\mu_1}}{r^3} \tag{I-89}$$

where r is the distance between nuclei; $\overleftarrow{r_0}$ is the unit vector in the direction from the first nucleus to the second.

If the nuclei are stationary relative to each other, then this field is steady and its magnitude and direction depend on the relative disposition of the nuclei. A change in $\overleftarrow{r}$ in value or direction leads to the appearance of variable components of the internal field.

The relation (I-89) could be used to calculate the functions $I(\omega)$ and $g(\omega_x)$ in the first approximation.

However, in a more accurate examination of the interaction between the magnetic moments of atomic nuclei it is necessary to take into account two more circumstances. The fact is that in undergoing a transition from the upper level to the lower and emitting a quantum of energy, an atomic nucleus may induce an energy transition of a neighboring nucleus of the same type. This means that pairs of nuclei of the same type may behave as coupled systems, a change in the energy state of which may require different resonance frequencies from the resonance frequency of a single nucleus. Moreover, by interacting with internal alternating fields, the electromagnetic radiation emitted by the nuclei may form combination frequencies, which coincide with the resonance frequency of the nuclei investigated.

Thus, in addition to the components of the energy spectrum in the region of the resonance frequency, some other frequencies may be effective. This also refers to the interaction of nuclei of different types.

These circumstances will be allowed for automatically if we examine the total energy of the spin system.

The simplest case of a spin system is a pair of nuclei of the same type. If this system is in an external field, then its total energy equals

$$E = -\vec{\mu}_1\vec{H}_0 - \vec{\mu}_2\vec{H}_0 - \frac{3\,(\vec{\mu}_1\vec{r}_0)(\vec{\mu}_2\vec{r}_0) - \vec{\mu}_1\vec{\mu}_2}{r^3} \qquad \text{(I-90)}$$

and the operator of this energy may be written in the form

$$\mathcal{H} = -\gamma H_0\,(\mathbf{I}_{z1} + \mathbf{I}_{z2}) - \gamma^2 r^{-3}\,[3\,(\mathbf{I}_1\vec{r}_0)\,(\mathbf{I}_2\vec{r}_0) - \mathbf{I}_1\mathbf{I}_2] \qquad \text{(I-91)}$$

Our problem consists of finding the eigenvalues of the energy and the probabilities of transitions between these values, which are connected with changes in the values r and $\vec{r}_0$ with time.

We will first assume that there is no interaction, i.e., the nuclei are sufficiently removed from each other. Then the last term in the expression for E may be neglected. In this case the measurement of the z projections of the spins may give four combinations of eigenvalues for each of them, which correspond to the four eigenvalues of the z projection of the total spin:

$$J_z = h \quad J_z = 0 \quad J_z = 0 \quad J_z = -h$$

Two different states of the pair examined correspond to the two zero eigenvalues of I_z. Therefore, there are four different eigenstates of our system.

It would be possible to use column matrices of four rows to describe these eigenstates. These matrices must also be functions of the eigenstates of each spin separately as the eigenstates of the system without interaction are combinations of the eigenstates of the elements forming it. Therefore, in the examination of a system of two spins the operators of the projections of the total spin may be represented by square matrices consisting of four rows and four columns.

However, another method of representing the eigenfunctions of a system of spins is usually used. In particular, the four eigenfunctions of the pair examined are represented in the form of symbolic products of the corresponding eigenfunctions of single spins:

$$\psi_1 = \begin{pmatrix} 1 \\ 0 \end{pmatrix}\begin{pmatrix} 1 \\ 0 \end{pmatrix} = \alpha\alpha \qquad \psi_3 = \begin{pmatrix} 0 \\ 1 \end{pmatrix}\begin{pmatrix} 1 \\ 0 \end{pmatrix} = \beta\alpha$$

$$\psi_2 = \begin{pmatrix} 1 \\ 0 \end{pmatrix}\begin{pmatrix} 0 \\ 1 \end{pmatrix} = \alpha\beta \qquad \psi_4 = \begin{pmatrix} 0 \\ 1 \end{pmatrix}\begin{pmatrix} 0 \\ 1 \end{pmatrix} = \beta\beta$$

(I-92)

The multiplication operation in these products is not matrix multiplication, but simply denotes the fact that each of the eigenstates of the system is a combination of the eigenstates of its elements.

With this choice of symbols of the eigenfunctions, the operators will be represented not by matrices, but by some other symbols, and their operation on the function of state will not be represented by matrix multiplication. Therefore, we will simply talk of the operation of the operator on the function of state, meaning only the final result of this operation.

For the functions (I-92) to be equivalent to the functions represented by the matrix columns it is necessary to stipulate the following rules.

1. In the expressions $\alpha\alpha$, $\alpha\beta$, $\beta\alpha$, and $\beta\beta$ the operator of a given spin operates only on its own function (operators belonging to the first spin operate on the first cofactor, while operators of the second spin operate on the second cofactor).

2. The result of the operation of the operators of each spin on the corresponding function is determined by the operation of the matrices (I-20) and (I-27) on the functions α and β, i.e.,

$$\mathbf{I}_{xi}(\alpha) = \frac{1}{2}\hbar\beta \quad \mathbf{I}_{yi}(\alpha) = \frac{i}{2}\hbar\beta \qquad \mathbf{I}_{zi}(\alpha) = \frac{1}{2}\hbar\alpha$$

$$\mathbf{I}_{xi}(\beta) = \frac{1}{2}\hbar\alpha \quad \mathbf{I}_{yi}(\beta) = -\frac{i}{2}\hbar\alpha \quad \mathbf{I}_{zi}(\beta) = -\frac{1}{2}\hbar\beta$$

(I-93)

$$\text{(where } i = 1, 2, \ldots)$$

$$\mathbf{I}_i^2(\alpha) = \frac{3}{4}\hbar^2\alpha \qquad \mathbf{I}_i^2(\beta) = \frac{3}{4}\hbar^2\beta$$

3. The cofactors in the products $\alpha\alpha$, $\alpha\beta$, $\beta\alpha$, and $\beta\beta$ cannot change places.

4. In accordance with the results of multiplication of the matrix columns with one element which differs from zero, we will assume that

$$(\alpha\alpha)(\alpha\alpha) = (\alpha\beta)(\alpha\beta) = (\beta\alpha)(\beta\alpha) = (\beta\beta)(\beta\beta) = 1 \qquad (I-94)$$

and that the products of the pairs of different functions equal zero,

$$(\alpha\alpha)(\alpha\beta) = 0 \quad (\alpha\alpha)(\beta\alpha) = 0 \text{ etc.} \qquad (I-95)$$

By using these rules it may be shown that the functions $\alpha\alpha$, $\alpha\beta$, $\beta\alpha$, and $\beta\beta$ are actually eigenfunctions of the z projection of the total spin, each one corresponding to one of its eigenvalues:

$$
\begin{array}{ll}
I_z(\alpha\alpha) = \hbar(\alpha\alpha) & I_z(\beta\alpha) = 0(\beta\alpha) \\
I_z(\alpha\beta) = 0(\alpha\beta) & I_z(\beta\beta) = -\hbar(\beta\beta)
\end{array}
\qquad (I-96)
$$

Moreover, the functions $\alpha\alpha$ and $\alpha\beta$ are eigenfunctions of the operator I_{z1} and both belong to its eigenvalue $\hbar/2$, while the functions $\beta\alpha$ and $\beta\beta$ belong to the corresponding eigenvalue $-\hbar/2$ of this operator.

Analogously, the pairs of functions $\alpha\alpha$, $\beta\alpha$ and $\alpha\beta$, $\beta\beta$ are eigenfunctions of the operator I_{z2} with the eigenvalue $-\hbar/2$ and $-\hbar/2$, respectively.

It was stated above that we did not know the form of the operators describing the system of spins, since we did not know the law of their operation on the functions $\alpha\alpha$, $\alpha\beta$, $\beta\alpha$, and $\beta\beta$. However, it is possible to find the matrix representations of these operators with which the functions of state will be represented by matrix columns, while the operation of the operator on them will be represented by matrix multiplication.

By means of relation (I-13) and the rules (I-93), we find, for example, the matrix representation of the operator I_z:

$$
I_z = \begin{pmatrix}
\hbar & 0 & 0 & 0 \\
0 & 0 & 0 & 0 \\
0 & 0 & 0 & 0 \\
0 & 0 & 0 & -\hbar
\end{pmatrix}
$$

Thus the eigenstates of the z projection of the total spin may be described conveniently by means of the functions $\alpha\alpha$, $\alpha\beta$, $\beta\alpha$, and $\beta\beta$.

In accordance with the fourth postulate (see page 62), any improper state of the system may be described by a function of the form

$$\psi = c_1(\alpha\alpha) + c_2(\alpha\beta) + c_3(\beta\alpha) + c_4(\beta\beta) \tag{I-97}$$

where the squares of the modules of the numbers c_1, c_2, c_3, and c_4 give the probabilities of finding the system in one of the eigenstates. Then the following equation must hold:

$$\Sigma (c_i)^2 = 1 \tag{I-98}$$

Let us examine the operator of the square of the module of the total spin

$$\mathbf{I}^2 = (\mathbf{I}_1 + \mathbf{I}_2)^2 = \mathbf{I}_1^2 + \mathbf{I}_2^2 + 2\mathbf{I}_1\mathbf{I}_2$$

In analogy with the classical relation for the scalar product of the operators $\mathbf{I}_1 \cdot \mathbf{I}_2$ we may write

$$\mathbf{I}_1\mathbf{I}_2 = \mathbf{I}_{x1}\mathbf{I}_{x2} + \mathbf{I}_{y1}\mathbf{I}_{y2} + \mathbf{I}_{z1}\mathbf{I}_{z2}$$

Therefore,

$$\mathbf{I}^2 = \mathbf{I}_1^2 + \mathbf{I}_2^2 + 2[\mathbf{I}_{x1}\mathbf{I}_{x2} + \mathbf{I}_{y1}\mathbf{I}_{y2} + \mathbf{I}_{z1}\mathbf{I}_{z2}]$$

By operating with the right-hand part of this expression on the functions $\alpha\alpha$, $\alpha\beta$, $\beta\alpha$, and $\beta\beta$ in accordance with the rules given above, we find

$$\mathbf{I}^2(\alpha\alpha) = 2\hbar^2(\alpha\alpha)$$
$$\mathbf{I}^2(\alpha\beta) = \hbar^2(\alpha\beta + \beta\alpha)$$
$$\mathbf{I}^2(\beta\alpha) = \hbar^2(\alpha\beta + \beta\alpha)$$
$$\mathbf{I}^2(\beta\beta) = 2\hbar^2(\beta\beta)$$

We see that only the functions $\alpha\alpha$ and $\beta\beta$ are eigenfunctions of the operator $\mathbf{I}^2$, and its eigenvalue in the states described by these functions equals $2\hbar^2$.

The other two functions are not its eigenfunctions, but if we multiply the second and third of these equations by any number a

and add them, then the function $a(\alpha\beta + \beta\alpha)$ will be an eigenfunction of the operator $\mathbf{I}^2$ with the eigenvalue $2\hbar^2$.

Analogously, the function $b(\alpha\beta - \beta\alpha)$ with any value of b will be an eigenfunction of this operator with an eigenvalue of zero.

By selecting in accordance with the normalization conditions (I-98),

$$a = b = \frac{1}{\sqrt{2}}$$

we obtain four eigenfunctions of the operator $\mathbf{I}^2$ in the form

$$\psi_1 = (\alpha\alpha) \quad \psi_2 = \frac{1}{\sqrt{2}}(\alpha\beta + \beta\alpha) \quad \psi_3 = \beta\beta$$

$$\psi_4 = \frac{1}{\sqrt{2}}(\alpha\beta - \beta\alpha) \tag{I-99}$$

Thus, we found that the square of the module of the total spin of a pair of atomic nuclei with a spin number $\frac{1}{2}$ may have two eigenvalues: $2\hbar^2$ and 0; in the states ψ_1, ψ_2, and ψ_3 the first of these values will be obtained and in the state ψ_4, the second.

By operating on the functions (I-99) with the operator $\mathbf{I}_z$ it may be demonstrated that they are also eigenfunctions of the z projection of the total spin with the eigenvalues $\hbar$, 0, $-\hbar$, and 0, respectively. With this selection of eigenfunctions the matrix representation of the operator $\mathbf{I}_z$ will have the form

$$\mathbf{I}_z = \begin{pmatrix} \hbar & 0 & 0 & 0 \\ 0 & 0 & 0 & 0 \\ 0 & 0 & -\hbar & 0 \\ 0 & 0 & 0 & 0 \end{pmatrix}$$

It follows from the above that a pair of noninteracting spins in the states ψ_1, ψ_2, and ψ_3 behaves as a single particle with a spin number 1, while in the state ψ_4 it behaves as a particle with a spin of 0.

The combination of the first three states is called the triplet state of the system, while the state ψ_4 is called the singlet state.

Let us now examine how the system will behave under the action of some perturbation, which is time-dependent. If the operator of the energy of the unperturbed system is $\mathscr{H}_0$, and the operator of the energy of perturbation is $\mathscr{H}'$, then the operator of the total energy of the system will be

$$\mathscr{H} = \mathscr{H}_0 + \mathscr{H}' \tag{I-100}$$

In the general case we will assume that the operator $\mathscr{H}_0$ has n eigenfunctions and n eigenvalues.

Our problem consists of finding the functions of state which describe the behavior of the system, i.e., solving the Schrödinger equation

$$\mathscr{H}\psi = j\hbar \frac{\partial\psi}{\partial t} \tag{I-101}$$

In the absence of perturbation the solution of this equation has the form $\psi_i e^{-jE_i t/\hbar}$ (where ψ_i represents the eigenfunctions of the operator $\mathscr{H}_0$, and E_i, its eigenvalues).

Let us expand ψ into a series with respect to the functions ψ_i:

$$\psi = \sum_{i=1}^{n} c_i \psi_i e^{-jE_i t/\hbar}$$

The coefficients of this expansion will depend on time, while the squares of their modules will be proportional to the probabilities of finding the system in the state ψ_i (i = 1, 2, . . . , n).

By substituting this series in the equation (I-101), we obtain

$$\sum_{i=1}^{n} c_i (\mathscr{H}_0 + \mathscr{H}') \psi_i e^{-jE_i t/\hbar} = j\hbar \sum_{i=1}^{n} \dot{c}_i \psi_i e^{-jE_i t/\hbar} + \sum_{i=1}^{n} E_i c_i \psi_i e^{-jE_i t/\hbar}$$

Since

$$c_i \mathscr{H}_0 \psi_i e^{-jE_i t/\hbar} = E_i \psi_i c_i e^{-jE_i t/\hbar}$$

then we will have

$$\sum_{i=1}^{n} c_i \mathscr{H}' \psi_i e^{-jE_i t/\hbar} = j\hbar \sum_{i=1}^{n} \dot{c}_i \psi_i e^{-jE_i t/\hbar}$$

We multiply both parts of the equation by any of the eigen-functions ψ_k^* of the operator $\mathscr{H}_0$. Then, because of the relations (I-8) and (I-9):

$$\psi_k^* \psi_i = 0 \qquad i \neq k$$
$$\psi_k^* \psi_k = 1$$

we obtain

$$j\hbar \dot{c}_k e^{-jE_k t/\hbar} = \sum_{i=1}^{n} c_i \left(\psi_k^* \mathscr{H}' \psi_i \right) e^{-jE_i t/\hbar} \tag{I-102}$$

These equations, which are written for all values of k, form a system of differential equations, which are equivalent to the initial Schrödinger equation. We will solve these equations approximately by assuming that at the initial moment the system was in the state ψ_j so that, for this moment, all the values $c_i = 0$, with the exception c_j, which equals unity. The approximation consists of replacing all the numbers c_i in equation (I-102) by their initial values. The solution thus obtained will be valid for moments of time which differ little from the initial moment. Thus, in the first approximation,

$$j\hbar \frac{dc_k}{dt} e^{-jE_k t/\hbar} = \left(\psi_k^* \mathscr{H}' \psi_j \right) e^{-jE_j t/\hbar}$$

or

$$c_k = (j\hbar)^{-1} \int_0^t \left(\psi_k^* \mathscr{H}' \psi_j \right) e^{-j\omega_{jk} t} \tag{I-103}$$

where

$$\omega_{jk} = \frac{E_j - E_k}{\hbar} \tag{I-104}$$

The square of the module of the number c_k evidently determines the probability of a transition of the system from the state ψ_j to the state ψ_k:

$$P_{j \to k} = |c_k|^2$$

We should note that under the integral there is the element of the k-th row and the j-th column of the matrix, which is the perturbation operator, which is obtained by means of the eigenfunctions of the operator $\mathcal{H}_0$.

As an example, let us examine transitions between levels of a noninteracting pair of spins under the action of an exciting external field:

$$H_x = H_1 \cos \omega t$$
$$H_y = - H_1 \sin \omega t$$

The operator of the perturbation energy in this case has the form

$$\mathcal{H}' = - \gamma H_1 [I_x \cos \omega t - I_y \sin \omega t]$$

where

$$I_x = I_{x1} + I_{x2} \qquad I_y = I_{y1} + I_{y2}$$

By calculating the matrix elements of this operator by means of the eigenfunctions (I-99) of the operator $\mathcal{H}_0$ in accordance with the rules presented above, we obtain

$$\mathcal{H}' = - \frac{\gamma \hbar H_1}{\sqrt{2}} \begin{pmatrix} 0 & e^{j\omega t} & 0 & 0 \\ e^{-j\omega t} & 0 & e^{j\omega t} & 0 \\ 0 & e^{-j\omega t} & 0 & 0 \\ 0 & 0 & 0 & 0 \end{pmatrix}$$

We see that the elements of the fourth column and the fourth row equal zero. This means that transitions to the state ψ_4 and from the state ψ_4 are impossible. As was stated, the function ψ_4 describes a singlet state of the system.

Transitions are possible only between levels of the triplet state, and since the elements of $\mathcal{H}'_{31}$ and $\mathcal{H}'_{13}$ equal zero a direct transition between the states ψ_1 and ψ_3 is impossible.

The mathematical system presented above may be used to describe interacting spins.

Since the functions $\alpha\alpha, (1/\sqrt{2})(\alpha\beta + \beta\alpha)$, and $(1/\sqrt{2})(\alpha\beta - \beta\alpha)$ are simultaneously eigenfunctions of the operators $\mathbf{I}_z$ and $\mathbf{I}^2$, it is convenient to use them to describe the interacting system.

As is evident from (I-91), the energy operator of such a system consists of two parts: the operator of a noninteracting pair

$$\mathscr{H}_0 = -\gamma H_0 (\mathbf{I}_{z1} + \mathbf{I}_{z2}) = -\gamma H_0 \mathbf{I}_z$$

for which the functions (I-99) are eigenfunctions, and the interaction operator:

$$\mathscr{H}_{int} = -\gamma^2 r^{-3} [3 (\mathbf{I}_1 \overleftarrow{r_0}) (\mathbf{I}_2 \overleftarrow{r_0}) - \mathbf{I}_1 \mathbf{I}_2]$$

The matrix representation of $\mathbf{I}_z$ was found above by means of the functions (I-99). Therefore, the matrix representation of the first parts of the energy operator of the system has the form

$$\mathscr{H}_0 = -\gamma H_0 h \begin{pmatrix} 1 & 0 & 0 & 0 \\ 0 & 0 & 0 & 0 \\ 0 & 0 & -1 & 0 \\ 0 & 0 & 0 & 0 \end{pmatrix}$$

We should remember that the fourth column and the fourth row of this matrix refer to the singlet state.

The matrix representation of the operator $\mathscr{H}_{int}$ may be found from the following considerations. In Cartesian coordinates we may write

$$\mathbf{I}_1 = \overleftarrow{i} \mathbf{I}_{x1} + \overleftarrow{j} \mathbf{I}_{y1} + \overleftarrow{k} \mathbf{I}_{z1}$$
$$\mathbf{I}_2 = \overleftarrow{i} \mathbf{I}_{x2} + \overleftarrow{j} \mathbf{I}_{y2} + \overleftarrow{k} \mathbf{I}_{z2} \qquad \text{(I-105)}$$
$$\overleftarrow{r_0} = \overleftarrow{i} \cos\alpha + \overleftarrow{j} \cos\beta + \overleftarrow{k} \cos\gamma$$

where $\overleftarrow{i}$, $\overleftarrow{j}$, and $\overleftarrow{k}$ are the unit vectors along the coordinate axes; α, β, and γ are the angles formed by the direction from one nucleus to the other and the x, y, and z coordinate axes, respectively.

These angles are conveniently expressed in terms of the polar and azimuthal angles of a spherical system of coordinates:

$$\cos \alpha = \sin \theta \cos \varphi$$
$$\cos \beta = \sin \theta \sin \varphi \qquad\qquad (I\text{-}106)$$
$$\cos \gamma = \cos \theta$$

By using relations (I-105) and (I-106) in the expression for the interaction operator and replacing $\cos \varphi$ by $\frac{1}{2}(e^{j\varphi} + e^{-j\varphi})$ and $\sin \varphi$ by $\frac{1}{2}j(e^{j\varphi} - e^{-j\varphi})$, we obtain after reduction of like terms

$$\mathscr{H}_{int} = [I_{z1}I_{z2} - \frac{1}{4}(I_1^- I_2^+ + I_1^+ I_2^-)]\, \gamma^2 Y_0 -$$
$$- \frac{3}{2}(I_1^+ I_{z2} + I_{z1}I_2^+)\, \gamma^2 Y_1 - \frac{3}{2}(I_1^- I_{z2} + I_{z1}I_2^-)\, \gamma^2 Y_1^* -$$
$$- \frac{3}{4} I_1^+ I_2^+ \gamma^2 Y_2 - \frac{3}{4} I_1^- I_2^- \gamma^2 Y_2^* \qquad (I\text{-}107)$$

where the operators I_i^- and I_i^+ are defined by the relations

$$I_i^{\mp} = (I_{x1} \mp jI_{yi})$$

and the values Y_0, Y_1, and Y_2 are functions of the coordinates:

$$\left.\begin{aligned} Y_0 &= r^{-3}(1 - 3\cos^2 \theta) \\ Y_1 &= r^{-3}\sin \theta \cos \theta e^{-j\varphi} \\ Y_2 &= r^{-3}\sin^2 \theta e^{-2j\varphi} \end{aligned}\right\} \qquad (I\text{-}108)$$

By operating on the functions (I-99) with the operator $\mathscr{H}_{int}$ and multiplying the results obtained by each of these functions, we find the matrix representation of the interaction operator:

$$\mathscr{H}_{int} = +\frac{1}{4}\gamma^2\hbar^2 \begin{pmatrix} Y_0 & -3\sqrt{2}Y_1 & -3Y_2 & 0 \\ -3\sqrt{2}Y_1^* & -2Y_0 & 3\sqrt{2}Y_1 & 0 \\ -3Y_2^* & 3\sqrt{2}Y_1^* & Y_0 & 0 \\ 0 & 0 & 0 & 0 \end{pmatrix} \qquad (I\text{-}109)$$

On comparing the operators $\mathscr{H}_0$ and $\mathscr{H}_{int}$ we see first of all that the singlet state is not affected by the interaction. This is understandable as zero total spin and, consequently, zero magnetic moment corresponds to the singlet state.

Then the interaction operator may be regarded as the sum of two operators, the first of which contains only the diagonal elements, while the second contains the other elements.

If the coordinates r, θ, and φ are constant, then the operation of the nondiagonal elements may be ignored in the first approximation, since the change in the level produced by them is small. In this case the matrix of the operator of the total energy is given by

$$\mathscr{H} = -\gamma\hbar \begin{pmatrix} H_0 - \frac{1}{4}\gamma\hbar Y_0 & 0 & 0 & 0 \\ 0 & +\frac{1}{2}\gamma\hbar Y_0 & 0 & 0 \\ 0 & 0 & -H_0 - \frac{1}{4}\gamma\hbar Y_0 & 0 \\ 0 & 0 & 0 & 0 \end{pmatrix} \qquad \text{(I-110)}$$

It is obvious that the initial levels are perturbed by the interaction. On the basis of previous reasoning we can readily conclude that under the action of an external, time-dependent perturbation transitions are possible only between the states ψ_1 and ψ_2 and between ψ_2 and ψ_3.

The first transition is accompanied by the absorption or emission of a quantum of energy:

$$\hbar\omega_1 = \gamma\hbar \left[H_0 + \frac{3}{4}\gamma\hbar r^{-3}(3\cos^2\theta - 1) \right] \qquad \text{(I-111)}$$

and the second is connected with the absorption or emission of a quantum:

$$\hbar\omega_2 = \gamma\hbar \left[H_0 - \frac{3}{4}\gamma\hbar r^{-3}(3\cos^2\theta - 1) \right] \qquad \text{(I-112)}$$

We should note that from the relation (I-89) it is possible to find the longitudinal component of the field produced by one nucleus at the point in space of the second. It is found to equal

$$h = \pm\frac{1}{2}\gamma\hbar r^{-3}(3\cos^2\theta - 1) \qquad \text{(I-113)}$$

i.e., less by a factor of one and a half than the effective value of the field produced by the interaction of like nuclei (I-111) and (I-112).

By repeating the calculations performed for pairs of spins with the spin number $\frac{1}{2}$ but with different gyromagnetic ratios (γ_1 and γ_2), we would obtain the resonance conditions in the form

$$\hbar\omega_{1,2} = \gamma_1 h \left[H_0 \pm \frac{1}{2}\gamma_2 r^{-3}h\,(3\cos^2\theta - 1) \right] \qquad \text{(I-114)}$$

Thus, a system of interacting spins may be described formally by the behavior of one spin in the magnetic fields of neighboring nuclei, by assuming that the magnetic moments of like neighboring nuclei are increased by a factor of one and a half.

If the coordinates r, θ, and φ change sufficiently rapidly with time, then the perturbation of the initial levels is reduced substantially and the resonance region is determined by the amplitudes of the zero or close-to-zero frequencies of the function Y_0. The non-diagonal matrix elements of the interaction operator then begin to play a part, causing transitions between energy levels. Thus, the matrix

$$\mathscr{H}'' = +\frac{1}{4}\gamma^2 h^2 \begin{pmatrix} 0 & -3\sqrt{2}Y_1 & -3Y_2 & 0 \\ -3\sqrt{2}Y_1^* & 0 & 3\sqrt{2}Y_1 & 0 \\ -3Y_2^* & 3\sqrt{2}Y_1^* & 0 & 0 \\ 0 & 0 & 0 & 0 \end{pmatrix} \qquad \text{(I-115)}$$

describes the relaxation mechanism.

The form of this matrix shows that transitions are possible between any levels of the triplet state under the influence of the interaction.

Since the matrix elements of the operator $\mathscr{H}''$ (i.e., the functions Y_1 and Y_2) are random functions of time, formula (I-103) may not be used directly for calculating the probabilities of transitions. These probabilities depend on the actual form of the functions Y_1 and Y_2. Therefore, in a given case, it is meaningful to speak only of the mean probabilities of transition.

Let us denote the mean probability of transition in unit time from the state ψ_i to the state ψ_j by Q_{ij}. By means of the theory of transitions presented above and the theory of random processes it may be shown [2] that this probability equals

$$Q_{ij} = \hbar^{-2} \int\limits_{-\infty}^{\infty} \overline{\mathcal{H}''_{ji}(t)\, \mathcal{H}''_{ji}(t+\tau)}\, e^{-j\omega_{ij}\tau}\, d\tau \qquad \text{(I-116)}$$

where $\mathcal{H}''_{ji}$ is the element of the j-th row and the i-th column of the interaction operator, and the line over the product indicates averaging with respect to time.

Let us examine a definite spin of an interacting pair and determine for it the probability of a transition w_{21} in unit time from the lower level to the upper. If the state of the system is described by the function ψ_3, then the selected spin is in the upper level with certainty; each of the two levels may correspond to the state ψ_2 with a probability of $\frac{1}{2}$. Then the transitions $\psi_3 \rightarrow \psi_1$ and $\psi_2 \rightarrow \psi_1$ convert the spin from the upper level to the lower. Moreover, the transition $\psi_3 \rightarrow \psi_2$ converts it to the lower level with a probability of $\frac{1}{2}$. Therefore, the probability sought (taking into account the fact that the spin is in state ψ_2 or ψ_3 with a probability of $\frac{1}{2}$) equals

$$w_{21} = \frac{1}{2}\left[Q_{31} + \frac{1}{2} Q_{21} + \frac{1}{2} Q_{32} \right] \qquad \text{(I-117)}$$

By substituting here the values of Q_{ij} calculated by means of formulas (I-115) and (I-116) we find

$$w_{21} = \frac{9}{16}\gamma^4\hbar^2\left[\int\limits_{-\infty}^{\infty} \overline{Y_1(t)Y_1^*(t+\tau)}e^{-j\omega_0\tau}d\tau + \frac{1}{2}\int\limits_{-\infty}^{\infty} \overline{Y_2(t)\,Y_2^*(t+\tau)}\,e^{-2j\omega_0\tau}\,d\tau \right] \text{(I-118)}$$

By definition (page 35) the values $\overline{Y_i(t)Y_i^*(t+\tau)}$ are autocorrelation functions of the random functions Y_1 and Y_2, and the integrals appearing in the expression obtained are the values of the energy spectra of these functions at the frequencies ω_0 and $2\omega_0$.

Therefore,

$$w_{21} = \frac{9}{16}\gamma^4\hbar^2\left[I_1(\omega_0) + \frac{1}{2}I_2(2\omega_0) \right] \qquad \text{(I-119)}$$

Chapter II

Recording and Interpretation
of NMR Spectra

1. Characteristics of NMR Spectra
of Liquids

Chemical Shift and Indirect Spin—Spin Interaction. The characteristics of the NMR spectra of liquids are
connected with the great freedom and the intensity of motion of the
molecules, as a result of which the internal steady fields due to the
indirect interaction of the magnetic moments of the atomic nuclei
(the so-called direct spin—spin interaction) are averaged out practically to zero, and the resonance lines become very narrow. As a
result of this it is possible to detect a small difference in the resonance frequencies of atomic nuclei of the same type in different
parts of a molecule or in different molecules.

In atoms and molecules in an external magnetic field there
arises an additional motion of the electrons, which is equivalent to
electrical currents flowing around a closed circuit in a plane perpendicular to the external field. The induced currents create an
additional field, which acts on the nucleus in a direction which is
always opposite to the direction of the external field.

The intensity of the induced currents is proportional to the
external field and therefore the magnetic field actually acting upon
the nucleus equals

$$H = H_0 (1 - \sigma) \qquad \text{(II-1)}$$

where σ is a value characterizing the relation between the external
and induced fields and is called the screening constant.

The value σ is usually about 10^{-6} for protons and depends on
the electronic environment of the nuclei.

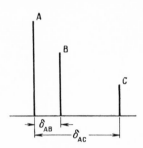

Fig. 3. Spectrum of the molecule A_3B_2C caused by the presence of chemical shifts.

For nuclei of a given type in different electronic environments, i.e., either in different molecules or in nonequivalent positions of one molecule, resonance will be observed at several different excitation frequencies of the field or at several different values of the external field if the excitation frequency is constant. This phenomenon is called the c h e m i c a l s h i f t of resonance. The chemical shift of a given nucleus or group of equivalent nuclei is usually measured relative to the resonance field of nuclei of the same type in some standard substance and is expressed in relative units:

$$\delta' = \frac{H_a - H_s}{H_s} \tag{II-2}$$

where H_a and H_s are the resonance values of the external field of the analysis and standard nuclei.

The absolute value of the shift in frequency units is sometimes given:

$$\frac{\delta}{2\pi} = \frac{\gamma}{2\pi}(H_a - H_s) \tag{II-3}$$

There is a relation between the relative shift and the screening constants of the analysis and standard nuclei:

$$\delta' = \sigma_s - \sigma_a \tag{II-4}$$

The chemical shift depends on the electronic structure of the molecules, the nature of the chemical bond, exchange processes, the concentration and temperature of the substance investigated, the character of the solvent, etc.

More or less successful attempts have been made at the theoretical calculation of magnetic screening, but in the main the chemical shift is measured experimentally.

If two or several nuclei in a molecule are screened identically, then they have the same shift, and their resonance is observed at the same value of the external field. Such nuclei usually occupy chemically equivalent positions in the molecules.

From what has been stated it follows that the nuclear resonance spectrum of a molecule caused by chemical shifts will contain as many resonance lines as there are chemically nonequivalent
groups of nuclei of a given type in it, and the intensity of each line
will be proportional to the number of nuclei in the corresponding
group (Fig. 3).

However, it is found that the nuclei of nonequivalent groups
may interact with each other. This interaction is explained by the
intereffect of the magnetic moments of the nonequivalent nuclei
transmitted from one nucleus to another by the paired valence electrons and is called i n d i r e c t   s p i n — s p i n   i n t e r a c t i o n.

This interaction is not averaged out by thermal motion as a
rule, does not appear in systems of equivalent nuclei, and does not
depend on the magnitude of the external field.

In this case the complication of the form of the spectra may
be explained if we assume that the energy of interaction is expressed by the relation

$$E_{ij} = -K_{ij}\overleftarrow{\mu}_i\overleftarrow{\mu}_j \tag{II-5}$$

where K_{ij} is a constant for a given molecule, which depends on its
electronic structure.

However, the indirect spin—spin interaction is usually characterized by the constant

$$J_{ij} = \gamma_i\gamma_j\hbar K_{ij} \tag{II-6}$$

which has the dimensions of rad · sec^{-1} or the constant $J_{ij}/2\pi$,
which has the dimensions of frequency (Hz).

The observed values of the spin—spin interaction constant
for nuclei of different types lies in the range from 1000 Hz to very
low values.

In the simplest cases, the effect of the indirect spin—spin interaction on the character of the spectrum may be described in the
following way.

For example, let the molecules of a given liquid contain two
groups of protons with a chemical shift δ. Let the first group con-

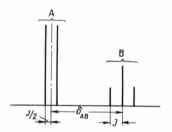

Fig. 4. Spectrum of the molecule A_2B_1 caused by the presence of a chemical shift and a weak spin—spin interaction between the nuclei of group A and group B.

tain two spins and the second, one. In the absence of spin—spin interaction the spectrum of the protons of the liquid would consist of two lines with a ratio of intensities of $2:1$, lying at a distance of $(\gamma/2\pi)(\sigma_1 - \partial_2)H_0$ Hz. The interaction results from the fact that at the points in space of the nuclei of one group there acts an additional field, which is determined by the orientation of the spins in the other.

At each given moment of time, in half (if we neglect the small difference due to the Boltzmann distribution) of the total number of molecules, the spin of the second group is oriented along the field and, in the other half, against the field. As a result, in half of all the molecules, on nuclei of the first group there will act an additional field of one sign and, in the other, the opposite sign, and the line of the protons of the first group is split into two lines of equal intensity with a separation of $J/2\pi$. On the other hand, the nuclei of the first group may form four equally probable orientations relative to the field:

Therefore, under the action of the protons of the first group the spectrum of the protons of the second is split into three lines with relative intensities of $1:2:1$ with a distance of $J/2\pi$ between neighboring lines (Fig. 4).

The picture described is found to hold for low values of the spin—spin interaction constant J in comparison with the chemical shift

$$\delta = \gamma\,(\sigma_1 - \sigma_2)\,H_0$$

With an increase in the energy of the spin—spin interaction, the nuclei in nonequivalent groups begin to behave as couple systems and the picture of the spectrum is still more complicated.

Thus, the analysis and interpretation of a spectrum in liquids with complex molecules are relatively simple with a small number of interacting groups of nonequivalent nuclei and with low ratios J/δ. In more complex cases the description of the spectra requires the use of quantum mechanical calculations.

The problem of calculating a spectrum usually consists of determining the eigenvalues of the energy of the system (from data on chemical shifts and the spin—spin interaction constant), calculating the resonance frequencies of the transitions between the energy levels found, and finding the probabilities of a transition under the action of an external exciting field.

Let us examine the method of calculating the spectrum on the example of a molecule containing two nonequivalent nuclei of the same type A and B with a spin number of $\frac{1}{2}$.

It is assumed that the molecule contains no other magnetic nuclei.

The first step in the calculation consists of formulating the operator of the total energy of the system. The energy of interaction of the nuclei A and B with the external field is given by

$$E_A = -\overleftarrow{\mu}_A \overleftarrow{H}_A = -\gamma(1-\sigma_A)H_0 J_{zA} = -\omega_A J_{zA}$$
$$E_B = -\overleftarrow{\mu}_B \overleftarrow{H}_B = -\gamma(1-\sigma_B)H_0 J_{zB} = -\omega_B J_{zB}$$

(II-7)

The values ω_A and ω_B are the resonance frequencies of the nuclei A and B in the field H_0 in the absence of interaction. The interaction energy equals

$$E_{AB} = -K\overleftarrow{\mu}_A\overleftarrow{\mu}_B = -\gamma^2 K \overleftarrow{J}_A \overleftarrow{J}_B = -J\hbar^{-1}\overleftarrow{J}_A\overleftarrow{J}_B$$

(II-8)

Therefore, the operator of the total energy of the interaction of the system will have the form

$$\mathscr{H} = -(\omega_A \mathbf{I}_{zA} + \omega_B \mathbf{I}_{zB} + J\hbar^{-1}\mathbf{I}_A\mathbf{I}_B)$$

(II-9)

The next step in the calculation consists of finding the eigenvalues and the eigenfunctions of the operator $\mathscr{H}$.

We saw above that to describe the states of a system of two spins it is possible to use the functions $\alpha\alpha$, $\alpha\beta$, $\beta\alpha$, and $\beta\beta$.

By operating with the operator $\mathcal{H}$ on these functions in accordance with the rules described previously (see page 61) we obtain

$$\left.\begin{aligned}
\mathcal{H}\,(\alpha\alpha) &= -\frac{1}{2}\,\hbar\left(\omega_A + \omega_B + \frac{1}{2}\,J\right)\alpha\alpha \\
\mathcal{H}\,(\alpha\beta) &= -\frac{1}{2}\,\hbar\left(\omega_A - \omega_B - \frac{1}{2}\,J\right)\alpha\beta - \frac{1}{2}\,\hbar J\beta\alpha \\
\mathcal{H}\,(\beta\alpha) &= -\frac{1}{2}\,\hbar\left(-\omega_A + \omega_B - \frac{1}{2}\,J\right)\beta\alpha - \frac{1}{2}\,\hbar J\alpha\beta \\
\mathcal{H}\,(\beta\beta) &= -\frac{1}{2}\,\hbar\left(-\omega_A - \omega_B + \frac{1}{2}\,J\right)\beta\beta
\end{aligned}\right\} \qquad \text{(II-10)}$$

We see that the functions $\alpha\alpha = \psi_1$ and $\beta\beta = \psi_4$ are already eigenfunctions of the operator $\mathcal{H}$ with the eigenvalues

$$\begin{aligned}
E_1 &= -\frac{1}{2}\,\hbar\left(\omega_A + \omega_B + \frac{1}{2}\,J\right) \\
E_4 &= -\frac{1}{2}\,\hbar\left(-\omega_A - \omega_B + \frac{1}{2}\,J\right)
\end{aligned} \qquad \text{(II-11)}$$

We seek two other eigenfunctions in the form of linear combinations of the functions $\alpha\beta$ and $\beta\alpha$:

$$\psi = a\cdot\alpha\beta + b\cdot\beta\alpha \qquad \text{(II-12)}$$

where

$$|a|^2 + |b|^2 = 1 \qquad \text{(II-13)}$$

Since ψ is an eigenfunction of $\mathcal{H}$, then it must be that

$$\mathcal{H}\psi = E\psi$$

where E is an eigenvalue of the operator $\mathcal{H}$.

By operating on ψ with the operator $\mathcal{H}$ and using the second and third relations from the system (II-10), we obtain

$$a\left[-\frac{1}{2}\,\hbar\left(\omega_A - \omega_B - \frac{1}{2}\,J\right)\alpha\beta - \frac{1}{2}\,\hbar J\beta\alpha\right] +$$
$$b\left[-\frac{1}{2}\,\hbar\left(-\omega_A + \omega_B - \frac{1}{2}\,J\right)\beta\alpha - \frac{1}{2}\,\hbar J\alpha\beta\right] = E\left[a\cdot\alpha\beta + b\cdot\beta\alpha\right]$$

By equating coefficients of like functions in the right- and left-hand parts of the equations we will have

$$a \left[\left(\omega_A - \omega_B - \frac{1}{2} J \right) + \frac{2E}{\hbar} \right] + bJ = 0 \atop aJ + b \left[\left(-\omega_A + \omega_B - \frac{1}{2} J \right) + \frac{2E}{\hbar} \right] = 0 \right\} \qquad \text{(II-14)}$$

This system of linear homogeneous equations has a solution if its determinant equals zero. Hence, we can find the eigenvalues of the energy operator sought:

$$E_2 = \frac{1}{2} \hbar \left(\frac{1}{2} J - \sqrt{\delta^2 + J^2} \right)$$
$$E_3 = \frac{1}{2} \hbar \left(\frac{1}{2} J + \sqrt{\delta^2 + J^2} \right) \qquad \text{(II-15)}$$

and then, from equations (II-14), taking into account the condition (II-13), we determine a, b, and the eigenfunctions:

$$\psi_2 = \frac{1}{\sqrt{1+Q^2}} (\alpha\beta + Q\beta\alpha)$$
$$\psi_3 = \frac{1}{\sqrt{1+Q^2}} (Q\alpha\beta - \beta\alpha) \qquad \text{(II-16)}$$

where

$$\delta = \omega_A - \omega_B \quad \text{and} \quad Q = \frac{J}{\delta + \sqrt{\delta^2 + J^2}} \qquad \text{(II-17)}$$

Thus, the system examined has four eigenstates, corresponding to the four possible values of the energy.

The last step in the calculation consists of calculating the probabilities of transitions between the different states under the action of the exciting field:

$$H_x = H_1 \cos \omega t; \qquad H_y = -H_1 \sin \omega t$$

By using the theory of perturbations presented previously (page 66), it may be shown [6] that the probability of the transition $\psi_j \rightarrow \psi_i$ in unit time, accompanied by the absorption of energy of this field, equals

$$Q_{ji} = \frac{1}{4} \, \omega_1^2 \, | \, \psi_i^* \mathbf{I}^- \psi_j \, |^2 \tag{II-18}$$

where

$$\mathbf{I}^- = (\mathbf{I}_{xA} + \mathbf{I}_{xB}) - j \, (\mathbf{I}_{yA} + \mathbf{I}_{yB}) \tag{II-19}$$

and transitions occur at frequencies of the exciting field

$$\omega_{ij} = \frac{E_i - E_j}{\hbar} \tag{II-20}$$

By operating with the operator $\mathbf{I}^-$ on the functions $\alpha\alpha$, ψ_2, ψ_3, and $\beta\beta$ and taking into account relations (II-16), we find

$$\mathbf{I}^- (\alpha\alpha) = \hbar \, (\alpha\beta + \beta\alpha) = \hbar \left(\frac{1+Q}{\sqrt{1+Q^2}} \, \psi_2 + \frac{Q-1}{\sqrt{1+Q^2}} \, \psi_3 \right)$$

$$\mathbf{I}^- (\psi_2) = \hbar \, \frac{Q+1}{\sqrt{1+Q^2}} \, \beta\beta$$

$$\mathbf{I}^- (\psi_3) = \hbar \, \frac{Q-1}{\sqrt{1+Q^2}} \, \beta\beta \tag{II-21}$$

$$\mathbf{I}^- (\beta\beta) = 0$$

By means of (II-18), from these expressions it is possible to find the probabilities of transitions sought.

The calculation of these probabilities makes it possible to establish which transitions will be possible in general, i.e., the frequencies at which resonance lines will be observed in a given field. Moreover, in most cases it may be assumed that the intensity of a line at a given resonance frequency is proportional to the probability of the corresponding transition.

The results of calculations for the system examined may be written in the form of the following table (for $R = \sqrt{\delta^2 + J^2}$):

Transition	Resonance frequency of exciting field	Intensity of absorption line
$\psi_1 \longrightarrow \psi_3$	$(\omega_A + \omega_B + R + J)/2$	$(Q-1)^2/4 \, (Q^2+1)$
$\psi_2 \longrightarrow \psi_4$	$(\omega_A + \omega_B + R - J)/2$	$(Q+1)^2/4 \, (Q^2+1)$
$\psi_1 \longrightarrow \psi_2$	$(\omega_A + \omega_B - R + J)/2$	$(Q+1)^2/4 \, (Q^2+1)$
$\psi_3 \longrightarrow \psi_4$	$(\omega_A + \omega_B - R - J)/2$	$(Q-1)^2/4 \, (Q^2+1)$

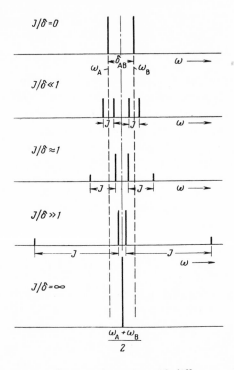

Fig. 5. Spectra of a two-spin system with different ratios J/δ.

In the absence of a spin—spin interaction ($J = 0$) the spectrum consists of two lines. With finite values of J it contains four lines, which are symmetrically placed relative to the mean value of the resonance frequencies of the nuclei A and B. With an increase in J (or a decrease in δ) the intensity of the outer lines falls, while the intensity of the middle lines increases. At the same time the middle lines get closer, while the outer lines move out. At very high values of J (or very small values of δ) the spectrum changes into one line (Fig. 5).

In principle, the calculation of spectra of more complex molecules is carried out analogously. Let the molecule have N nuclei with a spin number $\frac{1}{2}$, and let ω_j be the resonance frequency of the j-th nucleus and J_{jk}, the constant of the spin—spin interaction of the j-th nucleus with the nucleus k.

In such a system the energy operator has the form

$$\mathscr{H} = -\left[\sum_{j=1}^{N} \omega_j \mathbf{I}_{zj} + \sum_{j>k}\sum J_{jk} h^{-1} \mathbf{I}_j \mathbf{I}_k\right]$$

In the construction of the matrix representation of the operator $\mathscr{H}$ as the base functions we use product functions consisting of N factors. In these functions a definite factor corresponds to each nucleus. The operator of a given nucleus operates only on its factor. For a system of N nuclei with a spin number of $\frac{1}{2}$ the total of these functions will be $n = 2^N$. For example, for a system of three spins the base functions will have the form

$\alpha\alpha\alpha$ $\beta\alpha\alpha$

$\alpha\alpha\beta$ $\beta\alpha\beta$

$\alpha\beta\alpha$ $\beta\beta\alpha$

$\alpha\beta\beta$ $\beta\beta\beta$

To find the matrix representation we operate on each of the base functions with the operator. The result of this operation will always be a sum consisting of the same functions, multiplied by certain coefficients:

$$\mathscr{H}\psi_j = \sum_{k=1}^{n} a_{kj}\psi_k$$

By multiplying this expression from the left by any of the base functions, for example ψ_m^*, we obtain the matrix element of the operator, which lies on the m-th row and the j-th column:

$$\psi_m^* \mathscr{H}\psi_j = a_{mj}$$

It is assumed that products of the same base functions equal unity ($\psi_m^*\psi_m = 1$), while products of different functions equal zero ($\psi_m\psi_k = 0$ when $m \neq k$).

Thus, all the matrix elements of the energy operator are found. The eigenvalues of the energy operator are found from the equation (I-14).

After multiplying out the determinant (I-14) we obtain an algebraic equation of the power $n = 2^N$ with respect to E_i. It has n roots, which are the eigenvalues of the operator sought. In simple cases this equation may be broken down into a series of equations of lower powers. In more complex cases it is necessary to resort to calculating the spectra on computers.

To calculate the probabilities of transitions between levels, we use formula (II-18), in which the operator I^- equals the sum of the corresponding operators for each nucleus:

$$\mathbf{I}^- = \sum_{i=1}^{N} \mathbf{I}_i^- = \sum_{i=1}^{N} (\mathbf{I}_{xi} - j\mathbf{I}_{yi})$$

With the previous stipulations, the probabilities of the transitions give the relative intensities of the lines, while the differences in energy of pairs of levels between which a transition occurs, divided by the reduced Planck constant, give the resonance frequencies, i.e., the position of the lines in the spectrum.

From the above account it follows that the structure of the spectrum depends on the number of groups of nonequivalent nuclei, the numbers of equivalent nuclei in the groups, and the ratio of the spin—spin interaction constants to the chemical shifts for each pair of groups.

The following notation system is used for convenience in classifying different spectra. Groups of nonequivalent like nuclei, for which the chemical shift and the spin—spin interaction constants are of the same order of values, are denoted by the first letters of the alphabet with the number of nuclei in the group denoted by the first letters of the alphabet with the number of nuclei in the group denoted by a subscript with each letter (AB, A_2B, A_3B, A_2B_2, etc.). The symbols X, Y, and Z denote nuclei whose resonance frequency is far (in comparison with J) from the resonance frequency of the nucleus A either as a result of a relatively large chemical shift or because of a difference in the gyromagnetic ratios.

The spectra of typical systems (AB, A_2B, A_3B, A_2B_2, A_3B_2, etc.) have been calculated and data on them may be found in the literature [6].

Width of Resonance Lines and Relaxation Times. As has already been mentioned, because of the great freedom and intensity of the intramolecular motion in liquids, the internal steady magnetic fields are averaged out to zero. Because of this the assumptions made previously (see page 41) are valid, and the absorption signal may be described by a function v, which corresponds to the stationary solution of the Bloch equations (I-88).

Under conditions of weak saturation,

$$\omega_1^2 T_1 T_2 \ll 1$$

the signal is proportional to the function of the form of the line $g(\omega)$ which, as follows from the information given on page 47, is the normalized Lorentz function in this case:

$$g(\omega) = \frac{T_2}{\pi} \cdot \frac{1}{1 + (\Delta\omega T_2)^2} \qquad (\text{II-22})$$

The width of the resonance line $\delta\omega$, which is defined as the distance between the points at the half-height, equals

$$\delta\omega = \frac{2}{T_2} \qquad (\text{II-23})$$

For molecules which consist of a pair of like nuclei, the energy spectrum $I_z(\omega)$, on which T_2 depends, is determined by the fluctuations of the effective longitudinal component of the internal field:

$$h = \pm \frac{3}{4} \gamma \hbar z^3 Y_0 \qquad (\text{II-24})$$

The correlation function of this component may be written in the form

$$\varphi_z(\tau) = \frac{9}{16} \gamma^2 \hbar^2 \overline{Y_0(t) \, Y_0(t+\tau)} \qquad (\text{II-25})$$

If we regard a molecule as a rigid sphere, which executes random rotations in a medium with a viscosity η at a temperature T, it may be shown [5] that φ_z depends on τ in accordance with an exponential law:

$$\varphi_z(\tau) = \frac{9}{16} \gamma^2 \hbar^2 |Y(0)|^2 e^{-|\tau|/\tau_c} \qquad (\text{II-26})$$

where

$$\tau_c = \frac{4\pi\eta r^3}{3kT} \qquad (\text{II-27})$$

The value τ_c, which is called the correlation time of the stochastic process, is the time in which the internal statistical bonds which characterize the process are weakened by a factor of e.

By averaging $|Y_0|^2$ over all the values of the angle θ and calculating the energy spectrum $I(\omega)$ of the function (II-26) from the formula (I-50), we obtain

$$I_z(\omega) = \frac{9}{10}\hbar^2\gamma^2 r^{-6}\frac{\tau_c}{1+(\omega\tau_c)^2} \tag{II-28}$$

Therefore, for the width of the line due solely to the spin—spin interaction we obtain

$$\delta\omega = \gamma^2 I_z(0) = \frac{9}{10}\hbar^2\gamma^4 r^{-6}\tau_c \tag{II-29}$$

For example, for water at 20°C ($\eta = 10^{-2}$ poise and $\tau = 1.5 \cdot 10^{-8}$ sec), we find $\tau_c = 3 \cdot 10^{-12}$ sec and $\delta\omega = 0.15$ rad/sec (= 0.05 Hz).

When the broadening of the line due to dipole—dipole interaction is small it is necessary to take into account the second term in the expression (I-63) for T_2, which equals $\frac{1}{2}T_1$. Thus, the additional broadening of the line produced by spin—lattice relaxation is also a value of the order of $1/T_1$.

The relaxation time T_1 for a pair of interacting spins is determined by the relation (see also page 72)

$$\frac{1}{T_1} = 2w_{21} = \frac{9}{8}\gamma^4\hbar^2\left[I_1(\omega_0) + \frac{1}{2}I_2(2\omega_0)\right]$$

where $I_1(\omega)$ and $I_2(\omega)$ are the energy spectra of the functions Y_1 and Y_2.

In analogy with the previous case, the correlation functions of these random values may be represented in the form

$$\varphi_1(\tau) = \overline{|Y_1|^2}e^{-|\tau|/\tau_c}$$

$$\varphi_2(\tau) = \overline{|Y_2|^2}e^{-|\tau|/\tau_c}$$

and by averaging $|Y_1|^2$ and $|Y_2|^2$ over all the values of the angles θ and φ, and by calculating the energy spectra, we obtain

$$\frac{1}{T_1} = \frac{3\gamma^4 \hbar^2}{10r^6} \left[\frac{\tau_c}{1+(\omega_0 \tau_c)^2} + \frac{2\tau_c}{1+(2\omega_0 \tau_c)^2} \right] \qquad \text{(II-30)}$$

The relation of $1/T_1$ to the correlation time τ_c is shown in Fig. 6. At low values of τ_c the value of $1/T_1$ coincides with the width of the line $\delta\omega$, determined only by spin—spin interaction.

With large correlation times the assumptions made previously (see page 41) become doubtful, since the correlation functions (II-25) do not then fall sufficiently rapidly. In this case, for calculating the width of the line it is possible to use the following considerations. All the components of the spectrum $I_z(\omega)$ with frequencies close to zero participate in the broadening of the line. It may be assumed that the limits of the frequencies considered will be the frequencies approximately equal to the width of the line itself (expressed in frequency units). Therefore, we assume that the mean value of the square of the width of the line $(\delta H)^2$ in field strength units must equal the sum of the components of this spectrum over the frequency range $\pm \gamma \delta H/2$:

$$(\delta H)^2 = \int_{-\gamma \delta H/2}^{\gamma \delta H/2} I_z(\omega)\, d\omega \qquad \text{(II-31)}$$

By expressing the width of the line in frequency units, we obtain

$$(\delta\omega)^2 = \frac{9\gamma^4 \hbar^2}{10r^6} \int_{-\delta\omega/2}^{\delta\omega/2} \frac{\tau_c\, d\omega}{1+(\omega\tau_c)^2} = \frac{9\gamma^4 \hbar^2}{5r^6} \tan^{-1} \frac{\delta\omega \tau_c}{2} \qquad \text{(II-32)}$$

At small values of τ_c ($\delta\omega\tau_c \ll 1$) the expression obtained coincides with (II-29). However, with long correlation times (a high viscosity of the liquid investigated or low temperatures) the width of the line ceases to depend on the correlation time.

Naturally the form of the line remains indefinite in these discussions, but it may be described as previously by the Lorentz function by making $T_2 = 2/\delta\omega$.

Exchange Processes. In addition to the above factors, which limit the narrowing of the resonance lines and the possibility

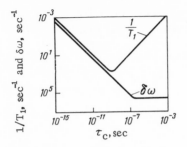

Fig. 6. Relation of the line width $\delta\omega$ and the reciprocal of the relaxation time $1/T_1$ to the correlation time τ_c.

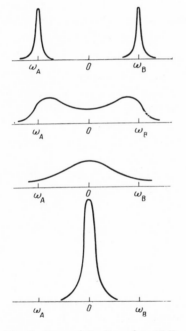

Fig. 7. Form of the line of an NMR spectrum with different rates of exchange of nuclei between two nonequivalent positions.

of observing the fine structure of spectra, distortion of spectra may occur as a result of exchange processes. Such processes may be exchange processes of like nuclei in chemically nonequivalent positions and the exchange of equivalent nuclei between identical molecules.

Let us assume that there is exchange of nuclei between positions A and B, and let τ_A and τ_B be the mean residence times of a nucleus in each of the states (lifetime). The probabilities of its residence in these states will be [7]

$$p_A = \frac{\tau_A}{\tau_A + \tau_B} \qquad p_B = \frac{\tau_B}{\tau_A + \tau_B} \qquad \text{(II-33)}$$

If the lifetimes τ_A and τ_B are long in comparison with $(\omega_A - \omega_B)^{-1}$, i.e., if the exchange proceeds slowly, then the spectrum consists of two narrow lines at the frequencies ω_A and ω_B.

With an increase in the exchange rate from low values, both signals are broadened and gradually approach. When

$$\frac{1}{\tau_A} + \frac{1}{\tau_B} = \frac{\omega_A - \omega_B}{\sqrt{2}} \qquad \text{(II-34)}$$

the signals merge into one broad signal (Fig. 7).

With a further increase in the exchange rate the line gradually becomes narrower.

If exchange proceeds rapidly, i.e., if the values τ_A and τ_B are small in comparison with the reciprocal of the chemical shift, then the spectrum merges into one line, which is observed at a frequency

$$\omega = p_A \omega_A + p_B \omega_B \qquad \text{(II-35)}$$

and has a width

$$\delta\omega = p_A \delta\omega_A + p_B \delta\omega_B \qquad \text{(II-36)}$$

An example of rapid exchange is the exchange of protons of H_2O molecules and the hydroxyl groups in CH_3COOH. The result of rapid exchange in this case is merging of the resonance lines of the protons of H_2O and the hydroxyl groups of CH_3COOH into one line and a linear relation between the shift of the resulting peak and the concentration of CH_3COOH. A second type of exchange process is the exchange of equivalent nuclei between identical molecules.

Let us assume that there is exchange of nuclei belonging to the groups B between molecules of the type A_2B. In the absence of exchange, the two nuclei of group A split the line of group B into three components, and the spectrum of this group consists of a triplet. With rapid exchange of the nuclei of group B they are attached to molecules which have an equal probability of any of the possible states of the spins in group A. As a result, the spectrum is merged into a singlet, which lies at the position of the middle line of the previous triplet. An example of this type of exchange process is the exchange of protons between OH groups in ethanol in the presence of a small amount of hydrochloric acid.

2. Characteristics of NMR Spectra

of Solids

In solids the freedom of thermal motion of the molecules is strongly limited, so that the exchange of energy between the spin system and the molecular system is hampered. Therefore, the longitudinal relaxation times T_1 must be very great and, as calculation shows, they must exceed the relaxation times for liquids by several orders. However, the experimentally observed relaxation times for crystalline solids are substantially less than the values which would be expected for a lattice of the given structure. This discrepancy may evidently be explained only by the presence of paramagnetic impurities. On the other hand, in solids containing reorienting groups of nuclei or reorienting molecules, these re-orientation processes explain quite satisfactorily the experimental relaxation times right down to very low temperatures.

As a rule, the width of the resonance lines for solids is very great, since the magnetic fields of dipole interaction are not aver-aged out in practice due to the limited freedom of motion of the molecules. Therefore, in many cases a solid may be regarded as a rigid system of immobile magnetic moments. The distribution of spins in this system will determine the form of the resonance line.

It is sometimes possible to obtain information on the struc-ture of the crystal lattice of the solid from this form.

The simplest case of calculation of the form of the resonance line in a solid is the case where the interacting magnetic moments form well-isolated pairs. As has been shown, the interaction is re-duced to one spin in each pair creating at the point in space of the other an additional internal field

$$h = \pm \frac{3}{4} \gamma h r^{-3} (3 \cos^2 \theta - 1) \qquad \text{(II-37)}$$

If the solid is a crystal in which the lines connecting the nu-clei in pairs form identical angles θ relative to the field, then the spectrum of this crystal will consist of two lines, broadened by in-termolecular interaction. If the frequency of the exciting field is constant $\omega = \omega_0$, then these lines are observed at values of the ex-ternal field H:

$$H = \frac{\omega_0}{\gamma} \pm \frac{3}{4} \gamma h r^{-3} (3 \cos^2 \theta - 1) \qquad \text{(II-38)}$$

The distance between the lines depends on the orientation of the crystal in the polarizing field. When $3 \cos^2 \theta = 1$, the lines merge into one, while when $\cos^2 \theta = 0$, the distance between the lines is maximal and equals $3 \gamma h r^{-3}$.

From this spectrum it is possible to determine the distance between the atomic nuclei.

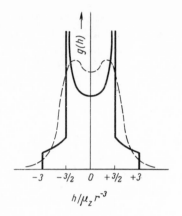

Fig. 8. Theoretical form of the line of an NMR spectrum caused by direct interaction of a pair of immobile spins. The broken line shows the experimental NMR line observed for a crystal of a substance with isolated pairs of nuclei.

If the solid examined is polycrystalline, i.e., if all possible directions from one nucleus to another are equally probable, then neglecting the intermolecular interaction, it is possible to calculate the form of the line.

Let us assume that all pairs of nuclei B_1—B_2 are arranged such that the nuclei B_1 are at the origin of the coordinates. Then the nuclei B_2 will be uniformly distributed over the surface of a sphere. If the number of nuclei is N, then on unit surface there will be $N(4\pi r^2)^{-1}$ nuclei. The number of nuclei B_2 (i.e., the number of pairs investigated) for which the angle θ lies within the range $\theta \pm d\theta/2$ will equal the area of a zone of the sphere $2\pi r^2 \sin\theta d\theta$, multiplied by the number of nuclei per unit surface:

$$dN = N (4\pi r^2)^{-1} 2\pi r^2 \sin\theta \, d\theta = \frac{1}{2} N \sin\theta \, d\theta$$

The relative fraction of these nuclei is

$$\frac{dN}{N} = \frac{1}{2} \sin\theta \, d\theta \tag{II-39}$$

On the other hand [from the definition of the function $g(h)$], this fraction must be $g(h)dh$. Therefore,

$$g(h) = \frac{1}{2} \sin\theta \frac{d\theta}{dh} = -\frac{1}{2} \frac{d\cos\theta}{dh} \tag{II-40}$$

By using equation (II-37), we find

$$\cos\theta = \sqrt{\frac{1}{3}\left(1 \pm \frac{h}{\frac{3}{2}\mu_z r^{-3}}\right)}$$

By differentiating the latter expression and substituting in equation (II-39), we obtain

$$g(h) = (6\sqrt{3}r^{-3})^{-1}\left(1 \pm \frac{h}{\frac{3}{2}\mu_z r^{-3}}\right)^{-\frac{1}{2}} \tag{II-41}$$

The form of this line is illustrated in Fig. 8. The distance between its maxima equals

$$\Delta h = 3\mu_z r^{-3}$$

In actual fact, the intermolecular interaction leads to smoothing out of this form and it will have the form of the curve denoted by a broken line in the figure.

Similarly it is possible to calculate the form of the line for the simplest combinations of three and four nuclei. However, in more general cases the calculations become too laborious. Therefore, for characterizing a resonance line we often use the moment of the second-order function describing the form of the resonance line; this is simply the mean value of the square of the strength of the internal field (for simplicity this concept is often called the "second moment of the line" and is denoted by ΔH_2^2).

In the general case the moment of the n-th order of the distribution function $g(h)$ is the expression

$$\Delta H_n^n = \int\limits_{-\infty}^{\infty} h^n g(h)\, dh \tag{II-42}$$

The second moment may be calculated most simply. Naturally it does not give the form of the line, but it is an important characteristic of the spectrum.

Let us assume that in a single crystal the magnetic moments form identical isolated groups of N nuclei. Neglecting the intermolecular interaction, we will take into account only the interaction inside the groups.

By definition, the second-order moment is the mean value of the square of the strength of the internal field acting on the nuclei of the group examined. The total field produced by all the other nuclei of the group acting on a nucleus j equals

$$h_j = \sum_{\substack{k=1 \\ k \neq j}}^{N} \frac{3}{2} \mu_{zh} r_{jk}^{-3} (3 \cos^2 \theta_{jk} - 1) \qquad (\text{II-43})$$

The mean value of the square of the strength of the field acting on the nucleus is given by

$$\bar{h}_j^2 = \left[\overline{\sum \frac{3}{2} \mu_{zh} r_{jk}^{-3} (3 \cos^2 \theta_{jk} - 1)} \right]^2 \qquad (\text{II-44})$$

where the line denotes averaging for all groups of nuclei.

The involution of the sum in brackets gives the squares of all the terms and all the possible products of them in pairs. The mean value of the sum of the pairs of products becomes zero on averaging for the different groups, since each of them will be encountered practically the same number of times with opposite signs. Consequently,

$$\bar{h}_j^2 = \sum_{\substack{k=1 \\ k \neq j}}^{N} \frac{9}{16} \gamma^2 \hbar^2 r_{jk}^{-6} (3 \cos^2 \theta_{jk} - 1)^2 \qquad (\text{II-45})$$

Finally, the mean value of the square of the strength of the internal field for all nuclei of the group, i.e., the second-order moment of the function $g(h)$ will be

$$\Delta H_2^2 = \frac{\sum_{j=1}^{N} \bar{h}_j^2}{N} = \sum_{j=1}^{N} \sum_{\substack{k=1 \\ k \neq j}}^{N} \frac{9 \gamma^2 \hbar^2 r_{jk}^{-6}}{16N} (3 \cos^2 \theta_{jk} - 1)^2 \qquad (\text{II-46})$$

The second-order moments have been calculated for many crystalline structures and data on them may be found in the appropriate literature.

In the case of polycrystalline structures the expression for ΔH_2^2 has to be averaged for all possible values of the angles θ_{jk}, bearing in mind the fact that, as was pointed out above, the number of pairs with the same angle will be proportional to $\frac{1}{2} \sin \theta d\theta$. Then

$$\overline{(3\cos^2\theta-1)^2} = \frac{1}{2}\int_0^{\pi} (3\cos^2\theta-1)^2 \sin\theta\, d\theta = \frac{4}{5} \qquad \text{(II-47)}$$

The second moment for a polycrystalline solid will equal

$$\Delta H_2^2 = \sum_{\substack{j=1 \\ }}^{N}\sum_{\substack{k=1 \\ k\neq j}}^{N} \frac{9\gamma^2\hbar^2 r_{jk}^{-6}}{20N} \qquad \text{(II-48)}$$

If the group examined contains, in addition to the N resonating nuclei, other atomic nuclei or magnetic moments of a different type (paramagnetic ions or radicals), then it is necessary to take into account their contribution to the second-order moment. For any magnetic moments this contribution may be calculated from the formula

$$\Delta H_2^2 = \frac{4}{15}N^{-1}\sum_{j=1}^{N}\sum_{f=1}^{M} \mu_f^2 r_{jf}^{-6} \qquad \text{(II-49)}$$

where μ_f^2 is the square of the length of the vector of the magnetic moment; M is the number of "foreign" magnetic moments.

The relations obtained are called Van Vleck's formulas and are used widely for practical calculations.

In addition to the limiting cases examined of identical and completely disordered orientations of the internuclear vectors, there may be intermediate structures in which the molecules are oriented along a certain axis (fiber) or in a plane (film). The calculation method presented may also be used in these cases.

It was shown above that the exact calculation of the form of the resonance line is practically impossible. However, it is sometimes convenient to approximate the observed curves by means of mathematical equations. The equations used most are those of Gaussian and Lorentzian curves, which have the form

$$g_G(h) = \frac{1}{b\sqrt{2\pi}}e^{-\frac{h^2}{2b^2}}; \qquad g_L(h) = \frac{1}{\pi b}\cdot\frac{1}{1+\left(\frac{h}{b}\right)^2}$$

The equations presented for both curves are normalized, i.e., the following condition holds for each:

$$\int_{-\infty}^{\infty} g(h)\,dh = 1$$

Both lines are bell-shaped curves, which are characterized by a single parameter b and have a maximum at h = 0.

For a Gaussian curve, the value b^2 simply equals the second-order moment

$$\Delta H_2^2 = \int_{-\infty}^{\infty} h^2 g_G(h)\,dh = b^2$$

For a Lorentzian curve the concept of a second-order moment is inapplicable, since the corresponding integral diverges. Then the parameter b is related by a simple equation to the width of the line, which is defined as the distance between the maxima of its derivative

$$\delta H_L = \frac{2}{\sqrt{3}}b$$

The width of a Gaussian line is expressed by means of its parameter equally simply:

$$\delta H_G = 2b$$

In many solids the molecules or groups of atoms may have several equivalent equilibrium positions corresponding to a potential energy minimum. A transition from one equilibrium position to another does not change the structure of the molecule, but requires the expenditure of a certain amount of energy. If the energy of thermal motion of the lattice is sufficiently great, then in such solids there will be reorientation of the molecules or groups of atoms between the equivalent equilibrium positions. The number of reorientations in unit time depends on the magnitude of the potential barrier E between the equilibrium positions and the temperature T, and is determined from the relation

$$f = f_\infty e^{-E/RT} \tag{II-50}$$

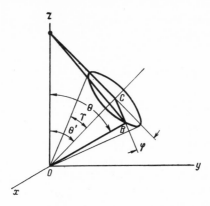

Fig. 9. Motion of the internuclear vector OB relative
to the reorientation axis OC. The angles θ' and γ are
constant, while the angles φ and θ change with time.

where f_∞ is the number of reorientations in unit time at a very high
temperature; R is the gas constant.

If the atomic nuclei for which the nuclear magnetic resonance
spectrum is observed are present in such reorienting groups, then
the spectrum will differ from the spectrum of a stationary structure.

Let us re-examine a pair of isolated nuclei, reorienting relative to a certain axis, which forms an angle θ' with the direction
of the field H_0. In this case the internuclear vector $\overleftarrow{r}$ has a constant length and forms a constant angle γ with the reorientation
axis, so that it moves over the surface of a cone during reorientation (Fig. 9).

During reorientation, the angle θ between the direction of the
field and the direction of the internuclear vector will change. Therefore, the component of the dipole—dipole interaction [see equations
(II-37)] will be a variable.

From the triangles AOB and ABC in Fig. 9, by using the
theorem of cones twice we obtain

$$\cos \theta = \cos \gamma \cos \theta' - \sin \gamma \sin \theta' \cos \varphi \qquad \text{(II-51)}$$

where φ is the angle of rotation of the internuclear vector.

By substituting this expression in (II-37), we find

$$h = \frac{3}{4} \mu_z r^{-3} [(3 \cos^2 \theta' - 1)(3 \cos^2 \gamma - 1) - 3 \sin 2\gamma \sin 2\theta' \cos \varphi$$

$$+ 3 \sin^2 \gamma \sin^2 \theta' \cos 2\varphi] \tag{II-52}$$

In the reorientations the angle φ is a random function of time, and therefore the zero and close-to-zero frequencies of the spectra of the last two terms of this expression may make a contribution to the value of h. If the reorientation occurs very rapidly, then these two terms are effectively averaged out and, therefore, the effect of these terms may be neglected. Then the spectrum will again consist of two lines,

$$h = \frac{3}{4} \mu_z r^{-3} (3 \cos^2 \theta' - 1)(3 \cos^2 \gamma - 1) \tag{II-53}$$

If the reorientation axis is perpendicular to the internuclear vector ($\gamma = 90°$), then the expression for h has the form

$$h = \frac{3}{4} \mu_z r^{-3} (3 \cos^2 \theta' - 1)$$

i.e., the spectrum is similar to the spectrum of a rigid structure, but has half the width.

For a single crystal containing identical isolated groups of molecules or atoms, which are reorienting relative to one axis, from the calculation (II-52) we find

$$h_j = \frac{3}{4} \mu_z (3 \cos^2 \theta' - 1) \left[\sum_{k=1}^{N} r_{jk}^{-3} (3 \cos^2 \gamma_{jk} - 1) \right]$$

and by carrying out an averaging process, analogous to the process in the derivation of formula (II-45), we obtain

$$\Delta H_2^2 = \frac{9}{64} \gamma^2 \hbar^2 N^{-1} (3 \cos^2 \theta' - 1)^2 \sum_{\substack{j=1}}^{N} \sum_{\substack{k=1 \\ k \neq j}}^{N} (3 \cos^2 \gamma_{ij} - 1)^2 r_{jk}^{-6} \tag{II-54}$$

It is evident that in comparison with a rigid structure, each term of the intramolecular interaction is changed by a factor of $\frac{1}{4}(3 \cos^2 \theta' - 1)^2$.

For a polycrystalline sample in which the angles θ' are distributed isotropically, the value $(3\cos^2\theta' - 1)$ must be replaced by its mean value $4/5$. Then

$$\Delta H_2^2 = \frac{9}{80}\gamma^2 h^2 N^{-1} \sum_{\substack{j=1}}^{N} \sum_{\substack{k=1 \\ k\neq j}}^{N} (3\cos^2\gamma_{jk} - 1)^2\, r_{jk}^{-6} \tag{II-55}$$

The relations presented are valid with sufficiently rapid reorientations, i.e., for samples at sufficiently high temperature.

With low rates of reorientation it is necessary to take into account the contribution to the value h from the components mentioned above. Then, in analogy with the calculation of the width of the line in liquids, it is possible to obtain the relation of the second-order moment to the reorientation rate (or to the temperature) in the form

$$(\Delta H_2^2)_f = (\Delta H_2^2)_\infty + [(\Delta H_2^2)_0 - (\Delta H_2^2)_\infty]\,\tan^{-1}\frac{\gamma\sqrt{(\Delta H_2^2)_f}}{2\pi f} \tag{II-56}$$

where $(\Delta H_2^2)_f$, $(\Delta H_2^2)_0$, and $(\Delta H_2^2)_\infty$ are the second moment of the line at the reorientation frequency, for a rigid structure, and for a very high reorientation rate, respectively.

3. Nonstationary Processes

In previous sections the main external factors determining NMR phenomena, namely the strength of the polarizing field H_0, the amplitude of the exciting field H_1, and its frequency ω, were assumed to be independent of time. If these values change sufficiently rapidly, then in the system of spins there arise transient processes which accompany the nuclear resonance phenomenon and are often used in the experimental technique. Processes of this type are primarily effects of passing rapidly through the resonance region, modulation effects, and spin echo.

The main task in an NMR experiment is the investigation of the form of the absorption line, i.e., recording the relation of the absorbed power to the field frequency difference. This is usually achieved by continuous measurement of the absorption signal v with a slow change in the field H_0 in the resonance region.

An analysis based on the solution of the Bloch equations
(I-88) for conditions of weak saturation ($\omega_1^2 T_1 T_2 \ll 1$) shows that
transient processes are absent in the system of spins and the observed signal corresponds to the true form of the line if the rate
of change of the field satisfies the inequality

$$\frac{dH_0}{dt} \ll \gamma (\delta H)^2$$

This inequality is called the condition of slow passage
of the resonance region.

If the slow passage condition does not hold, then the observed
signal is distorted by transient processes, and this appears as low-
frequency damped oscillations (wiggles) with a falling period,
which arise immediately after the center of the resonance region,
and whose form may be described approximately by the relation

$$v = v_0 e^{-\frac{t}{T_2}} \cos \left(\frac{1}{2} \gamma \frac{dH_0}{dt} t^2 \right)$$

Very interesting effects arise if an alternating low-frequency
sinusoidal component is superimposed on the steady polarizing
field. (This method is often used to increase the sensitivity and
accuracy of apparatus for recording NMR spectra and is called
modulation of the polarizing field.) If the frequency of the modulating component substantially exceeds the width of the resonance
line, expressed in frequency units, then there arise modulation effects, in which, for a given frequency ω of the exciting field, resonance may occur not at a single value of the polarizing field, equal
to ω/γ, but at a series of values, which are defined by the relation

$$H = \frac{\omega \pm k\omega_M}{\gamma}$$
$$(k = 0, 1, 2, \ldots)$$

where ω_M is the frequency of the modulating component.

The physical essence of these effects is that if the spin system is in a polarizing field H_0, which is not the resonance field
with respect to the frequency ω of the exciting field H_1 (i.e., if
$\omega \neq \gamma H_0$), then it is an oscillatory system with a natural oscilla-
tion frequency close to $|\omega - \gamma H_0|$. The superimposition of a modu-

lating component with a frequency ω_M means a periodic change in the natural oscillation frequency of the spin system and may produce resonance if the frequency ω_M is close to the frequency difference of the field $\omega - \gamma H_0$. This phenomenon is analogous to the so-called parametric excitation of resonance in oscillatory circuits. If in the Bloch equations we substitute

$$H_z = H_0 + H_M \cos \omega_M t$$

then the solution of this equation, for example, for a component of the magnetization vector M_x, will have the form

$$M_x = A(t) \cos \omega t + B(t) \sin \omega t$$

where $A(t)$ and $B(t)$ are found from the relation

$$B(t)+jA(t) = -\omega_1 M_0 T_2 \sum_{k=-\infty}^{\infty} \sum_{n=-\infty}^{\infty} \frac{J_k(\beta)[1-j(\Delta\omega+k\omega_M)T_2]J_n(\beta)e^{j(k-n)\omega_M t}}{1+(\Delta\omega+k\omega_M)^2 T_2^2 + \omega_1^2 T_1 T_2 J_k^2(\beta)} \qquad \text{(II-57)}$$

in which $J_k(\beta)$ is the Bessel function of the first kind of the parameter

$$\beta = \frac{\gamma H_M}{\omega_M}$$

This expression, which is valid for a modulation frequency greater than $1/T_1$, $1/T_2$, and ω_1, shows that the envelopes $A(t)$ and $B(t)$ of the high-frequency oscillations of the transverse components of the magnetization vector contain components which are harmonics of the modulation frequency, and their amplitudes are the absorption and dispersion signals for points of the field $H = (1/\gamma) \cdot (\omega \pm k\omega_M)$.

The normal resonance conditions are obtained when $k = 0$. The signals corresponding to $k = 0$ are called the c e n t r a l s i g - n a l s. For central signals with low modulation indices ($\beta \ll 1$), taking into account the fact that $J_1(\beta) = -J_{-1}(\beta)$, we obtain

$$B_0(t) = -\frac{\omega_1 M_0 T_2 J_0(\beta)}{1+(\Delta\omega T_2)^2 + \omega_1^2 T_1 T_2 J_0^2(\beta)} [J_0(\beta) - 2J_1(\beta)\Delta\omega T_2 \sin \omega_M t] \qquad \text{(II-58)}$$

$$A_0(t) = \frac{\omega_1 M_0 T_2 J_0(\beta)}{1+(\Delta\omega T_2)^2 + \omega_1^2 T_1 T_2 J_0^2(\beta)} [J_0 \Delta\omega T_2 - 2J_1(\beta) \sin \omega_M t] \qquad \text{(II-59)}$$

With these conditions the constant components of the envelopes $B(t)$ and $A(t)$ are, respectively, the absorption and dispersion signals, corresponding to the effective exciting field $\omega_1 J_0(\beta)$ with amplitudes increased by a factor $J_0(\beta)$. When $\beta = 0$, these expressions revert to the normal functions of v and u [see expression (I-88)]. The coefficients of the alternating components are also absorption and dispersion signals, but with amplitudes changed by a factor of $2J_1(\beta)$. We should note that both the envelopes of the alternating components have a 90° phase shift relative to the modulating field $(H_M \cos \omega_M t)$.

The signals corresponding to the value $k = 1$ or $k = -1$ are called the first lower or first upper side signals, respectively. Let us look at the alternating components of these signals at low values of β, when it is possible to consider only the first harmonics of the modulating frequency.

For $k = +1$ we obtain

$$B_1(t) = -\frac{\omega_1 M_0 T_2 J_1(\beta)}{1 + (\Delta\omega + \omega_M)^2 T_2^2 + \omega_1^2 T_1 T_2 J_1^2(\beta)} \times$$
$$[(J_0 + J_2) \cos \omega_M t + (\Delta\omega + \omega_M) T_2 (J_0 - J_2) \sin \omega_M t]$$

(II-60)

$$A_1(t) = \frac{\omega_1 M_0 T_2 J_1(\beta)}{1 + (\Delta\omega + \omega_M)^2 T_2^2 + \omega_1^2 T_1 T_2 J_1^2(\beta)} \times$$
$$[(\Delta\omega + \omega_M) T_2 (J_0 + J_2) \cos \omega_M t - (J_0 - J_2) \sin \omega_M t]$$

The amplitudes of the first harmonics of the modulating frequency are again either absorption signals or dispersion signals, which are observed at values of the external field

$$H_0 = \frac{\omega + \omega_M}{\gamma}$$

The effective excitation field for these signals equals $\omega_1 J_1(\beta)$, so that the saturation condition for them has the form

$$\omega_1^2 T_1 T_2 J_1^2(\beta) \gg 1$$

In the same envelope the absorption and dispersion signals correspond to different phases of the first harmonic of the modulating frequency, so that, by the use of appropriate electronic methods, they may be recorded independently. The upper side signals

have an analogous form. Side signals are practically absent at
very low values of β. With an increase in β, their intensity increases. At considerable values of β, second side signals begin
to appear (k = ±2).

The phenomenon describes is used in NMR equipment for
stabilization of the resonance conditions and for calibration of high-
resolution spectra. In this case the distance between the central
and side signals, which equals the modulation frequency, is used
as the scale for measuring the distances between the lines of the
spectrum.

Finally, the phenomenon of spin echo is of very great interest for practical applications.

The phenomenon of spin echo is observed as a sequence of
several short but intense pulses when the exciting field is switched
on with a constant strength of the polarizing field H_0.

Let us assume that at the initial moment of time the system
of spins is in the equilibrium state in a field H_0 with an inhomogeneity ΔH. At this moment the magnetization vector equals its
equilibrium value M_0 and is directed along the z axis (the direction
of the field H_0). If we now switch on the exciting field with the resonance frequency

$$H_x = H_1 \cos \omega_0 t \qquad H_y = -H_1 \sin \omega_0 t$$

then in the initial moment of time, which is short in comparison
with T_2, the movement of the vector $\overleftarrow{M}$ may be regarded as independent of the relaxation time and described by the equation

$$\frac{d\overleftarrow{M}}{dt} = \gamma \, [\overleftarrow{M}\overleftarrow{H}]$$

The solution of this equation shows that after the field has
been switched on, the vector $\overleftarrow{M}$ begins to deviate from the z axis
with a constant angular velocity $\omega_1 = \gamma H_1$ and will simultaneously
precess about this axis with a frequency $\omega_0 = \gamma H_0$. The movement
of the vector $\overleftarrow{M}$ is conveniently represented in a rotating system
of coordinates. Let the z' axis of this system coincide with the z
axis of the stationary coordinates and the x' axis with the direction
of the vector $\overleftarrow{H}_1$. In this system the motion of the vector in the initial moments after the field H_1 has been switched on will be rep-

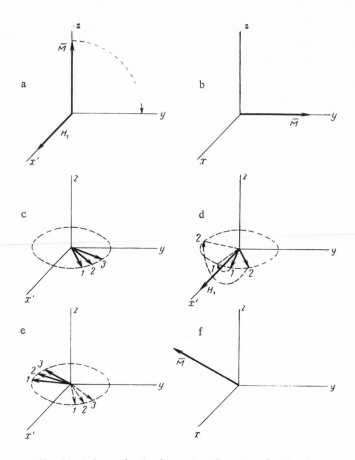

Fig. 10. Scheme for the formation of a spin-echo signal.

resented by steady rotation about the x' axis in the zy' plane (Fig. 10a).

At the moment of time $t_w = \pi/2\omega_1$, the vector $\overleftarrow{M}$ deviates from the z axis by 90° and will rotate in the xy plane of the stationary system of coordinates. In the rotating system of coordinates its direction at this moment of time coincides with the y' axis (Fig. 10b).

Let us assume that at the moment t_w the exciting field is switched off, so that the duration of the first pulse equals $\pi/2\omega_1$.

We will assume that the inhomogeneity of the field ΔH considerably exceeds the value $(\gamma T_2)^{-1}$. Then after the first pulse the movement of the vector $\overleftarrow{M}$ will be determined solely by the inhomogeneity. Then after a time of the order of $(\gamma \Delta H)^{-1}$ the precessing magnetic moments of separate sections of the sample (isochromats), which constitute the vector $\overleftarrow{M}$, get out of phase, and the total magnetic moment reverts to zero. However, the precession in each individual section is still preserved (since it is determined by the time T_2), and the magnetic moments of all the sections will rotate in the xy plane of the stationary system of coordinates. In the rotating system of coordinates they form a fan of vectors directed in different directions. Figure 10c shows three adjacent isochromats. The vector 1 preceded the vector 2, while the vector 3 lagged behind it.

Now let the field again be switched on at the moment τ ($\tau <$ T_2). As previously, it may be assumed that the motion of each isochromat is described by the equation

$$\frac{d\overleftarrow{M}_i}{dt} = \gamma \, [\overleftarrow{M}_i \overleftarrow{H}_i]$$

This means that while precessing about the z axis, each isochromat will change its position in space, performing nutation about the direction of the field H_1. In the rotating system of coordinates this motion is represented by precession about the x' axis with an angular velocity $\omega_1 = \gamma H_1$ (Fig. 10d). It is clear that after a time $\pi/\omega_1 = 2t_w$ all the isochromats have executed half a turn in this motion and are again in the x'y' plane of the rotating system of coordinates (Fig. 10e), but their relative motion is reversed. In Fig. 10c, the vector 1 preceded the vector 2 because its precession frequency was greater than the precession frequency of the vector 2. In Fig. 10e, because of the completed half-turn in the nutational motion, the vector 1 was found behind the vector 2. Since the precession frequency of the vector 1 remained as before, it now overtakes the vector 2. Therefore, while the phases of the isochromats diverged after the first pulse, after the second, they begin to converge at the same rate. At the moment 2τ their directions coincide (Fig. 10e) and in the receiver coil there arises a strong spin echo signal. At this moment the vector $\overleftarrow{M}$ will be rotating in the xy plane of the stationary system of coordinates.

In the scheme described, two successive pulses of the high-frequency field are applied to the sample for observing the spin echo. The first pulse is called the 90-degree (it rotates the vector by 90° in the rotating system of coordinates) and the second, the 180-degree pulse.

The pulses may also be applied in other ways with an increase in the number of them. This gives a more complex spin-echo picture.

The width of the spin-echo signal depends on the rate of divergence and convergence of the isochromats. This rate is higher and, consequently, the signal width is less, the greater is the homogeneity of the field. From the condition $t_w < T_2$ it follows that the intensity of the pulses must be greater, the shorter is the relaxation time T_2.

The intensity of the spin-echo signal depends on self-diffusion processes. With weak diffusion, i.e., when the molecules remain for a long time at the same points in the sample, the magnitude of the echo signal equals $M_0 e^{-2\tau/T_2}$. In this case, by comparing the intensity of the echo at different values of τ it is possible to measure the relaxation time. When there is diffusion, the relation of the signal intensity to τ is more complex, and for a sequence of 90- and 180-degree pulses it has the form

$$M_0 e^{-\frac{2\tau}{T_2} - \frac{1}{12} \gamma^2 G^2 D (2\tau)^3}$$

where G is the mean value of the magnetic field gradient in the volume of the sample, and D is the diffusion coefficient.

4. Apparatus for Observing and

Recording NMR Spectra

Special instruments called NMR spectrometers are used for observing and recording NMR spectra. At the present time NMR spectrometers are divided into two classes, namely, high-resolution spectrometers, which are designed for recording nuclear spectra of pure liquids, and broad-line spectrometers. This division is explained by the fact that the sharp difference in the character and intensity of the resonance lines in the spectra of liquids and solids causes such a great difference in the techniques for recording the corresponding spectra.

Because most practical problems are connected with proton resonance, in recent years there have appeared high-resolution spectrometers, whose possibilities are limited to the recording of proton resonance spectra. These spectrometers are sometimes called analyzer spectrometers.

Because of the conditions under which the phenomenon of nuclear resonance arises, any NMR spectrometer must include the following equipment:

1) a magnetic system for energizing the polarizing field H_0;

2) a device for exciting and receiving nuclear signals;

3) a device for stabilization of the polarizing field;

4) a device for sweeping the spectra.

Magnet Systems. The magnet systems for NMR spectrometers are usually electromagnets or permanent magnets with an iron magnetic circuit, in the air gap of which is produced the field H_0.

Since the sensitivity of the NMR method is proportional to the square of the strength of the polarizing field, and also because the chemical shifts are proportional to this field, while the splittings of second-order lines, which are connected with spin—spin interaction, are reduced with an increase in H_0, the refinement of NMR spectrometers is associated with a considerable increase in the working strength of the field in the spectrometer magnets.

At the present time practically the standard strength for high-resolution instruments is a strength of about 14,100 Oe, corresponding to a proton resonance frequency of 60 MHz, and in some countries there is mass production of instruments with a proton working frequency of 100 MHz (about 25,000 Oe). There is also intensive development of instruments with magnets in which superconductivity is used and with which it is already possible to observe the spectra of protons at a frequency of 200 MHz (about 50,000 Oe).

Electromagnets with an iron yoke, which give a field strength of 15,000-25,000 Oe, are heavy instruments which require water-cooling; their mass may be from several hundred kilograms to several tons.

The main characteristic of an NMR spectrometer is the resolving power of its magnet. The resolving power of the magnet is the ratio of the maximum value of the field H_0 inside the magnet gap to the width ΔH of the distribution of this field through the volume of a sample placed in the gap: $R = H_0/\Delta H$. Obviously, the concept of the resolving power remains indefinite until we have specified the volume of the sample and its form. In contemporary instruments of high resolution the resolving power is usually determined for a cylindrical sample about 5 mm in diameter with a volume of about 20 mm^3. In addition, the resolving power naturally depends on the point in the gap at which the sample lies. It is therefore assumed that the sample lies in the region of greatest homogeneity of the field where the value ΔH is minimal. (This region usually coincides with the region of maximum field H_0.)

In describing the distribution of the magnetic field in the gap of the spectrometer magnet it is usually assumed that the external field H_0 is horizontal and that the direction of this field coincides with the z coordinate axis, while the y coordinate axis is vertical. We assume that the sample is at the center of the air gap of the magnet and that this point coincides with the origin of the coordinates. As a result of the edge effect, i.e., the bulging of the magnetic field from the gap into the surrounding space, the field will not be the same either in magnitude or direction at different points inside the gap. However, the transverse components H_x and H_y, which characterize the deviation of the direction of the field vector from the z axis, are small at the center of the gap in comparison with the magnitude of the longitudinal component H_z. We will therefore consider only the latter component.

The longitudinal component at different points of the air gap close to its center may be represented as a Taylor series

$$H_z(x, y, z) = H_0 + \left(\frac{\partial H_z}{\partial x}\right)_0 x + \left(\frac{\partial H_z}{\partial y}\right)_0 y + \left(\frac{\partial H_z}{\partial z}\right)_0 z +$$
$$+ \frac{1}{2}\left(\frac{\partial^2 H_z}{\partial x^2}\right)_0 x^2 + \frac{1}{2}\left(\frac{\partial^2 H_z}{\partial y^2}\right)_0 y^2 + \frac{1}{2}\left(\frac{\partial^2 H_z}{\partial z^2}\right)_0 z^2 + \ldots$$

The coefficients of this series are called the gradients of the field. If the geometric dimensions of the sample are small in com-

parison with the diameter of the pole pieces, then we need consider
only the first- and second-order gradients. We should note that
the second-order gradients of the field are not independent, but
are related by the Laplace equation

$$\frac{\partial^2 H_z}{\partial x^2} + \frac{\partial^2 H_z}{\partial y^2} + \frac{\partial^2 H_z}{\partial z^2} = 0$$

The values of the gradients determine the resolving power
of the instrument. The magnets are usually designed so that these
gradients are minimal. This is achieved primarily by selecting a
sufficiently large diameter of the pole pieces with a given gap width
and ensuring that they are coaxial and as parallel as possible.
The homogeneity of the field will also be affected by the quality of
the material of the pole pieces.

By means of a series of design and technological methods in
the construction of the magnet, it is possible to reduce the gradients of the fields to values at which it is possible to record broad
lines quite accurately. However, relatively small gradients always
remain even with well-designed magnets and, moreover, they change
somewhat with the passage of time, depending on the temperature
of the magnet and the method of switching it on. As a result of this,
special devices and methods have to be used to increase the resolving power above 10^{-7} with sample volumes of the order of 20 mm^3.

The natural width of the resonance lines of protons in pure
liquids is often less than 10^{-4} Oe. For the instrument to be able
to record the true form of the line with $H_0 = 15,000$ Oe, the resolving power of the instrument required is

$$R = \frac{1.5 \cdot 10^4}{10^{-4}} = 1.5 \cdot 10^8$$

i.e., the field should be homogeneous over the whole volume of the
sample with an accuracy of one two-hundred millionth.

It is a difficult technical problem to obtain such a resolving
power. It is solved in high-resolution spectrometers by position-
ing the sample in the most homogeneous region of the field, by re-
ducing the volume of the sample to a few tens of cubic millimeters,
by using special coils for correction of the homogeneity of the field
(electrical s h i m m i n g), by mechanical rotation of the sample at
a sufficiently high rate, and by cycling the electromagnet.

The most important of these methods are the last three. As has already been mentioned, the first two will give a resolving power of 10^7. A further decrease in the sample leads to a decrease in the sensitivity of the apparatus and, therefore, it is hardly used in practice.

Electrical shims, which are designed to compensate for linear field gradients, consist of a system of coils, the z component of the field of which contains only uneven gradients along the corresponding axis (x, y, or z). Their geometric dimensions are such that the third-order gradient reverts to zero at all values of the shimming current. The first-order gradient is used to compensate for the corresponding gradient of the magnet field by selection of an appropriate current direction and magnitude. The fifth-, seventh-, etc. order gradients remain uncompensated.

In particular, a system of shims for compensation of the z gradient consists of a pair of flat round coaxial coils, the currents in which are in opposite directions, and whose common axis coincides with the z axis of the magnet. The distance between the coils is limited by the dimensions of the gap. This distance and the radius of the coils are selected on the basis of the considerations given above.

The y gradient is compensated by two pairs of flat rectangular coils, lying in pairs in planes parallel to the planes of the pole pieces, which are symmetrical relative to the xy plane of the magnet system of coordinates.

The second-order gradients are compensated by means of two pairs of circular coaxial coils, whose axis coincides with the z axis of the magnet. The ratio of the currents in the coils and their geometric dimensions are chosen so that the constant component and the fourth-order gradient of the z field of these coils revert to zero.

By means of the shimming devices it is possible to improve substantially (by approximately half an order) the resolving power of a magnet and, in particular, to compensate for those gradients which are not eliminated by mechanical rotation of the sample (see below).

In practically all spectrometer designs the z axis is horizontal, and therefore we will assume that the rotation of the sample is relative to the vertical y axis.

Let us examine a small element of volume of the sample. When the sample is rotated, this element describes a circle in a plane parallel to the xz plane. If the field is not homogeneous around this circle, then the element considered will be periodically in a somewhat stronger and then in a somewhat weaker field. This is equivalent to the sample being in a homogeneous field with the superposition of an additional field

$$H'_z = \frac{\Delta H}{2} \cos \omega_T t$$

where ΔH is the difference between the highest and lowest values of the field; ω_T is the angular velocity of rotation of the sample.

It is clear that for the element considered the inhomogeneity essentially does not exist, but instead of it there appear side signals, which are at a distance from the main signal of $\pm \omega_T$.

By increasing the rate of rotation of the sample it is possible, on the one hand, to move these side signals (satellites) away from the main one and, on the other hand, to reduce their intensity to practically zero. As was shown previously (see page 101), the essential condition for this will be

$$\frac{\gamma \frac{\Delta H}{2}}{\omega_T} \ll 1$$

or

$$\omega_T \gg \frac{1}{2} \gamma \Delta H$$

From the above account it is obvious that rotation does not eliminate gradients along the axis of rotation.

So-called cycling of the magnet is used to reduce the second-order gradients. Cycling consists of a certain increase in the magnet current relative to the current at which the resonance field is obtained. The increased current is maintained for several minutes, and then the current is returned to the resonance value. The cycling effect is individual for each given magnet. In some spectrometers cycling is carried out automatically, and the duration of the cycle is selected experimentally in the factory adjustment of the spectrometer.

Another equally important characteristic of the magnetic system of a spectrometer is the stability of the polarizing field H_0 with time. This requirement is particularly rigid again in high-resolution spectrometers. The point is that to obtain the form of the line undistorted by transient processes the time for passage through the resonance region should not be substantially less than $1/\delta\omega$, as was shown in the previous section. If the width of the line in field units $\delta H = 10^{-4}$ Oe, then this corresponds to $\delta\omega = \gamma\,\delta H = 2.67$. Thus, the rate of sweep of the spectrum in the spectrometer should not be greater than

$$\delta H : \frac{1}{\delta\omega} = \gamma\,(\delta H)^2 = 10^{-4} : \frac{1}{2.67} = 2.67 \cdot 10^{-4} \text{ Oe/sec}$$

For uncontrollable drifts of the polarizing field not to distort the sweep rate they should be at least an order less than this value. Thus, the rate of slow drifts of the field should be no greater than $4 \cdot 10^{-6}$ Oe/sec and in relative units with a field of 14,000 this is $3 \cdot 10^{-10}$ sec^{-1}. It is clear that this stability should be maintained during the whole time of recording of the spectrum, which is an average of 10-15 min.

The magnitude of sudden jumps in the field should be substantially less (again about an order) than the width of the resonance line, i.e., it should be no more than $5 \cdot 10^{-10}$ of the magnitude of the field.

In this connection we distinguish between the long-term and short-term stability of the magnetic field in the magnetic system of a spectrometer.

The main sources of the instability of the magnet field are the instability of the current which feeds the magnet, the presence of external, randomly varying magnetic fields (from contact circuits of electrical transport, electrical machines, etc.), the movement of metal objects close to the magnet, and the temperature instability of the magnet.

By technical means it is possible to reduce the current instability to relative values of the order of 10^{-7}. For this purpose special power sources are used for the magnets of NMR spectrometers, which include stabilizers of the alternating voltage of the power supply, electronic stabilizers of the rectified voltage, and current stabilizers.

Since, as a rule, the other sources of instability of the field introduce a relative instability which is greater than 10^{-7}, there is no point in further improvement of the stability of the current. Therefore, we also use special stabilizers of the magnetic flux and also stabilizing devices which are based on the use of nuclear resonance itself.

A magnetic flux stabilizer consists of two coils, a receiving and an operating coil, which are placed on the magnet pole pieces, and also an integrator and an amplifier, which are connected between these coils.

Any change in the magnetic flux for any of the above reasons induces in the receiving coil of the stabilizer an emf which is proportional to the rate of change of the flux. After integration and amplification, this emf produces in the operating coil currents which eliminate the change in the magnetic flux in the gap.

The stabilization factor of these devices is about 10^2, so that the short-term instabilities of the order of 10^{-7} are reduced to almost the desired value. However, these systems are almost insensitive to a slow drift of the field.

The above stability, which is guaranteed by stabilizers of the magnet power supply, is usually quite adequate for broad-line spectrometers. It is only with difficult external conditions that it may be profitable to use a flux stabilizer. However, since the development of methods of increasing the sensitivity of instruments of this type leads to systems with so-called information storage, stabilizing devices for these instruments will undoubtedly be refined.

In high-resolution spectrometers the stability (particularly the long-term stability) provided by flux stabilizers is inadequate. Therefore, these instruments are fitted with special proton stabilizers, whose principles of operation will be described below.

Excitation and Reception of Nuclear Signals. For the excitation of an alternating field H_1 in a sample, the latter is placed in a coil, which is introduced into the magnet gap and arranged so that its axis is perpendicular to the direction of the field. The coil is fed with a current from a special high-frequency generator. This current produces inside the coil an alternating magnetic field, whose vector coincides in direction with the axis of the coil, but changes in magnitude and sign in accordance with a sinu-

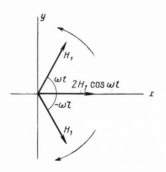

Fig. 11. Resolution of a linear polarized field into two rotating components.

soidal law. This linearly polarized field may be regarded as the sum of two components, which rotate in different directions in a plane perpendicular to the direction of the polarizing field (Fig. 11). The component rotating clockwise is the exciting field:

$$H_x = H_1 \cos \omega t$$
$$H_y = - H_1 \sin \omega t$$

The other component has only an insignificant effect on the behavior of the spin system.

Since in all relations describing NMR the magnitude of the polarizing field H_0 and the frequency ω of the exciting field appear in the form of the difference $\gamma H_0 - \omega$, the requirements for the stability of the frequency ω are exactly the same as the above requirements for the stability of the polarizing field.

The frequency of the oscillations of the exciting generator in NMR spectrometers is usually stabilized by means of quartz resonators, and this, on the one hand, guarantees the above requirements but, on the other hand, it almost completely excludes the possibility of changing this frequency for selecting the resonance conditions or for sweeping spectra. Therefore, NMR spectrometers are fitted with sources of one or several quartz-stabilized frequencies, while the resonance conditions are selected and the spectra swept by changing the magnitude of the polarizing field.

The reception of nuclear resonance signals is based on the electromagnetic principle mentioned above (see page 57). The transverse component of the magnetization vector excited M_x [see equation (I-87)] induces in the exciting coil an emf (see also page 57),

$$e_x = - n \, 4\pi S \frac{dM_x}{dt} \cdot 10^{-8}$$

where n is the number of turns; S is the cross section of the coil, cm^2.

Thus, the exciting coil is simultaneously the receiving coil for nuclear resonance signals. We should note that the frequency of nuclear resonance signals coincides with the frequency of the exciting field H_1.

To be able to distinguish the comparatively small voltages (of the order of hundredths or tenths of microvolts), induced in the coil by precession of the magnetization vector, from the considerable voltages of the same frequency (of the order of several tens of millivolts), which are fed to the coil to create the exciting field, and also to pick out from the two components of M_x the absorption component, the coil is connected to the circuit of an rf bridge.

A very simple bridge circuit is shown in Fig. 12. If we feed to the input of the bridge a voltage $u_1 = u_{1max} \cos \omega t$, then, in the general case, the voltage at its output terminal will be

$$u_2 = u'_{2\,max} \cos \omega t + u''_{2\,max} \sin \omega t$$

i.e., the output voltage differs from the input voltage in both amplitude and in phase. By selection of the tuning elements of the bridge (in this case the resistance R and the capacitance C) it is possible to reduce to zero the amplitude of any of the components of the output voltage with a constant amplitude of the input voltage. Compensation of the component which coincides in phase with the input voltage is called amplitude balance of the bridge, while compensation of the component which is in quadrature with this voltage ($u''_{2\,max} \sin \omega t$) is called phase balance of the bridge.

If the voltage of the nuclear resonance signal is now induced in the coil of the bridge, then the voltage at the output of the bridge changes and will have the form

$$u_2 = (u'_{2\,max} + kv) \cos \omega t + (u''_{2max} + ku) \sin \omega t$$

where k is the proportionality coefficient, which depends on the parameters of the bridge.

Since the voltages kv and ku are not put in from the input of the bridge, but from its internal elements (coil), the process of balancing the bridge produces only a slight change in these components in the output voltage.

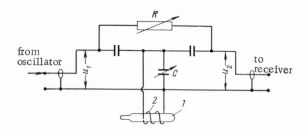

Fig. 12. Basic circuit of the simplest receiver bridge. 1) Ampoule with substance investigated; 2) receiver coil.

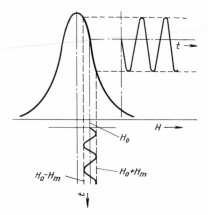

Fig. 13. High-frequency voltage envelope at the output of the bridge with sinusoidal modulation of the polarizing field.

It would be possible to balance the bridge completely, i.e., to adjust it so that $u'_2{}_{max} = u''_2{}_{max} = 0$. However, in this case, a mixture of absorption and dispersion signals is obtained at the output of the bridge. By balancing the bridge with respect to the phase $(u''_2{}_{max} = 0)$ and leaving some amplitude balance $(u'_2{}_{max} \gg kv)$, we find that the amplitude of the output voltage of the bridge will change in accordance with the law

$$A_2 = \sqrt{(u'_2{}_{max} + kv)^2 + (ku)^2} \approx$$
$$u'_2{}_{max} + kv$$

i.e., it will depend practically solely on the absorption signal interesting us.

The high-frequency output voltage envelope of the bridge is separated by means of a so-called amplitude detector, which is usually preceded by a high-frequency voltage amplifier. After amplitude detection and some further amplification, the signals obtained may be used to feed a recording apparatus (oscillograph or recording potentiometer).

A modulation method is used in NMR spectrometers for increasing the sensitivity and the accuracy of the apparatus, and for this purpose an alternating field $H_M \cos \omega_M t$ is superimposed on

the polarizing field by means of special modulating coils. In broad-
line spectrometers the modulation frequency is substantially less
than the width of the resonance line (usually 30-50 Hz) to avoid distortion of the form of the line. In addition, the amplitude of the
modulating component selected is considerably less than the width
of the line in field units. Under these conditions the modulation
process will be slow in comparison with transient processes in the
spin system, and the envelope of the high-frequency oscillations at
the output of the bridge with accurate tuning will have the form of
a sinusoidal curve with an amplitude which is proportional to the
slope of the resonance curve (Fig. 13). Thus, by changing the field
H_0 in the resonance region it is possible to obtain signals which
are proportional to the derivative of the function describing the
resonance curve.

In high-resolution spectrometers the modulation frequency is
greater than the whole width of the spectrum (usually about 2 kHz).
The modulation amplitude is selected (see page 101) so as to avoid
saturation and to guarantee sufficient signal intensity. The alternating components of the central or first side signals are used for
recording.

In both of the cases examined the envelopes of the high-frequency components of the output voltage of the bridge will no longer
be signals of v or u, changing together with the sweep, but sinusoidal oscillations, whose frequency equals the modulation frequency and whose amplitudes are proportional to the signals v or u or
values derived from them (in broad-line spectrometers), which also change together with the sweep.

It is obvious from the above that with phase balance of the
bridge the output voltage of the amplitude detector in these cases
will have the form

$$u_{ad} = A \cos \omega_M t + B \sin \omega_M t$$

where either A and B are the signals v or u or B = 0, while A is
the derivative of the signal v.

The signals A and B interesting us are separated in spectrometers by means of a so-called cophasal detector, whose operation consists of multiplying the signal detected u_{ad} by a voltage of
the form $\cos [\omega_M t + \varphi]$, controlled by the detector, whose phase φ

(relative to the phase of the modulating field) may be chosen arbitrarily. When $\varphi = 0$, the effect of the reference voltage on the signal detected gives

$$u_{ad} \cos \omega_M t = \frac{1}{2} A + \frac{1}{2} A \cos 2\omega_M t + \frac{1}{2} B \sin 2\omega_M t \qquad (\text{II-61})$$

To suppress oscillations with twice the modulation frequency, the cophasal detector contains an integrating circuit, which consists of a resistor R and a capacitor C. The time constant RC of this circuit is chosen so that $RC2\omega_M \gg 1$. In this case, the output voltage of the cophasal detector (after the integrating circuit) will be proportional to the signal interesting us, A.

When $\varphi = 90°$, at the output of the cophasal detector we obtain a voltage which is proportional to the signal B.

Simultaneously with suppression of the second harmonic of the modulation, the integrating circuit will also suppress relatively high-frequency components of the noise, which is inevitably present in radio circuits and is comparable with weak NMR signals and sometimes exceeds the level of the effective signal. The greater the constant RC, the more the low-frequency components of the noise will be suppressed effectively by the circuit and the better will be the signal-to-noise ratio at the output of the instrument.

The method described is at present the main means of improving the signal-to-noise ratio in broad-line spectrometers.

So-called autodyne receivers are often used to receive nuclear signals in broad-line spectrometers. An autodyne receiver consists of a tube rf generator , whose oscillation circuit includes the coil containing the sample investigated. The absorption of energy by the spin system produces a decrease in the amplitude of the oscillations in the circuit of this receiver, which is detected by the method described above. The sensitivity of an autodyne receiver is particularly high when it is operating under conditions close to breakdown of the oscillations. However, the sensitivity may be varied by changing the rf voltage on the circuit of the receiver (to establish the desired H_1). Therefore, the autodyne receiver is fitted with a sensitivity calibrator, which simulates a nuclear signal of standard intensity.

Autodyne circuits are simple and convenient to operate. However, they have a series of substantial drawbacks. The main ones of these are as follows: the low stability of the frequency (about 10^{-4}), the impossibility of stable operation at very low levels of oscillations, and the effect of the natural oscillations of the spin system on the frequency of the autodyne receiver, which leads to distortion of the form of the lines with quite narrow lines.

Proton Field Stabilizers. Stabilizers, whose action is based on the use of the NMR phenomenon, are used to increase the stability of the magnetic field in high-resolution spectrometers. For this purpose an auxiliary sample, which consists of water with a small amount of paramagnetic salts added, is introduced into the magnet gap. The signals of the auxiliary (or reference) sample are excited with the same rf source as for exciting the sample investigated and are received by the modulation method described above with receiver devices which form the circuit of the stabilizer receiver.

By varying the tuning of the receiver bridge and the phase of the reference voltage, the receiver is tuned to the dispersion signal and after amplification, the output voltage of the receiver is fed to actuating coils on the pole pieces of the magnet. With the appropriate choice of the direction of the current in these coils, the stable point of such a closed-circuit system will be the center of the dispersion signal of the auxiliary sample. Thus, the resonance conditions $\gamma H_0 = \omega$ (or $\gamma H_0 = \omega \pm \omega_M$, if the system operates on side signals) are maintained continuously for the auxiliary sample.

Since the auxiliary sample must be at some distance from the main sample, as a result of which the resonance conditions for the two samples are not the same, an additional compensating field is applied to the auxiliary sample by means of special small coils, which are called bias coils.

The system described is used in combination with a current stabilizer. It gives extremely good results, but it is not the best, since the field is stabilized essentially at the position of the auxiliary sample. Therefore, there is a tendency to use the analysis sample itself as the reference sample. For this purpose we add to the analysis sample a substance whose line is not present in the spectrum of the sample (for example, tetramethylsilane). The resonance signal of this substance is used for stabilization. In comparison with the two-sample system, this method gives substantially better results, but the construction of the stabilizer is very complicated.

Stabilizers with one sample are used in spectrometers manufactured in USA and Japan with an operating frequency (for protons) of 100 MHz.

Another form of stabilizing device, which is used in systems with both two samples and with one, is the so-called side-band spin generator.

It was shown above (page 100) that when there is modulation with a frequency ω_M, which considerably exceeds the width of the line, the side signals observed at fields H = $(\omega \pm \omega_M)/\gamma$ contain components with the modulation frequency at the output of the amplitude detector. Therefore, if we remove the modulation and connect the output of the amplitude detector to the modulating coils of the reference sample with detuning of the field substantially greater than the width of the resonance line of the reference sample, then it may be calculated that in such a system with sufficiently high amplification of the receiver circuit there arise autooscillations with a frequency which is determined by the detuning of the field:

$$\omega_g = \omega - \gamma H_0$$

These autooscillations actually arise, and with a change in the frequency or the field, the generation frequency in the system will change, i.e., the system will "follow" the field (or ω). If the oscillations generated by such a spin generator on the auxiliary sample are used for modulation of the main sample, then for the latter we may write the resonance conditions in the form

$$\omega = \gamma H_0 + \gamma \Delta H + \gamma \Delta H' + \omega_g$$

where ΔH is the distance between the field of the main and the auxiliary samples; $\Delta H'$ is the field created by the bias coils.

It is obvious that when $\Delta H + \Delta H' = 0$ this condition becomes an identity, i.e., with a suitable choice of bias field applied to the main sample it is possible to guarantee for it continuous resonance conditions with any changes in H_0 or ω.

In actual fact, the system described is not so ideal. A change in the generation frequency is accompanied by a change in the phase shift of this frequency in the receiving device, and this results in different regions of the reference sample participating in the formation of autooscillations. Since the width of the line of the reference sample is usually substantially greater than the width of the line of the main sample (even though it is due to the natural non-uniformity of the field), the change in the generation frequency will not follow changes in the field H_0 (or the frequency ω) completely exactly. This system therefore requires the creation of a receiver circuit with a weak dependence of the phase shifts on the frequency ω_g. To make it easier to fulfill this requirement, additional feedback is introduced into the system to correct for the considerable departures of the generation frequency from some set mean value.

Systems for Sweeping Spectra. In principle, spectra may be swept by changing any of the parameters which determine the resonance conditions, namely, the magnitude of the polarizing field, the frequency of the exciting field, and the modulation frequency (in systems using the modulation method). The first method is used mainly in NMR spectrometers, but the last method is also used sometimes.

In broad-line spectrometers the required sweep width is usually several tens of oersteds. Sweeping is achieved with additional coils on the magnet pole pieces, which are fed with a linearly changing current.

Sweep rates are from tenths to thousandths of an oersted per second.

Both mechanical and electronic systems are used as master units for providing a linearly changing current.

In high-resolution spectrometers the sweep width is usually a few parts of an oersted. The sweep system is closely connected with the field stabilization system.

In spectrometers with only current stabilization the sweeping is usually achieved by applying a direct voltage to the receiver coils of the stabilization system. The integrating unit of the system converts this voltage into a linearly increasing current in the actuating coils. The sweep rate is proportional to the magnitude of the applied voltage.

In spectrometers with proton stabilization through an auxiliary sample, the sweep field is provided by a current in small coils (sweep coils) arranged about the auxiliary or about the main (analysis) sample. In both cases the design of the sweep coils must be such that their field has no appreciable gradients at the point of the main sample. For this reason, in the first case these coils sometimes have the form of a solenoid, inside which lies the auxiliary sample (NMR-5535 spectrometer). In the second case, they consist of coils whose radius equals the distance between them (Helmholtz coils), as in the S-60 spectrometer (Japan), or they have a complex form, as in the A-60 spectrometer (USA).

In spectrometers with proton stabilization through the reference line of the main sample, the only sweep method is to change the modulation frequency.

NMR Spectrometer 5535 (TsLA).* In conclusion, we will explain the intereffect of the units of a spectrometer on the example of an NMR spectrometer 5535, which is produced by the Central Automation Laboratory of the Ministry of Ferrous Metallurgy.

A block diagram of the instrument is shown in Fig. 14. The exciting coils 1 of the electromagnet are fed from a special source, which contains a stabilizer for the alternating voltage 2, a rectifier 3, a stabilizer for the rectified voltage 4, a unit for adjusting the current 5, and a current stabilizer 6. The field in the gap is also stabilized with a proton-current stabilizer, which contains a head with an auxiliary sample 8, receiver coils 9 for detecting the instability of the flux, actuating coils 10 for compensating for the instabilities, and electronic units 11.

When the sweep generator 12 is switched on, a linearly changing voltage is fed to the stabilization system. The stabilization system then changes the field in the magnet gap linearly.

Through an attenuator 14 (controlling the amplitude of the rf voltage) a quartz rf oscillator 13 feeds the receiver coil 23, which is an element in the RF bridge. The receiving part of the spectrometer consists of a superheterodyne radio receiver 15.

---

*High-quality spectrometers with a working frequency of 60 MHz are being developed at the present time in the USSR and will be mass-produced.

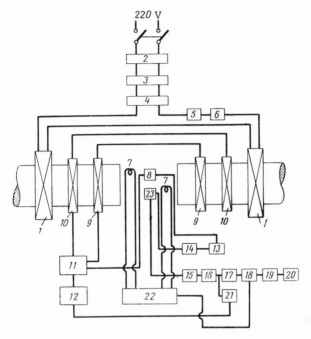

Fig. 14. Block diagram of NMR-5535 spectrometer. 1,7,9,10,23) Coils; 2,4,6) stabilizers; 3) rectifier; 5) current adjusting unit; 8) head; 11) electronic units; 12,13,22) oscillators; 14) attenuator; 15) radio receiver; 16,18) detectors; 17) amplifier; 19) integrating circuit; 20) potentiometer; 21) oscillograph.

After the amplitude detector 16, which picks out the envelope of the rf oscillations in the receiver, there is an output to an oscillograph 21 for visual observation of the spectra. In visual observation the horizontal sweep of the oscillograph is synchronized with the operation of the sweep generator.

In accordance with the considerations given above, the modulation method is used for recording the spectra in the NMR spectrometer 5535. Modulation is achieved by means of the coils 7, fed with an alternating current from the modulation oscillator 22. The modulation frequency is 2 kHz. With modulation the nuclear signal at the output of the first detector consists of oscillations with a frequency equal to the modulation frequency and with an amplitude which is proportional to the power absorbed. Therefore, after the first detector there is a low-frequency amplifier 17 and a cophasal detector 18, which makes it possible to obtain a signal that is proportional to the power absorbed. An integrating circuit 19 is connected to the output of the cophasal detector for suppressing noise. The signal is recorded on the chart of a high-speed potentiometer 20.

The main controls of the spectrometer are as follows: a knob for switching on the current supply to the spectrometer and a knob for switching on the magnet smoothly; knobs for adjusting the current*; switches for the rf level and the modulation level; a switch for the sweep rate; a knob for starting the sweep; a switch for the time constant of the integrating circuit of the cophasal detector; a knob for adjusting the phase of the reference signal of the cophasal detector; knobs for controlling the current of the correction coils.

5. Recording and Interpretation

of Spectra

In the recording and interpretation of nuclear magnetic resonance spectra, it must be borne in mind above all that the theoretically accurate form of the spectrum may be obtained only with an infinitely low amplitude of the exciting field and an infinitely low rate of passage through the resonance region. Since these values are necessarily finite in practice, and since there is also a series of factors which may distort the spectrum (nonuniformity and instability of the polarizing field, instability of the frequency and amplitude of the exciting field, the presence of modulation, the tuning of the receiving devices of the spectrometer, the finite value of the time constants of the cophasal detector, noise from electronic circuits, etc.), then the spectrum recorded will be only a good or bad approximation to the actual spectrum. The degree of this approximation depends on how accurately the investigator is able to select the experimental parameters which he is able to control and to allow for those which he cannot control.

The former include the amplitude of the exciting field, the amplitude of the modulating field, the rate of passage through resonance, the tuning of the receiver devices of the spectrometer, the

---

* The signal of the reference sample is controlled by adjusting the current to the proton resonance level; the stabilization system is switched on automatically.

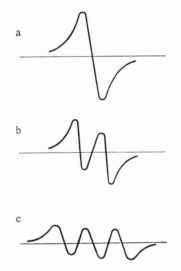

Fig. 15. Form of recording of derived function describing an absorption line. a) Singlet; b) doublet; c) triplet.

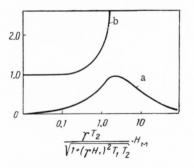

Fig. 16. Relation of the relative intensity of the signal v_1/v_{max} (a) and the measured line width $\delta H/\delta H_{HM}$ (b) to the modulation amplitude for the Lorentzian form of an NMR line.

magnitude of the time constant of the cophasal detector, the phase of the reference signal of the cophasal detector, and the uniformity of the magnetic field.

For controlling these values, each spectrometer is fitted with appropriate knobs and controlling instruments. However, the main difficulty lies in the fact that the correct selection of some of these values depends on the character of the spectrum, which is unknown before the experiment.

The process of recording a spectrum therefore consists of a series of recordings, accompanied by a change in some of these parameters until the optimal result is obtained.

In addition to the factors given, the spectrum recorded may often be distorted by the presence of undesirable impurities in the sample such as ferromagnetic impurities in solid samples and paramagnetic salts and oxygen in liquids.

In many cases careful preparation of the sample is essential to obtain an accurate spectrum.

Recording and Interpretation of Broad Resonance Lines. For broad lines it is not the resonance line itself which is recorded on the chart of the spectrometer, but a mathematically derived function, which describes it.

Figure 15 shows a typical recording obtained if the line has a simple bell shape (a), and also the appearance of the spectra when the form of the line is determined by the presence of well-isolated pairs of nuclei (b, doublet) and triplets of nuclei (c, triplet). In the first case we observe one positive maximum and one negative; in the second case, two positive and two negative maxima; and in the third case, three maxima of each sign.

The width of the resonance line equals the distance between the extrema of its derivative provided that the spectrum is recorded accurately.

To obtain the undistorted form of a line it is necessary to observe the following conditions:

1. The nonuniformity ΔH of the polarizing field in the sample should be considerably less than the width of the line recorded δH:

$$\Delta H \ll \delta H \qquad\qquad (\text{II-}62)$$

2. The modulation frequency should be substantially less than the width of the line, expressed in angular frequency units:

$$\omega_M \ll \gamma \delta H \qquad\qquad (\text{II-}63)$$

3. The modulation amplitude should be substantially less than the line width:

$$H_M \ll \delta H \qquad\qquad (\text{II-}64)$$

4. There should be no saturation and for this the following inequality must hold:

$$H_1 \ll \sqrt{\frac{\delta H}{\gamma T_1}} \qquad\qquad (\text{II-}65)$$

5. The sweep rate should satisfy the condition for steady-state passage:

$$\frac{dH_0}{dt} \ll \gamma (\delta H)^2 \qquad\qquad (\text{II-}66)$$

6. The time of passage through the resonance region should be greater than the time constant of the cophasal detector:

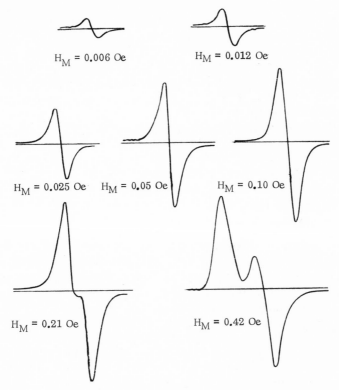

Fig. 17. Effect of increasing the modulation amplitude on the intensity and form of the signal.

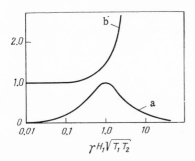

Fig. 18. Relation of the relative intensity of the signal (a) and the relative line width (b) to the exciting field H_1 for the Lorentzian form of the NMR line.

$$\frac{\delta H}{dH_0/dt} \gg RC \qquad \text{(II-67)}$$

All these conditions are determined by the line width, which is unknown before the experiment.

As regards the first two conditions, in a broad-line spectrometer with a given resolving power $R = H_0/\Delta H$ and a definite modulation frequency ω_M, they are uncontrollable parameters. However, if the width of the resonance line recorded is found to be close to the value RH or ω_M/γ, then it may be assumed that the form of the line and its width are determined by the instrument factors R and ω_M.

If the modulation amplitude is small, then the signal recorded is weak and may be strongly distorted by noise. An increase in the modulation amplitude leads first to an increase in the signal without a substantial change in its form or width. A further increase in the modulation amplitude begins to increase the width of the line and to distort its form, and then the intensity of the signal falls (Fig. 16).

Figure 17 shows recordings of the spectra of protons in vulcanized rubber with different modulation amplitudes. An increase in the modulation amplitude from 0.006 Oe to 0.025 Oe gave a practically proportional increase in the signal from 19 to 71 relative units without an appreciable increase in the line width.

A further increase in the modulation amplitude to 0.10 Oe led to an increase in the intensity to 181 units, but produced an increase in the line width (0.17 Oe). When $H_M = 0.21$ Oe, distortion of the form of the line appeared. In the case of a complex spectrum an extremely high modulation amplitude leads to obliteration of the structure of the spectrum.

Thus, an increase in the modulation amplitude up to certain levels leads practically only to an increase in the signal-to-noise ratio, but beyond these levels distortion of the spectrum begins.

An increase in the strength of the exciting field H_1 also leads at first to an increase in the signal recorded, and then to a decrease in it. There are two reasons for this decrease. In actual fact, in a spectrometer we do not measure the power absorbed by the spin system, but the change in the rf voltage in the receiver coil induced by this absorption. The energy of the external field H_1 absorbed by the spin system is proportional to the amplitude of this field and the magnitude of the change in the rf voltage. Under saturation conditions (at high values of H_1) the power absorbed does not increase with an increase in H_1 and, consequently, there is no increase in the work done by the exciting field.

Consequently, under these conditions an increase in H_1 must produce a decrease in the change in rf voltage induced by the absorption, so that the products of H_1 and the magnitude of this change remains constant.

The other reason for the decrease in the intensity of the signal recorded with an increase in H_1 is line broadening. The physical reasons for broadening were examined in Chapter I (page 52). With a constant modulation amplitude broadening of the line leads to the ratio $H_M / \delta H$, as a result of which the intensity of the signal recorded falls.

Figure 18 shows the relations of the intensity of the recorded signal (a) and the line width (b) to H_1 for the Lorentzian form of the line, while Fig. 19 shows oscillograms of the spectrum of protons in vulcanized rubber with different strengths of the exciting field H_1. With an increase in H_1 there is a gradual increase in the intensity of the signal at first without an appreciable change in the line width (up to $H_1 = 9.6$ mOe), and then the width of the line begins to increase, while the signal recorded is weakened and, finally, the form of the line is distorted.

It was shown above (see page 116) that an increase in the time constant of the cophasal detector leads to the appearance of noise at the output of the instrument. However, since a change in H_0, i.e., sweeping of the field, is achieved at a finite rate, the term $A/2$ of relation (II-61) will contain frequencies which are not zero.

An excessive increase in the time constant with a given sweep rate may lead to suppression of part of the spectral components of the effective signal, which produces a distortion of its form that consists of the appearance of asymmetry of the positive and negative halfwaves.

Figure 20 illustrates the effect of suppressing the noise with the correctly selected time constant and the effect of distortion of the form of the signal with an excessive increase in RC of the cophasal detector. (When $\tau = 1$ sec and $\tau = 4$ sec there is no distortion and noise is suppressed. When $\tau = 12$ sec and $\tau = 30$ sec, further suppression of the noise is accompanied by distortion of the form of the signal. It is necessary to reduce the field sweep rate to eliminate distortions.)

The incorrect selection of the phase of the reference signal of the cophasal detector leads only to a decrease in the intensity of the signal relative to the noise without producing further distortions of the spectrum.

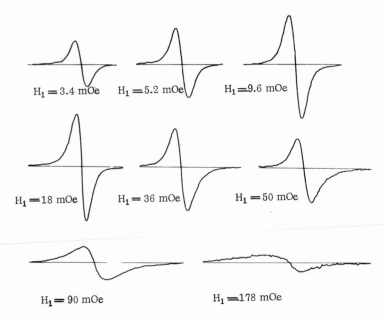

$H_1 = 3.4$ mOe $H_1 = 5.2$ mOe $H_1 = 9.6$ mOe

$H_1 = 18$ mOe $H_1 = 36$ mOe $H_1 = 50$ mOe

$H_1 = 90$ mOe $H_1 = 178$ mOe

Fig. 19. Effect of the strength of the exciting field on the form of the NMR signal.

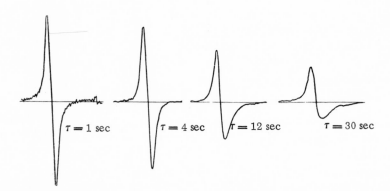

$\tau = 1$ sec $\tau = 4$ sec $\tau = 12$ sec $\tau = 30$ sec

Fig. 20. Effect of the magnitude of the time constant of the cophasal detector on the form of the NMR signal.

Let us examine another factor which may produce distortion of the form of the spectrum. The absorption of energy by the spin

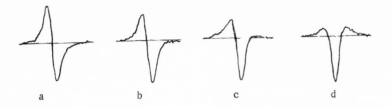

Fig. 21. Effect of the bridge balance on the form of the NMR signal recorded.
a) Purely phase balance (the derived function of the absorption is recorded);
b,c) very slight and substantial imbalances of the bridge (a "mixture" of absorption and dispersion signals is recorded); d) purely amplitude balance of the
bridge with imbalance of the phase (the derived function of the dispersion is
recorded).

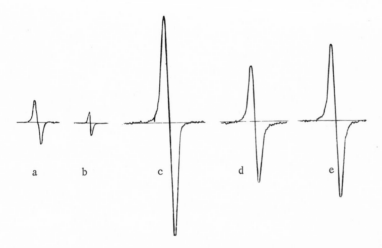

Fig. 22. An example of the selection of the conditions in recording a relatively narrow NMR signal (protons of rubber at room temperature).

system actually results in the appearance in the receiver coil of
two rf electromotive forces, which have a phase shift of 90°.

If the bridge is balanced with respect to phase, then detection gives us the required signal v.

If the bridge is balanced with respect to amplitude but not
with respect to phase, detection will give the signal u. In inter-

mediate cases of balance the signal observed will be a mixture of the signals u and v and the recorded absorption spectrum is distorted.

The effect of the balance of the bridge on the form of the signal recorded is illustrated in Fig. 21.

In work with autodyne units the absorption signal is selected automatically. However, in this case there may also be distortions of the spectrum if the width of the line recorded is small.

To illustrate the principles presented above we give two examples of the selection of the conditions for plotting spectra. The spectra were recorded on an NMR spectrometer 5535.

Example I

S a m p l e .   Natural rubber at room temperature. The spectrum of the protons is recorded. A narrow line is expected.

A p p a r a t u s .   A bridge is set up on the coordinate system of the spectrometer. It may be assumed that an autodyne unit would distort the spectrum observed.

T e s t   r e c o r d i n g . H_1 = 5.2 mOe, H_M = 0.21 Oe. A line of complex form is obtained (Fig. 22a). Overmodulation is suspected.

S e c o n d   r e c o r d i n g . H_1 = 5.2 mOe, H_M = 0.025 Oe. The signal has the right form, but its intensity is low (Fig. 22b).

T h i r d   r e c o r d i n g . H_1 = 5.2 mOe, H_M = 0.025 Oe. The amplification is increased by a factor of eight (Fig. 22c). An intense signal is recorded.

F o u r t h   r e c o r d i n g .   As a check we reduce the modulation amplitude. H_M = 0.012 Oe (Fig. 22d). The intensity of the signal falls, while the width of the line remains as before.

F i f t h   r e c o r d i n g .   As a check we increase the intensity of the exciting field, H_1 = 9.6 mOe. The amplitude of the signal increases (Fig. 22e), while the width of the line remains as before.

C o n c l u s i o n s .   It may be assumed that the conditions used for the third recording are correct and the
true width of the line $\delta H = 0.069$ Oe.

C o m m e n t s .   The resolving power of the instrument
$R = 10^{+6}$ when $H_0 = 5000$. Therefore, there is no reason to assume that the observed line width is an instru-
mental factor.

Example II

S a m p l e .   Polymethyl methacrylate at a temperature
of 77°K (−196°C). The spectrum of the protons is recorded. A broad line is expected.

A p p a r a t u s .   The autodyne unit is set up on the coordinate system of the spectrometer. (It is simpler to
work with it with broad lines.)

T e s t   r e c o r d i n g . $H_M = 3.1$ Oe; $dH_0/dt = 58$
mOe/sec; RC = 30 sec. A broad line is recorded (Fig.
23a). We can surmise that the structure is distorted
due to overmodulation.

S e c o n d   r e c o r d i n g . $H_M = 0.42$ Oe with the other
conditions the same. The structure is clearly expressed (Fig. 23b), but the signal is asymmetric.

T h i r d   r e c o r d i n g .   We reduce the time constant
RC = 4 sec, and reduce the amplification somewhat.
A symmetrical signal is observed (Fig. 23c) with good
separation of the lines, but it is somewhat noisy.

F o u r t h   r e c o r d i n g .   We increase the time constant RC = 12 sec, and reduce the scanning rate $dH_0/dt =$
29 mOe/sec. A spectrum suitable for interpretation
is obtained (Fig. 23d).

The interpretation of nuclear magnetic resonance spectra in
the case of broad lines is reduced mainly to determining the line
width, calculating the second-order moment, and explaining the observed form of the line.

It is most convenient to take as the line width the distance
between its points of greatest slope relative to the abscissa axis.

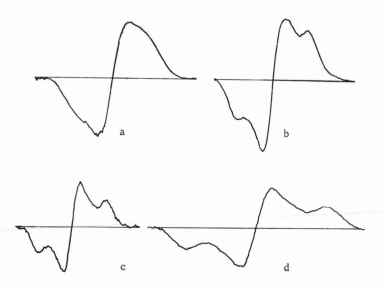

Fig. 23. Example of the selection of conditions in recording a broad line
(polymethyl methacrylate at −196°C).

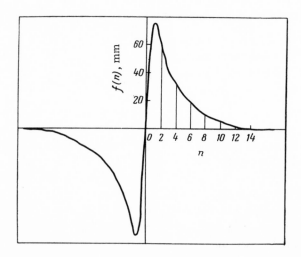

Fig. 24. Graphical calculation of the second moment of a broad line.

The measurement of the line width from the recording on the spectrometer is not difficult, since these points correspond to the two
extrema of the derivative recorded by the spectrometer.

The second-order moment of the resonance line, according
to equation (II-42), is defined by the relation

$$\Delta H_2^2 = \int\limits_{-\infty}^{\infty} h^2 g\,(h)\,dh \qquad\qquad \text{(II-68)}$$

where h is the deviation of the external field from the resonance
value $H_0 = \omega_0/\gamma$ (ω_0 is the excitation frequency) with the condition

$$\int\limits_{-\infty}^{\infty} g\,(h)\,dh = 1 \qquad\qquad \text{(II-69)}$$

The line actually recorded f (h) is the derivative of g(h), obtained on some arbitrary scale, i.e.,

$$f\,(h) = A g'\,(h)$$

where A is an unknown coefficient.

By integrating the right-hand part of equation (II-68) by parts
we find on condition that g(h) falls more rapidly than h^{-3}:

$$\Delta H_2^2 = -\frac{1}{3} \int\limits_{-\infty}^{\infty} h^3 g'\,(h)\,dh = -\frac{1}{3A} \int\limits_{-\infty}^{\infty} h^3 f\,(h)\,dh$$

Analogously, by integrating the left-hand part of equation
(II-69) by parts we obtain

$$1 = \int\limits_{-\infty}^{\infty} g\,(h)\,dh = -\int\limits_{-\infty}^{\infty} hg'\,(h)\,dh = -\frac{1}{A} \int\limits_{-\infty}^{\infty} hf\,(h)\,dh$$

and, consequently,

$$A = -\int\limits_{-\infty}^{\infty} hf\,(h)\,dh$$

Thus, we will have

$$\Delta H_2^2 = \frac{1}{3} \frac{\int\limits_{-\infty}^{\infty} h^3 f\,(h)\,dh}{\int\limits_{-\infty}^{\infty} hf\,(h)\,dh} \tag{II-70}$$

where $f\,(h)$ is the derivative of the resonance line recorded by the spectrometer.

The calculation of ΔH_2^2 from the experimental curve may be carried out graphically. For this purpose we divide the abscissa axis of the curve $f\,(h)$ into equal parts. Let the size of a division equal c oersteds.

Then, by replacing integration by summation we may write

$$\Delta H_2^2 = \frac{1}{3} \frac{\sum (nc)^3 f\,(nc)\cdot c}{\sum (nc) f\,(nc)\cdot c} = \frac{c^2}{3}\cdot \frac{\sum\limits_{-\infty}^{\infty} n^3 f_n}{\sum\limits_{-\infty}^{\infty} n f_n}$$

where n is the number of the section on the abscissa axis; f_n is the ordinate of the experimental curve, corresponding to this number.

We will illustrate the method of calculating ΔH_2^2 on the example of an experimental curve (Fig. 24). The calculation is carried out for the right-hand half of the curve (the curve is symmetrical).

The abscissa axis is divided into equal sections, each equal to 0.7 Oe. At each division point we construct the ordinates of the curve and measure their length in millimeters.

The results of the calculation are given as a table, shown on page 135.

The second-order moment obtained experimentally is actually somewhat high because of the finite values of the modulation amplitude, the sweep rate, the time constant of the cophasal detector, and the amplitude of the exciting field.

n	n^3	f_n	nf_n	$n^3 f_n$
1	1	71	71	71
2	8	45	90	360
3	27	38	114	1016
4	64	30	120	1920
5	125	23	115	2875
6	216	18	108	3888
7	343	13	91	4459
8	512	10	80	5120
9	729	7	63	5103
10	1000	5	50	5000
11	1331	3	33	3993
12	1728	2	24	3456
13	2197	1	13	2197

$$\sum_{n=1}^{n=13} nf_n = 972 \qquad \sum_{n=1}^{n=13} n^3 f_n = 39\,458$$

$$(\Delta H_2^2)_{\exp} = \frac{c^2}{3}\,40.4 = 6.6\ \text{Oe}^2$$

To take into account the effect of these factors on the second-order moment, we can introduce corrections:

a) for the finite modulation amplitude [4]:

$$(\Delta H_2^2)_{\text{true}} = (\Delta H_2^2)_{\exp} - \frac{H_M^2}{4} \tag{II-71}$$

b) for the finite rate of passage through resonance with a time constant of the cophasal detector which is not zero [1]:

$$(\Delta H_2^2)_{\text{true}} = (\Delta H_2^2)_{\exp} - \left(RC \cdot \frac{dH_0}{dt}\right)^2 \tag{II-72}$$

For the error in the use of this formula not to exceed 0.5%, the sweep rate should be such that

$$\frac{dH_0}{dt} \leqslant \frac{\sqrt{\Delta H_2^2}}{RC} \tag{II-73}$$

To allow for the effect of the finite amplitude of the exciting field on the magnitude of the second moment of the line we can plot the spectrum with different values of H_1, and by constructing the relation of ΔH_2^2 to H_1 we can extrapolate this value to zero amplitude of the exciting field.

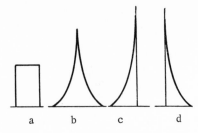

Fig. 25. Theoretical form of signals from a single narrow line with different field gradients. a) The presence of only a y gradient; b) only an x or only a z gradient; c,d) a quadratic gradient of negative and positive sign (the field strength increases from left to right).

Recording and Interpretation of High-Resolution Spectra. Because of the severe requirements as regards the stability and resolving power of high-resolution spectrometers, the process of recording spectra often is not a trivial task.

The operating state of the spectrometer is reached after a more or less prolonged warmup of the whole apparatus. The warmup time may be 1-2 h and is usually indicated in the instructions for operating the instrument. However, even after the appropriate warmup it is still impossible to guarantee that the required resolution will be obtained with the knobs for correcting the gradients in the same position as when the instrument was switched off from the day before. Therefore, the first operation after the warmup of the instrument is to adjust it to the required resolution. This adjustment is usually carried out with a control sample, and at the present time this is generally ethylbenzene or, more accurately speaking, the spectrum of its methyl triplet.

In most instruments produced previously the adjustment was carried out by mechanical movement of the head through the three coordinate axes in the magnet gap by means of a special coordinate mechanism.

In contemporary spectrometers the shimming devices described above are used for this purpose, though some instruments are also fitted with mechanisms for moving the heads.

In the adjustment of the instrument to the required resolution, the signal of the control sample is observed on the oscillograph screen. Apparently in only one of the mass-produced instruments (A-60, USA) is the adjustment carried out through the intensity of the control signal, which is recorded by the usual measuring instruments.

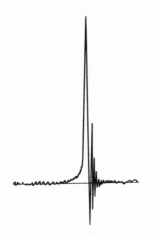

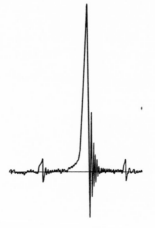

Fig. 26. A single narrow NMR line with rapid passage through resonance.

Fig. 27. The appearance of satellites in an NMR spectrum with insufficiently rapid rotation of the turbine.

With unsatisfactory uniformity of the field the observed signal is broad and of low intensity and may have different forms, which are determined by the combined effect of all the gradients. The following factors should be taken into account in adjusting the instrument.

Since the sample is usually in the form of a cylinder with the axis lying along the y axis, when there is one y gradient the form of the signal should approach a rectangle (Fig. 25a). When there is one x gradient or one z gradient, the signal has a symmetrical form with a sharp point and sloping sides (Fig. 25b). With quadratic gradients the signal has an unsymmetrical form: it either rises slowly and falls rapidly or rises rapidly and falls slowly (Fig. 25c, d).

A rapid rise and a slow fall of the signal with a sweep in the direction from low to high fields indicates that the polarizing field falls from the center of the sample to the edges, i.e., has a convex character. The opposite picture indicates that the field in the center of the sample is less than at the edges.

The form of the signal observed on the oscillograph screen is complicated both by the combined effect of all the gradients and

by transient processes in the spin system with rapid passage
through the resonance region. These transient processes are accompanied by the appearance of oscillations (arising after passage
through the central resonance region), i.e., wiggles.

With unsatisfactory uniformity and not too high a rate of passage through resonance, the wiggles are usually absent.

By compensation for the field gradients it is possible to obtain a resolving power of the instrument of the order of 10^7. When
this resolving power is reached, the intensity of the control signal
increases markedly, and with sweep rates greater than 50 mOe/sec
wiggles appear after the signal (Fig. 26).

Rotation of the sample is used to average out the remaining
gradients. We turn to rotation when good resolution has been obtained with a stationary sample.

The rotation rate of the turbine should be several hundred
revolutions per minute. Too high a rotation rate may lead to
spreading of the liquid over the walls of the ampoule. An insufficiently high rotation rate leads to the appearance of satellites, i.e.,
side signals lying on each side of the main signal at distances corresponding to the rotation frequency of the turbine (Fig. 27). As
has been pointed out, the appearance of satellites is explained by
the fact that the rotation of any point of the sample in a field with
a nonuniformity ΔH is equivalent to the application at this point of
an alternating longitudinal component of the field with an amplitude
$\Delta H/2$ and a frequency equal to the rotation frequency ω_T.

This leads to the appearance of side bands. However, if

$$\frac{\gamma \Delta H}{\omega_T} \ll 1$$

then the intensity of all side bands is negligibly small. Consequently, for the satellites to disappear, the rotation rate of the turbine
(f) should satisfy the relation

$$f \gg \frac{\gamma \Delta H}{2\pi}$$

When $H_0/\Delta H = 10^7$ and $H_0 = 15,000$, we find $f \gg 5$ rev/sec.

Therefore, we may take, for example, $f = 40$ rev/sec =
2400 rev/min.

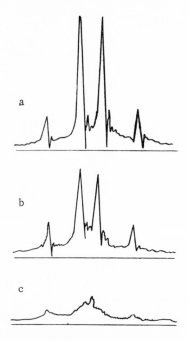

Fig. 28. Effect of saturation on the form of NMR signals. a) With a small amplitude of the rf field; b) the amplitude of the field is increased, while the amplification factor is reduced by a factor of 3.4; c) the amplitude is increased by a further factor of 3.4.

It was shown above that rotation of the sample does not compensate for field gradients along the axis of rotation. Therefore, when the turbine has been switched on it is necessary to change the currents in the correction circuits for the y gradients and the quadratic gradient, and sometimes it is also necessary to change slightly the positions of the knobs for compensation of the other gradients (or to move the head in the gap).

To establish that the optimal resolution has been reached from the picture of the spectrum observed on the oscillograph screen naturally requires experience, since this picture is usually strongly distorted by wiggles.

In addition to the main conditions, the successful recording of a high-resolution spectrum (reaching the maximum resolving power of the instrument) requires that saturation should be absent, and that the rate of passage through resonance should not be too high.

As was stated above, an excessive increase in the strength of the exciting field may lead to broadening of the line and a decrease in the signal intensity. In the plotting of a complex spectrum, which consists of a series of close lines, this would mean a deterioration in the resolving power and a fall in the sensitivity of the instrument (Fig. 28).

In exactly the same way, the spectrum recorded may be distorted if the condition for slow passage through resonance is not satisfied:

$$\frac{dH_0}{dt} \ll \gamma \, (\delta H)^2$$

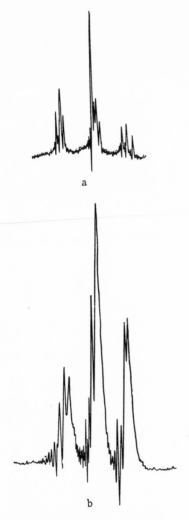

In the recording of very narrow lines in liquids, this condition is often difficult to fulfill, since the required sweep rate becomes comparable with the field drifts in some instruments. When the steady-state conditions break down, wiggles may distort the intensity of the line and change the position of its neighboring lines in the spectrum.

Figure 29a shows the spectrum of the CH_3 group of ethylbenzene, obtained on an NMR spectrometer 5535 with very slow passage through resonance. For comparison, Fig. 29b shows the same spectrum with a high rate of passage.

In work with a high-resolution spectrometer it is often necessary to check the stability of the magnetic field. For this purpose, spectrometers are usually provided with appropriate instruments and control sockets.

For accurate assessment of the magnitude and character of the instabilities we can recommend the following method, which is used frequently nowadays. In the main head of the spectrometer is placed a control sample, which gives a single and quite intense

Fig. 29. Spectrum of the CH_3 group of ethylbenzene with slow passage (a) and distorted by transient processes (b) because of relatively rapid passage (the sweep direction was changed).

proton line with a width of about 10 Hz. The sample used may be water with a small amount of a paramagnetic salt to broaden the line to the given value. The width of this line may be measured beforehand or determined directly during the measurement of the instability by recording with a calibration signal or on a calibrated chart.

With these conditions and a resolution of about 10^7, the intensity of this line is sufficient to give a deflection of the recording instrument of almost the whole scale (without appreciable noise in the absence of the signal) with the optimal value of the exciting field.

After determination of the width of the line, it is recorded without the calibration signal. Then the sweep of the instrument is switched off and with the field shift knob, the pen of the recorder is moved to the side slope of the line at a height corresponding to the maximum tangent to the line. If the spectrometer were absolutely stable, then when the sweep was switched off, the pen would remain at the initial height for an indefinite time. With any changes in the field (transient or slow) the pen will fluctuate, and the range of the fluctuations will be proportional to the range of fluctuations in the field, naturally, provided that the fluctuations of the pen do not go beyond the more or less linear section of the slope of the line.

The picture is particularly clear if the movement of the chart or the carriage of the recorder is switched on without the sweep switched on. In this case it is easy to observe the whole character of the instability of the instrument, i.e., transient kicks and slow drifts of the field.

It is obvious that knowing the width of the line, it is easy to calculate the numerical magnitude of the instability (Fig. 30).

Analysis of high-resolution spectra may give information on the chemical shifts of equivalent nuclei in a molecule relative to the signal of a standard substance, the values of the constants of spin—spin interaction between nonequivalent groups, and the relative numbers of equivalent nuclei in each group. From these data it is possible to draw conclusions on the constitution and the chemical structure of the substance studied.

Hence, the interpretation of the spectra consists of measuring the distances between the lines of the spectrum, measuring the positions of these lines relative to the resonance line of a standard, and measuring the relative intensities of the lines.

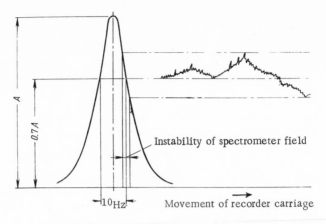

Fig. 30. Measurement of the instability of a spectrometer.

In contemporary spectrometers the spectrum is recorded on calibrated charts, so that it is possible to read off the distances directly in millionths of the strength of the polarizing field or in frequency units.

In some spectrometers the spectrum recorded is calibrated by modulation of the polarizing field with a low-frequency component. The distance between the side bands of the spectrum, which equals the modulation frequency, provides a scale for determining the distances between the lines of the spectrum.

In the determination of chemical shifts of H^1, the standards used are tetramethylsilane, benzene, water, acetic acid, etc.

To obtain the signal of the standard substance simultaneously with the spectrum of the substance investigated, the two substances are sometimes simply mixed. This method is called the internal standard method. If mixing of the analysis and standard substances is undesirable, then the latter is placed in an isolated compartment (for example, in a separate capillary) in the ampoule containing the substance investigated.

In using the external standard method, a correction must be introduced into the measured value of the chemical shift to allow for the difference in the magnetic susceptibilities of the analysis substance and the standard. This correction may be very considerable and reaches $5 \cdot 10^{-7}$.

In determining the numbers of equivalent nuclei in groups it must be remembered that these numbers are proportional to the integral intensities of the resonance lines and not the amplitudes. Integrators are incorporated into spectrometers for this reason.

In the simplest cases (weak spin—spin interaction) the interpretation of the results obtained presents no great difficulty. Additional methods are used in the analysis of more complex spectra. One of them, which makes it possible to distinguish a chemical shift from splitting due to spin—spin interaction, consists of repeating the recording of the spectrum at two different values of the frequency of the exciting field (and at the two corresponding polarizing field strengths). The method makes use of the fact that the chemical shift is proportional to the polarizing field strength, while the spin—spin interaction constant is independent of it.

The second method used in the analysis of spectra is the isotopic substitution method. It is used widely in the study of NMR in polymers and will be described below.

Finally, the double nuclear resonance method is yet another means of facilitating the analysis of complex spectra, which is beginning to find increasingly wide use with the appearance of special apparatus. This method consists of applying two rf fields simultaneously to the analysis sample, one field with a high intensity and a frequency corresponding to the resonance of one group of nuclei and the other field with a low intensity and a frequency equal to the resonance frequency of the other group. The resonance of the group excited by the weak field is observed during the experiment. Under these conditions the multiplet splitting of the line of this group, produced by spin—spin interaction of its nuclei with the nuclei of the group excited by the strong field, is almost completely eliminated and the spectrum is simplified.

This method is applicable in the case of weak spin—spin interaction, i.e., when the constant J is much less than the magnitude of the chemical shift δ.

Let us examine, for example, a molecule of the type AB, containing two spins with a spin number $\frac{1}{2}$ and characterized by a chemical shift δ and a spin—spin interaction constant J.

When $J \ll \delta$, the spectrum of this molecule consists of two pairs of lines of the same intensity and in each pair the lines are

at a distance J (page 81). This spectrum may be obtained if resonance is excited with an alternating field at constant frequency ω
and the polarizing field is changed so that it gradually approaches
the resonance region. Exactly the same spectrum is obtained if
the polarizing field is left constant, while the frequency of the exciting field is changed continuously.

Let us now assume that under the conditions of this experiment the sample is acted upon by yet another alternating field with
an intensity H_{1A} and a frequency ω_A, which equals the resonance
frequency of protons of group A. Calculations show that when the
frequency of the first exciting field reaches the resonance region
of the nuclei B, then instead of the previous two lines, corresponding to these nuclei, we obtain three lines lying at the frequencies
$\omega_B + 2a$; ω_B; $\omega_B - 2a$ and having the intensities $\cos^2\theta$; $2\sin^2\theta$;
$\cos^2\theta$. In these relations,

$$a = \frac{1}{2} \sqrt{\frac{1}{4}J^2 + \gamma^2 H_{1A}^2} \; ; \qquad \tan\theta = \frac{2\gamma H_{1A}}{J}$$

With low intensities of the additional field or, more accurately, when $2\gamma H_{1A} \ll J$, the intensity of the central line is very
low, while the outer lines are at practically the same points at
which they would be observed in the absence of the field H_{1A}.

With an increase in the amplitude of H_{1A} of the additional
field the intensity of the central line increases, while the intensity
of the outer lines falls and they move to the side.

When $\gamma H_{1A} \gg J/2$, practically only the one central line remains. We should note that the condition for disappearance of the
multiplet structure $(\gamma H_{1A}) \gg J/2$ is often more complicated than
the condition for normal saturation of resonance $\gamma H_{1A}\sqrt{T_1 T_2} \gg 1$.

Literature Cited

1.   Aleksandrov, N. M., and Moskalev, V. V., Vestnik LGU,
      10(2):55 (1960).
2.   Pople, J. A., Schneider, W., and Bernstein, H., High-Resolu-
      tion Nuclear Magnetic Resonance, McGraw-Hill, New York
      (1960).
3.   Schiff, L. I., Quantum Mechanics, McGraw-Hill, New York
      (1956).

4.   Andrew, E., Phys. Rev., 91 : 425 (1953).
5.   Bloembergen, N., Nuclear Magnetic Relaxation, Thesis,
       Leiden (1948).
6.   Corio, P. L., Chem. Rev., 6 : 363 (1960).
7.   McConnell, J., Chem. Phys., 28 : 430 (1958).

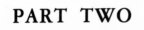

PART TWO

Chapter III

Study of the Structure of Polymers by the NMR Method

The structure of a solid is understood to pertain, in the broadest sense, not only to its crystal lattice and atomic orientation, but also in polymeric macromolecules, to its chain morphology and stereoregularity.

The NMR method has been used repeatedly for studying the crystal structure (see [10]) and is a valuable addition to the x-ray method. It has been shown theoretically [471] that by studying the relation of the second moment of an NMR line to the angle of rotation of a single crystal in a magnetic field about four axes, it is possible to calculate 15 structural parameters and to determine the dimensions of the elementary cell and the coordinates of three nuclei. However, due to the complexity of the structure and the difficulty of obtaining perfect single crystals of polymers, this direct approach to the study of their sturcture is practically impossible. A change in the form, width, and second moment of a line in relation to the position of the sample in the magnetic field is observed only for oriented polymers (fibers and films) and may be used to obtain data on the character of the orientation. For an unoriented polymer in bulk, information may be obtained on the structure from the form of the line and also from the second moment at low temperatures when molecular motions are restrained.

The most complete data on the chemical and stereochemical structure of macromolecules are provided by a study of the high-resolution NMR spectra of polymer solutions, but some conclusions may be drawn from the temperature dependence of the NMR line of a polymer in bulk.

1. Form of the Lines of NMR Spectra of Polymers [110]

The NMR spectrum of a polymer in bulk (i.e., in the form of a condensed system) generally consists of one bell-shaped line, and this is explained by dipole—dipole interaction of a large number of nuclei. However, the line sometimes assumes a more complex form with a more or less clearly expressed structure.

The complication of the line may be for different reasons, such as the presence of weakly interacting groups of atoms in the polymer, the effect of a chemical shift, and the presence of two phases in the system.

The exact quantum mechanical calculation of the form of the NMR line of such a complex system as a solid polymer is probably impossible at the present time. A qualitative description and systematization of the experimental data are apparently expedient.

Presence of Weakly Interacting Groups of Atoms in the Polymer. The lines of the NMR spectrum of isolated groups of atoms consisting of two or three nuclei (for example, CH_2, CF_2, CH_3, CF_3) have a complex form, which may be calculated theoretically (see page 89). These groups may be present in a polymer and, in cases where the distance between these groups in the bulk polymer is considerably greater than the distance between the protons or fluorine atoms in the group, the line broadening produced by the interaction of the groups is so small that the form of the line is distorted insignificantly, and the characteristic structure of the line is preserved.

Thus, the NMR line of a methyl group consists of a triplet. As Powles and Gutowsky [564] showed, even with additional Gaussian broadening of 8 Oe [2], there are clearly expressed extrema on the curve of the derived absorption function at a distance of ±8 Oe from the center of the line.

Figure 31 shows the NMR lines of polymers containing methyl groups (polymethyl methacrylate, polycarbonate, and butyl rubber) and a low-molecular substance, namely, methylchloroform CH_3CCl_3, determined at low temperature.

It is obvious that the form of the line in the spectra of all these substances is complex and the width between the outer extrema is about 16 Oe.

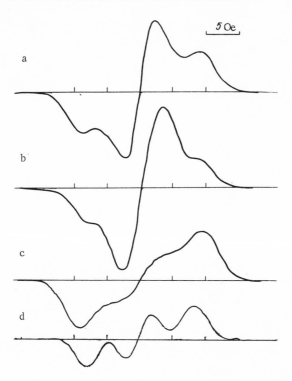

Fig. 31. Lines of the NMR spectra of substances containing CH_3 groups [110].
a) Polymethyl methacrylate at $-196°C$:   b) polycarbonate at $-196°C$:

$$\cdots-\overset{\displaystyle CH_3}{\underset{\displaystyle COOCH_3}{C}}-CH_2-\cdots$$

$$\cdots-\overset{}{\underset{\displaystyle O}{C}}-O-\langle\!=\!\rangle-\overset{\displaystyle CH_3}{\underset{\displaystyle CH_3}{C}}-\langle\!=\!\rangle-O-\cdots$$

c) butyl rubber at $-189°C$:   d) methylchloroform at $-196°C$.

$$\cdots-\overset{\displaystyle CH_3}{\underset{\displaystyle CH_3}{C}}-CH_2-\overset{\displaystyle CH_3}{\underset{\displaystyle CH_3}{C}}-CH_2-CH_2-\overset{\displaystyle CH_3}{C}=CH-CH_2-\cdots$$

A methylene group gives a doublet in the spectrum with a distance between the outer extrema of about 10 Oe. In the spectrum of polyethylene at $-196°C$ there is only a hint of a doublet structure of the line (Fig. 32); the strong interaction between the CH_2 groups in the chain and in neighboring chains evidently has an effect. This is also confirmed by the considerable increase in the second moment of the line (up to 28 Oe^2), while the contribution of an isolated CH_2 group to the second moment equals 11.2 Oe^2. When chlorine

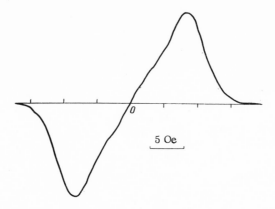

Fig. 32. Line of the NMR spectrum of low-density
polyethylene at −196°C [110].

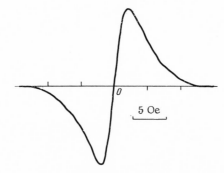

Fig. 33. Line of the NMR spectrum of polymethyl
methacrylate at room temperature [110].

atoms are introduced into the polymer chain, the distance between
the CH_2 groups is increased and this promotes the appearance of a
better expressed doublet structure to the line.

In polyethylene terephthalate (PET),

$$\cdots -O-C-\langle\!\!\!\langle\underline{\quad}\rangle\!\!\!\rangle- C-O-CH_2-CH_2-\cdots$$
$$\qquad\quad\; \overset{\|}{O} \qquad\qquad\quad \overset{\|}{O}$$

there are protons of methylene groups and phenyl nuclei. If three
hydrogen atoms in the nuclei are replaced by deuterium, the one

remaining proton interacts weakly with the methylene protons and a complex NMR line is obtained [706].

Orientation of the macromolecules promotes the appearance of the form of the line of the CH_2 group in the general contour of the NMR line of the polymer, as is shown by the spectra of monoaxially and biaxially drawn films of PET (see page 174).

As a rule, the complex structure of an NMR line due to the presence of weakly interacting groups appears only in spectra obtained at low temperature when the molecular motions are restrained. This is obvious from a comparison of the lines of polymethyl methacrylate and polycarbonate at −196°C (see Fig. 31) and at room temperature (Figs. 33 and 34). The amount of information obtainable from a spectrum is increased by observing a complex NMR line. Thus, the change in the form of the line of polycarbonate led to conclusions on the mechanism of thermal decomposition of the polymer, which were confirmed by the data of other authors [65] (see Chapter V).

Manifestation of a Chemical Shift in the NMR Spectrum of a Polymer in Bulk. The width of the line of the NMR spectra of polymers in bulk (in crystalline, vitreous, and highly elastic states) is 0.1-20 Oe.

The chemical shift for hydrogen at the magnetic field strengths normally used does not exceed hundredths of an oersted and does not appear in practice in the spectra of polymers.

For fluorine nuclei the chemical shift is an order greater than for protons, and if the NMR line is sufficiently narrow, it may have a complex form due to the chemical shift. Lyubimov and his co-workers [80] observed a chemical shift from resonance of CF_2 and CF_3 groups in the NMR spectrum of a copolymer of vinylidene fluoride and hexafluoropropylene at 90°C. When the spectrum of this copolymer was plotted by means of a bridge system, the doublet structure of the line was evident even at 60°C (Fig. 35).

A chemical shift was also observed from the resonance of fluorine in the NMR spectrum of a crystalline copolymer of tetrafluoroethylene and hexafluoropropylene [716] and fluoro rubbers [11].

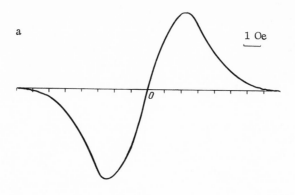

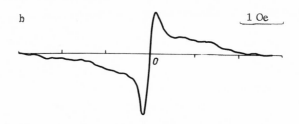

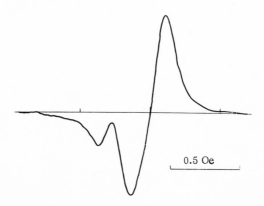

Fig. 34. Line of the NMR spectrum of polycarbonate [110].
a) At room temperature; b) at 176°C.

Fig. 35. Fluorine line in the NMR spectrum of a copolymer
of vinylidene fluoride with hexafluoropropylene at 60°C [110].

Two-Phase Systems. The appearance of an NMR line of complex form may be due to the atoms of the same element being present in two phases in the system with the molecular structure or the mobility in the two phases being different.

Nuclei in the phase in which the interatomic distances are greater or the mobility is greater give a narrow line; a broad line from the second phase is superimposed on it. An NMR spectrum is obtained which consists of two components, one broad and one narrow. NMR lines of this form are characteristic of polymers containing traces of low-molecular-weight substances such as moisture (see Chapter VI), polymers in which there are regions with different cross-link densities (see Chapter V), and polymers in which there are amorphous and crystalline regions.

As a rule, the complex structure of the NMR line in two-phase systems appears only at high temperature, when the molecular motion in one of the phases is sufficiently intense to give a narrow line. As the temperature is lowered, the structure disappears, as is shown by a comparison of the spectra of polyformaldehyde obtained at −60 and −10°C (see page 157).

The same polymer may give a complex NMR line at both low and high temperatures; this is due to different reasons. Thus, at low temperature, polycarbonate gives a complex NMR line, which is characteristic of a methyl group (Fig. 31b); at room temperature it gives a simple line (Fig. 34a), while at an elevated temperature it gives a complex line again (Fig. 34b), but this time due to increased molecular motion in the amorphous regions of the polymer.

In some cases the same spectrum may show line structure for two reasons. For example, in the NMR spectrum of oriented PET, a narrow line characteristic of amorphous regions of the polymer is superimposed on the doublet form of the line, characteristic of the CH_2 group (see page 174).

Second Derivative of the Absorption Function [113]. The characteristics of the form of the line appear very clearly when we record the second derivative of the absorption function. If the line has a simple bell shape, then from the second derivative we may decide whether the function of the form of the line is close to Lorentzian or Gaussian. For this purpose it is

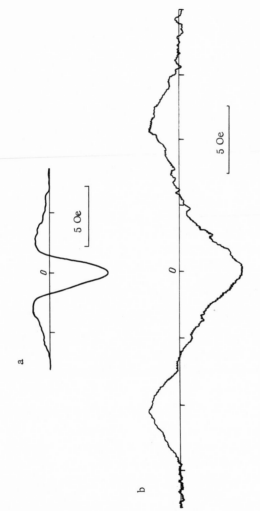

Fig. 36. Form of the second derivative of the line of the NMR spectrum [113]. a) Polymethyl metha-crylate at room temperature; b) isotactic polypropylene at −196°C.

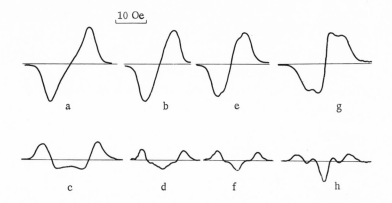

Fig. 37. Form of the first (a, c, e, g) and second (b, d, f, h) derivatives of the
line of the NMR spectrum of polyformaldehyde at different temperatures [113].
a,b) At −60°C; c,d) at −25°C; e,f) at −20°C; g,h) at −10°C.

sufficient to determine the ratio of the height of the side maxima
to the height of the central minimum. For lines of Lorentzian form
this ratio equals 0.25, while for those of Gaussian form it equals
$2/e^{3/2} \approx 0.45$. Thus, for the NMR line of polymethyl methacrylate
obtained at room temperature (Fig. 36), the ratio of the heights of
the extrema equal 0.29, while for the NMR line of isotactic polypropylene at −196°C (77°K) this ratio equals 0.44. It is obvious
that, in the first case, the form of the line is close to Lorentzian
and in the second it is close to Gaussian.

The advantages of studying the NMR in polymers using the
second derivative of the function may be illustrated on the example
of recording the absorption line of polyformaldehyde (Fig. 37).

At low temperatures the NMR line of polyformaldehyde has
a doublet structure which is due to CH_2 groups. With a rise in
temperature the line has a two-component form. This is explained
by the fact that under these conditions there is an increase in the
mobility of segments of amorphous regions of the polymer, while
the crystalline regions remain rigid.

At −60°C, on the curve of the first derivative (Fig. 37a),
there is only a hint of a doublet, while on the curve of the second
derivative (Fig. 37b), there is a clear doublet.

At −25°C, on the curve of the first derivative there are no
signs of a complex structure (Fig. 37c), while on the curve of the

second derivative (Fig. 37d) the presence of two components is quite clear.

A clearly expressed complex structure with additional maxima is observed at −20°C when the second derivative is plotted (Fig. 37f), while it appears only at −10°C when the curve of the first derivative is recorded (Fig. 37g).

2. Study of Crystallinity of Polymers

Reflection of Crystallinity in NMR Spectra of Polymers.   Even in the first studies of the NMR spectra of polymers it was observed that the magnitude and the form of the signal changes markedly, depending on the degree of crystallinity of the polymer. Holroyd et al. [357] observed that with rapid cooling of natural rubber to −35°C the amplitude of the NMR signal remains large, since crystallization of the polymer cannot occur, but when a sample is kept at a temperature of −25°C, which is close to the temperature where the crystallization rate is maximal, there is broadening and weakening of the line. A study was made recently [640] of the NMR spectrum of a sample of natural rubber which had been stored at low temperature for 25 years and had a degree of crystallinity (estimated from the density) of about 30%. It was found that for this sample the temperature of contraction of the line was considerably higher than for normal rubber, while the line width over the range from −60 to −40°C was substantially greater. Powles and Kail [569] observed a marked difference between the curves of the temperature dependence of the line width for a "native" (highly crystalline)* preparation of polytetrafluoroethylene and sintered polymer, and explained the great scatter of literature data on NMR of polytetrafluoroethylene by differences in the degree of crystallinity of the samples. With native polytetrafluoroethylene the NMR line narrows at a higher temperature than with sintered polymer. Analogous relations are observed for polyethylene terephthalate [205]: heating low-molecular prepolymer in vacuum gave powdered polyethylene terephthalate with a very well developed crystal lattice. In the NMR spectrum of this polymer the narrow component appears at a temperature 25° higher than in the NMR spectrum of unoriented film of the same polymer, obtained by melt extrusion.

---

*"Native" means a sample which has not been heated after polymerization.

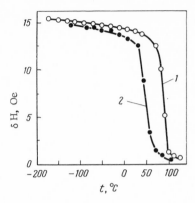

Fig. 38. Relation of the width of the line of the NMR spectrum
(δ H) of polyethylene crystals to the temperature [336]. 1) Fresh-
ly obtained crystals; 2) after 15 days.

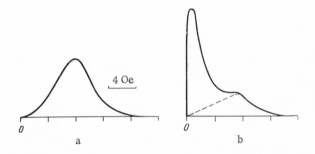

Fig. 39. Line of the NMR spectrum of polyethylene at different
temperatures [720] (half curves are shown in the figure). a) At
$-183°C$; b) at $3.5°C$.

Study of Polymer Single Crystals. The NMR
method has been used to study the structure of so-called "single
crystals" of polyethylene, obtained by cooling dilute solutions of
the polymer in xylene or other solvents. Slichter [632] showed
that a "single crystal" of polyethylene, which had not been subjected to thermal treatment, gives a simple NMR line (the spectrum was obtained at room temperature). The NMR line of the
preparation, after it had been heated to 120-140°C and subsequently
cooled, consisted of two components, namely, a broad and a narrow
component and the intensity of the narrow component was greater,
the higher the heating temperature. A few minutes heating of the

sample was sufficient for the appearance of the narrow component in the NMR spectrum, but the equilibrium intensity of the narrow component was established after 30 min of thermal treatment. Peterlin and Pirkmajer [553] also observed a narrow component in the NMR line of a single-crystal sample of polyethylene after it had been heated above 70°C. It was concluded that the narrow component of the NMR signal corresponds to regions with a defect crystal structure. The same conclusion was drawn by Odajima, Sauer, and Woodward [527], who studied the NMR of crystals of polyethylene and a series of normal paraffins. Thus, lattice defects which give a narrow line in the NMR spectrum were observed in crystals in n-$C_{94}H_{190}$ as well as in polyethylene. Under certain conditions the thermal treatment of single crystals of polyethylene leads to "healing" of the lattice defects, as is confirmed by an irreversible increase in the width and second moment of the NMR line [290].

Herring and Smith [349] found that in linear polyethylene "Marlex 50," which had been recrystallized from trichloroethylene, at a temperature from −33 to −3°C there is a transition, which is connected with the movement of disordered sections of the chains, consisting of six CH_2 groups.

By studying the temperature dependence of the width and the form of the NMR line it was possible to observe [315, 687] that the structure of a single crystal of polyethylene changes gradually during storage. Even after 15 days the temperature of narrowing of the line is shifted from +87 to +47°C (Fig. 38), while after several months, judging by the form of the NMR line, polymer obtained from solution is close in crystal structure to polymer obtained from the melt.

Determination of the Degree of Crystallinity. The problem of determining the degree of crystallinity of polymers from the form of the NMR line has attracted great attention. Even in 1953, Wilson and Pake [720, 721] showed that over a definite temperature range the NMR line of ^1H in polyethylene and ^{19}F in polytetrafluoroethylene has a structure which is characteristic of a two-phase system (Fig. 39). At a temperature above 0°C, the line consists of two components, namely, a narrow and broad component (as is shown by a broken line in Fig. 39). Wilson and Pake put forward the hypothesis that the narrow line corresponds to the amorphous part of the sample and the broad line to the crystalline part.

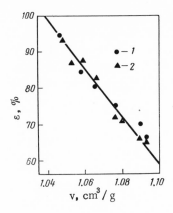

Fig. 40. Relation of the degree of crystallinity (ε) and the specific volume (v) for polyolefins [654]. 1) X-ray method; 2) NMR method.

In the absence of saturation effects the area under the curve of the signal is proportional to the number of nuclei. Therefore, it may be assumed that the ratio of the areas under the curves bounding the broad and narrow components of the signal (or the ratio of the first moments of the lines when the derivative of the absorption function is plotted) equals the ratio of the numbers of nuclei in the crystalline and amorphous parts of the sample.

The use of this method to determine the degree of crystallinity of polyethylene and other polyolefins at room temperature has been very successful. For a series of polyethylenes [245] with a degree of crystallinity ε equal to 59-93%, the discrepancy between the values of ε determined by the x-ray method and the NMR method did not exceed 1.8% and, as was pointed out by the author, the NMR method has the advantages of better reproducibility and insensitivity to orientation. Smith [654] used the NMR method to study a series of polyethylenes and copolymers of ethylene with propylene and butene-1. On a graph (Fig. 40) of the degree of crystallinity against the specific volume of the polymer, the points obtained by the NMR method and by the x-ray method lay on a single straight line. Infrared spectroscopy also gives similar values of ε. Good agreement between values of the degree of crystallinity determined by the NMR method and from the density for a series of samples of polyethylene with different degrees of branching has also been reported by Fuschillo and Sauer [307].

However, attempts to extend the method of determining crystallinity of Wilson and Pake to other polymers and to use the method over a wide range of temperatures did not give satisfactory results. For polyethylene and nylon over a wide range of temperatures the degree of crystallinity determined from the relative magnitude of the broad component of the NMR signal did not coincide with that determined by the x-ray method. This is confirmed by measurements on irradiated polymers. Despite the fact that irradiation reduces the crystallinity to zero, the high value of the broad component of the NMR line is retained, since cross-links are formed [306]. For polybutene and polypropylene [603] and also for irradiated polyethylene [647] a considerable discrepancy has been found between the results of measuring the degree of crystallinity by Wilson and Pake's method and by other methods, namely, by density and by the x-ray method (see also [337]).

Slichter and McCall [650] and Fuschillo and his co-workers [305] put forward a reason for these discrepancies. Different methods of determining the degree of crystallinity, namely, x-ray, specific volume, and NMR methods, measure different physical values. Analysis of x-ray diffraction patterns makes it possible to establish the degree of order in the sample; the specific volume depends on the packing density; the form of the NMR signal is determined not only by the position of the chains and groups in the polymer, but also largely by the character of the intermolecular motion in it (see Chapter I). If there are mobile groups in a solid, the NMR line narrows because of averaging of the local magnetic fields. Narrowing begins when the correlation time τ_c is less than 10^{-4}-10^{-5} sec. If a sample contains regions (crystalline) with $\tau_{c,cr} \gg 10^{-4}$ sec and regions (amorphous) with $\tau_{c,am} \ll 10^{-4}$ sec, then it may be regarded as a two-phase system. Then an NMR curve of complex structure is obtained and the degree of crystallinity may be determined from the ratio of the areas of the components.

Two other cases are also possible, and these are represented schematically in Fig. 41, which shows the change in the form of the derived absorption function with successively decreasing correlation times (rising temperature). At a low temperature (case a), the two phases give a broad line. With a rise in temperature (b), in the amorphous region there begins movement of groups or segments, which appears as a kink in the curve. With a further rise in temperature (case c), the amorphous phase gives a narrow line and the crystalline phase, a broad line. At high temperature (case d), movement is considerable in both phases, even before the beginning of melting and one narrow line is obtained.

It is obvious that values of the degree of crystallinity ε_{NMR} obtained by the NMR method must depend on the temperature at which the spectrum is obtained. Slichter and McCall [650] found that the values of ε_{NMR} of linear polyethylene are independent of

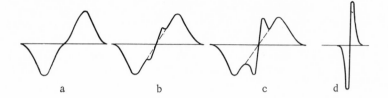

Fig. 41. Effect of the correlation times in two phases on the form of the line of the NMR spectrum [650]. a) $\tau_{c,am} > 10^{-4}$ sec, $\tau_{c,cr} > 10^{-4}$ sec; b) $\tau_{c,am} \approx 10^{-4}$ sec, $\tau_{c,cr} > 10^{-4}$ sec; c) $\tau_{c,cr} \gg 10^{-4}$ sec $\gg \tau_{c,am}$; d) $\tau_{c,am} \ll 10^{-4}$ sec, $\tau_{c,cr} < 10^{-4}$ sec.

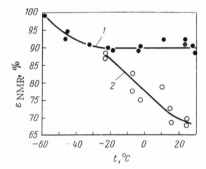

Fig. 42. Relation of the degree of crystallinity of polyethylene determined by the NMR method (ε_{NMR}) to the temperature [650]. 1) Linear polyethylene; 2) polyethylene with branched chains.

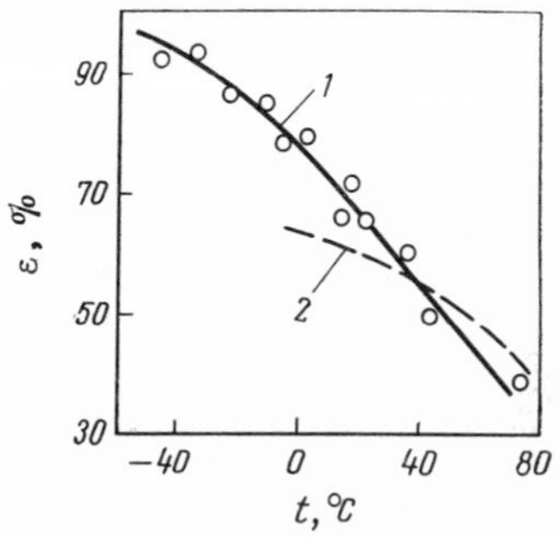

Fig. 43. Comparison of data [305] on the temperature dependence of the degree of crystallinity of polyethylene obtained by the NMR method (curve 1) and from density (curve 2).

temperature and coincide with data obtained by other methods over the range from −33 to +27°C, while ε_{NMR} of branched polyethylene over the same temperature range falls rapidly as the temperature rises (Fig. 42).   Fuschillo, Rhian, and Sauer [305] compared ε_{NMR} of polyethylene with the ratio $CH_3/CH_2 \approx 0.2$ and the number average molecular weight $M_n \approx 25,000$ (Fig. 43, curve 1) with the degree of crystallinity determined from the density (curve 2). The two methods give the same value of ε only over a narrow temperature range. For polybutadiene [603], $\varepsilon_{NMR} = 65\%$ at 0°C and falls to 45% at 27°C, while the degree of crystallinity determined from the specific volume is constant over this temperature range, and equals 32%.

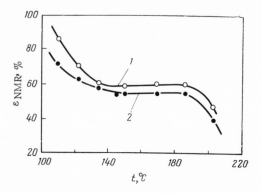

Fig. 44. Temperature dependence of the degree of crystallinity (ε_{NMR}) [464]. 1) 6,6 polyamide; 2) 6, 10 polyamide.

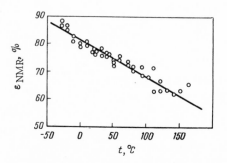

Fig. 45. Temperature dependence of the degree of crystallinity (ε_{NMR}) of polyformaldehyde [132].

The two-component form of the line caused by molecular motion in amorphous regions appears only at a temperature above the glass transition point. For polyethylene terephthalate [282], a two-component signal is observed at a temperature above 110°C. The ratio of the areas of the narrow and broad components rises with temperature, and at 180°C it reaches a constant value, corresponding to ε_{NMR} approximately equal to 70% for unoriented and 80% for oriented polymer. The degree of crystallinity determined by the x-ray method is close to 60% and changes little over the temperature range from room temperature to 180°C.

With 6,12 polyamide (from hexamethylenediamine and decamethylenedicarboxylic acid) an NMR signal of complex form is observed at a temperature above 70°C [373]. With 6,6 and 6,10 polyamides, the separation of the NMR line into two components is possible only over the range between 110 and 190°C [463]; curves of the relation of ε_{NMR} of a polyamide to the measurement temperature (Fig. 44) have a horizontal section in the region of 140-180°C.

In our investigation of the NMR spectrum of polyformaldehyde [132] it was also shown that a complex line is obtained at a temperature above −30°C and that the degree of crystallinity ε_{NMR} falls with a rise in temperature (Fig. 45). Analogous results were obtained recently by Chudesima and Kakigzahi [232], who determined ε_{NMR} of polyformaldehyde samples (Delrin) obtained by crystallization from the melt and from solution. The fraction of mobile protons rises with temperature and depends strongly on the thermal history of the sample.

For some polymers, for example for natural rubber, the condition $\tau_{c,cr} \gg 10^{-4}$ sec $\gg \tau_{c,am}$ apparently does not hold at any temperature and no complex NMR line is obtained at all [338].

Thus, available literature data on the relation of ε_{NMR} to the temperature at which the spectrum is obtained qualitatively confirm the above arguments of Slichter, McCall, and Fuschillo on the relation of the form of the NMR line of the crystalline polymer to the correlation times in the crystalline and amorphous regions ($\tau_{c,cr}$ and $\tau_{c,am}$).

A more strict quantitative theory, which takes into account the presence of a distribution (spectrum) of correlation times, has been developed recently by Chujo [233]. Together with the concept of the static degree of crystallinity ε_{stat}, i.e., the ratio of the number of segments in crystalline regions to the total number of segments in the polymer sample, the concept of the dynamic degree of crystallinity ε_{dyn} is introduced. The value of ε_{dyn} is determined from NMR experiments and also from the mechanical relaxation and dielectric relaxation in the polymer, and is the ratio of the number of segments which move with a frequency less than some critical value to the total number of segments. For NMR, the critical frequency is the square root of the second moment of the line (expressed in frequency units). To derive a relation between ε_{dyn} and ε_{stat}, Chujo starts from the relaxation theory of Kubo and Tomita [424] and Miyake [483]. It is assumed that the spectrum of correlation times of a partly crystalline polymer consists of the sum of the contributions of the crystalline and amorphous regions and that it is possible to neglect the effect of lattice defects and regions of intermediate order. It is shown that in this case over a certain middle range of temperatures $\varepsilon_{dyn} \approx \varepsilon_{stat}$, at low temperatures $\varepsilon_{dyn} > \varepsilon_{stat}$, and at high temperatures $\varepsilon_{dyn} < \varepsilon_{stat}$. If we know the form of the spectrum of correlation times in the amorphous regions I $(\tau_{c,am})$, then from the value of ε_{dyn} measured in the region of low temperatures it is possible to determine the fraction of nuclei in the amorphous regions which make a contribution to the broad component of the line and to calculate ε_{stat}. From the known values of ε_{dyn} and ε_{stat} determined by two independent methods at several temperatures it is possible to draw conclusions on the width of the spectrum of correlation times and to estimate the activation energy of molecular motions in the amorphous regions E. In the first approximation it may be assumed that in amorphous regions of the polymer there is a rectangular logarithmic distribution of correlation times with a lower limit τ_a and an upper limit τ_b [518]:

$$I(\tau) = \frac{1}{\ln(\tau_b/\tau_a)} \quad \text{when} \quad \tau_a \leqslant \tau \leqslant \tau_b$$
$$I(\tau) = 0 \quad \text{when } \tau < \tau_a \text{ and when } \tau > \tau_b \tag{III-1}$$

while the temperature dependence of the correlation time obeys the Arrhenius equation:

$$\tau = \tau_0 e^{\frac{E}{kT}} \tag{III-2}$$

Then we obtain the simple relation

$$\varepsilon_{dyn} = \varepsilon_{stat} + \frac{1 - \varepsilon_{stat}}{\ln(\tau_b/\tau_a)} \left\{ \frac{E}{kT} + \ln \sigma_0 \tau_{0b} \right\} \tag{III-3}$$

where σ_0 is the square root of the second moment of the line.

As is obvious from equation (III-3), the experimental values give a straight line on a graph of ε_{dyn} versus $1/T$. From the slope of the line relative to the ordinate axis for $1/T = 0$ it is possible to determine τ_b/τ_a and E. From the experimental data of Fuschillo et al. [305] (see Fig. 43) for polyethylene, by substituting the value $\tau_{0b} = 10^{-13}$ sec and $\sigma_0 = 10^4$ sec^{-1}, Chujo calculated the width of the spectrum of correlation time $\tau_b/\tau_a \approx 10^3$ and the activation energy $E \approx 10$ kcal/mole.

The case of more complicated spectra of correlation times and the relation $\tau(T)$ is also examined in Chujo's work.

In determining the degree of crystallinity by the NMR method it is also necessary to take into account difficulties caused by the very nature of the crystalline state of high polymers, namely, the presence of regions of intermediate order. The existence of such regions has also been detected by NMR observations.

Hyndman and Origlio [371] plotted the NMR spectrum of a bundle of cold-drawn polyethylene fibers, oriented at different angles to the field vector (Fig. 46). In addition to the narrow and broad lines, there is a third component, which gives the middle maximum in Fig. 46a. The third maximum disappears after annealing. The authors consider that the third component corresponds to regions which are amorphous, but are oriented during drawing.

By a series of methods, including NMR, Miller [478] showed that pouring a melt of polypropylene into cold water gives a sample with properties which are intermediate between the properties of crystalline and amorphous polymer. He calls this polymer "noncrystalline polypropylene" and considers that the spiral structure of the chains is retained in it, but not the strict order in a transverse direction.

A detailed study of the NMR of a series of samples of polytetrafluoroethylene [296, 383] showed that at room temperature the NMR signal consists of three superimposed lines with widths δH = 1.2, 3.9, and 8 Oe. These lines correspond to amorphous regions of the polymer, regions in which crystallization occurred only perpendicular to the c axis, and regions which crystallized both perpendicular and parallel to the c axis. A third "intermediate" component of the NMR lines was observed by Fischer and Peterlin [290] for single crystals of polyethylene obtained from solution. Structures intermediate between amorphous and crystalline have also been observed by Schlichter [630] in samples of polyethylene after cold drawing. Analysis of x-ray data and the temperature dependence of the second moment of the NMR line shows that in these samples there are regions in which the chains are packed in a pseudohexagonal lattice, but retain the disorder produced by random rotation of the zigzag chains about their axes. These structures are analogous to the "gas-crystalline" state of polymers described by Kitaigorodskii [62, 64].

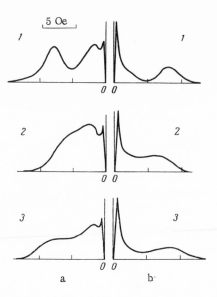

Fig. 46. NMR spectrum of a bundle of polyethylene fibers
after cold drawing (a) and after annealing (b) with the
fibers lying at different angles to the field vector [371].
1) 0°; 2) 45°; 3) 90°.

By the NMR method it is also possible to detect defects
formed in the crystal lattice of polyethylene by irradiation [647]
and by thermal treatment [632]. A comparative study of polypropylene and polybutene by the NMR method and by the dilatometric
method over a wide range of temperatures [602] shows that the
crystalline sections of polybutene are less well developed than in
polypropylene. Thus, in a number of cases we cannot characterize the degree of order of the structure by one value, namely, the
degree of crystallinity. Some authors have suggested that the ratio
of the area of the broad component to the area of the signal should
not be called the degree of crystallinity, but the index of high-frequency rigidity of the polymer N_{rig}/N (where N_{rig} and N are the
number of nuclei giving the broad component of the signal and the
total number of nuclei in the sample) [306, 308]. If this index is
high, the polymer is rigid and glass-like. This is true regardless
of whether the high value of N_{rig}/N is due to high crystallinity,
cross-links, low temperature, or, finally, the effects of hydrogen
bonds or dipole interactions. The value of N_{rig}/N shows the degree

to which thermal activation of the movement of the chains may overcome the intra- and intermolecular hindrance to motion. McCall and Anderson [462] rightly point out that this value is a more important characteristic of a polymer than the x-ray degree of crystallinity, since the chemical and physical properties of polymeric materials are affected by precisely their rigidity and not by the presence of crystallites.

To calculate the degree of crystallinity of a polymer from the form of the NMR line it is necessary to separate the complex line into two components, namely the broad and narrow components. The form of the NMR line of a two-phase system depends on the ratio of numbers of nuclei in the two phases and the ratio of the width of the line of one phase to the width of the line of the other phase. With certain values of these ratios [111] we obtain a curve of the derivative of the absorption function with two maxima, as shown in Fig. 41c (see page 163). In this case the line may be divided visually without a large error. If there is no obvious division into narrow and broad lines on the curve, special methods are used.

The method of linear anamorphoses [124] was used in [90] to separate the NMR curve of polypropylene into two components. For separation into components, other authors [349, 588] have assumed that the broad component retains at all temperatures the same form which the line has at low temperature, when molecular motions are restrained.

The derivative of the line of the dispersion signal has a more clearly expressed structure, and it may be separated into components more readily than when the derivative of the absorption function is recorded [688].

The "two-phase" character of a partly crystalline polymer appears not only in the form of the line, but in the presence of two spin—lattice relaxation times T_1, which may be observed by the spin-echo method or through the saturation curve. This circumstance should be borne in mind in measuring the ratio of the intensities of the components of a complex line. The narrow line of the amorphous phase is saturated more strongly than the broad line of the crystalline phase and with too high a strength of the rf field H_1 we obtain a high value of ε.

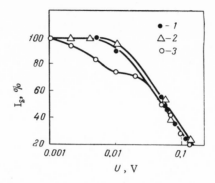

Fig. 47. NMR saturation curves of fluorine in polytetrafluoroethylene at different temperatures [719]. 1) At
−182°C; 2) at −138°C; 3) at −103°C. (U is the voltage
at the output of the oscillator; the signal intensity I_s
without saturation is taken as 100%.)

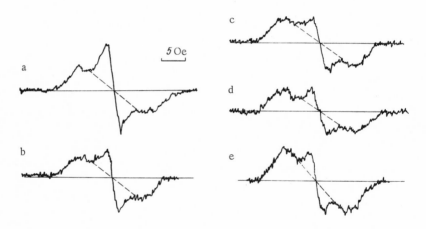

Fig. 48. Effect of time of crystallization from the melt at 127°C on the form of the line
of NMR spectrum for polyethylene samples (the spectra were obtained at room temperature) [554]. a) 10 min; b) 90 min; c) 150 min; d) 300 min; e) 1050 min.

McCall and Anderson [462, 463] recommend extrapolation of
the experimental values to $H_1 = 0$. The degree of crystallinity may
be estimated from the form of the saturation curve. Thus, for
polytetrafluoroethylene [545, 719] the curve of the relation of the
relative intensity of the signal to the strength of the rf field (Fig.
47) at −103°C has a step form. About 28% of all the ^{19}F nuclei

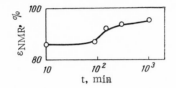

Fig. 49. Relation of the degree of crystallinity (ε_{NMR}) of polyethylene to the time of crystallization from the melt at 127°C [554].

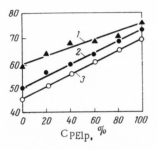

Fig. 50. Comparison of data on the relation of the degree of crystallinity (ε) of a mixture of high- and low-pressure polyethylenes to the composition of the mixture, determined by different methods (C_{PElp} is the content of low-pressure polyethylene in the mixture) [303]. 1) NMR method; 2) infrared spectroscopy; 3) from density.

present in the sample are saturated first and then the other 72%. For the narrow line (amorphous part), for which $T_2 = 5 \cdot 10^{-5}$ sec, saturation is reached at a lower strength of the rf field than for the broad line with $T_2 = 0.9 \cdot 10^{-5}$ sec.

The ratio of the area of the narrow component to the total area of the NMR curve is also close to 30%, i.e., the degree of crystallinity equals approximately 70%. With a fall in temperature the values of the relaxation times of the two phases become equal and the normal saturation curve is obtained at −138°C.

Nonstationary processes, for example, observing the fall in the nuclear signal after a 90° pulse of the rf field (see Chapter II), are also used to determine the dynamic degree of crystallinity. With linear polyethylene this method gives the same value of ε as analysis of the form of the NMR line [343, 466].

Determination of the degree of crystallinity from the form of the NMR line may be used to study the kinetics of crystallization processes in a polymer. Collins [247] used this method to study the kinetics of the crystallization of polyethylene. A detailed investigation of the recrystallization of polyethylene was carried out in 1963 by Peterlin [554]. Figure 48 shows the change in the form of the NMR line during the crystallization of polyethylene. It shows how the intensity of the narrow component of the line gradually falls.

Figure 49 gives the curve of the rise in ε_{NMR} of polyethylene in relation to the crystallization time. A combination of the

NMR and dilatometric methods made it possible to draw conclusions on the mechanism of the process.

The pulse method is particularly convenient for studying crystallization kinetics as the determination requires only a few seconds in this case [343].

Fujiwara and Narasaki [303] studied the crystallization of mixtures of high- and low-pressure polyethylenes by NMR, spectroscopic, and dilatometric methods. As Fig. 50 shows, additivity of the values of ε was observed.

The NMR method has been used to study crystallization of polypropylene [508] and for characterizing the structure of samples of polypropylene with different degrees of crystallinity [88]. It has also been used to study the effect of process conditions on the degree of crystallinity of polyvinyl chloride films [301] and to determine the amorphous fraction of viscose fibers [7].

Ermilova, Urman, and Slonim [42] used the NMR method to show that during the granulation of powdered polypropylene and its conversion into film by extrusion and blowing, the dynamic degree of crystallinity falls. Moreover, it was shown that the introduction of an effective stabilizer reduces the fall in the degree of crystallinity and improves the mechanical properties of the film obtained.

It should be borne in mind that the crystallization of a polymer may occur during the actual determination of the NMR spectrum. In such cases (for example, for polyethylene terephthalate and aromatic polyamides [84]), the relation $\Delta H_2^2 = F(T)$ has a minimum, i.e., the second moment of the line falls at the glass transition point of the initially amorphous polymer and then increases because of crystallization of the sample in the spectrometer.

3. Nuclear Magnetic Resonance
in Oriented Polymers

A change in the orientation of the molecules of a polymer such as in the formation of fibers has a great effect on the NMR spectrum. Thus, Ward [706] observed an increase in the second moment of the NMR line with an increase in the degree of orientation of polyethylene terephthalate and a number of other polyesters.

Molecular motion is apparently substantially reduced in drawn oriented films of polymer.

An analogous effect (an increase in the width of the line) has also been observed during the cold-drawing of linear polyethylene [630], while there is no substantial difference between the NMR signals of unoriented and drawn samples of high-pressure branched polyethylene.

During the drawing of polymethyl methacrylate, ΔH_2^2 first increases until the degree of drawing exceeds 135%, and this is explained by a decrease in molecular mobility and then, when the degree of drawing is greater than 300%, the second moment of the line (ΔH_2^2) falls as a result of breakdown of the structure [71].

When fibers of 6,6 polyamide are drawn there is an increase in the width of the wide component of the NMR line and a sharp fall in the intensity of the narrow component. From this it may be concluded that the rigidity of the lattice in an oriented polymer is greater than in an unoriented polymer, despite the fact that the degree of crystallinity determined by the x-ray method falls during the drawing of a fiber [661].

Relation of the Form of the NMR Line to the Position of the Sample in the Magnetic Field. There are great possibilities for studying the structure of polymer fibers and films by observing the change in the NMR spectrum of the polymer in relation to the angle between their axis of orientation and the magnetic-field vector.

For the form of the NMR line to change during rotation of the polymer sample in the magnetic field the following conditions should hold simultaneously. The structure of the polymer must contain weakly interacting groups of atoms (in other words, in general, a simple bell-shaped line should be obtained, see page 150). The form of the NMR line of these groups should depend substantially on their position in the field. A considerable part of these groups must have the same orientation in the sample. In practice, this combination is found only in oriented polymers containing CH_2 groups.

The expression for doublet splitting for a pair of protons is

$$\Delta H = \frac{3\mu}{r^3} (3 \cos^2 \theta - 1)$$

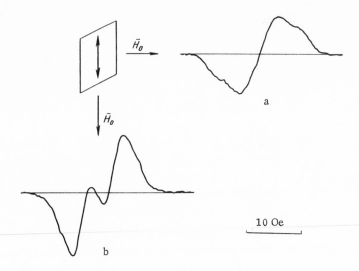

Fig. 51. NMR spectra of monoaxially drawn film of polyethylene tere-phthalate at −196°C [133]. a) Drawing direction perpendicular; b) parallel to the field vector.

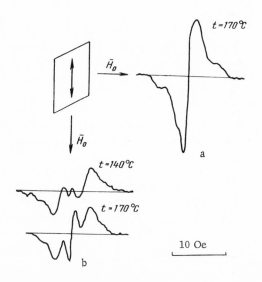

Fig. 52. NMR spectra of monoaxially drawn film of polyethylene terephthalate at 140 and 170°C [133]. a) Drawing direction perpendicular; b) parallel to the field vector.

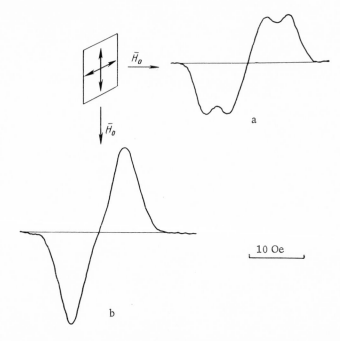

Fig. 53. NMR spectra of biaxially drawn film of polyethylene terephthalate at −196°C [133]. a) Field vector perpendicular to the plane of
the film; b) field vector parallel to the direction of drawing.

where μ is the magnetic moment of a proton, which equals 14.10 ·
10^{-24} erg/Oe; r is the distance between the protons; θ is the angle
of the internuclear vector with the field vector.

For protons of a CH_2 group, r = 1.78 · 10^{-8} cm and $3\mu/r^3 \approx$
7.5 Oe. If the internuclear vectors of H—H are directed along the
field vector, then θ = 0° and the splitting equals 15 Oe. If the
H—H vectors are perpendicular to the field vector, then θ = 90°
and the splitting equals 7.5 Oe. In polyethylene terephthalate,

$$\cdots -O-\underset{O}{\overset{\|}{C}}-\left\langle\underline{}\right\rangle-\underset{O}{\overset{\|}{C}}-O-CH_2-CH_2-\cdots$$

the protons of the methylene groups are remote from the protons
of the benzene rings and neighboring chains. In a monoaxially
drawn film of polyethylene terephthalate a large part of the chains
lie along the direction of drawing and the internuclear vectors of

H—H are perpendicular to this direction (shown by an arrow in Fig. 51). If the magnetic field is directed along the axis of orientation, then we obtain the doublet signal which is characteristic of a methylene group, but with a splitting equal to only 3.5 Oe due to the broadening produced by dipole—dipole interaction with the remaining protons [133]. The same form of NMR line has been observed for a film of polyethylene terephthalate in [205] and for drawn fiber of the polymer in [372].

With a rise in temperature to 170°C, another narrow component is superimposed on the doublet line of the CH_2 group, corresponding to mobile protons in the amorphous regions of the polymer (Fig. 52).

In a biaxially drawn film of polyethylene terephthalate the orientation of the crystallites is such that a considerable part of the H—H vectors of the methylene groups are perpendicular to the plane of the film [138]. When the film lies perpendicular to the magnetic field vector, for many CH_2 groups the angle $\theta = 0°$ and the NMR line (Fig. 53) shows a broad component with a distance between the extrema greater than 15 Oe. When the film is turned along the field, for a considerable part of the CH_2 groups $\theta \approx 90°$ and the line narrows.

An analogous change in the form of the NMR line is observed when a drawn film of polyformaldehyde [736] $\cdots - CH_2 - O - \cdots$ is rotated in the magnetic field.

In oriented polyethylene fiber a change has been observed in the form of the line in relation to the angle between the fiber and the field vector (see page 168).

The pairs of protons of water molecules sorbed by the fibrillar protein of stretched tendon, also give a complex NMR line, whose form varies with rotation of the sample [173].

In fluorine-containing oriented polymers it is possible to observe the distortion of the form of the line caused by the magnetic anisotropy of the C—F bond [296, 717].

Calculation of the Relation of the Second Moment of the NMR Line to the Position of the Sample in the Magnetic Field. According to Van Vleck's formula (see Chapter I), the second moment, ΔH_2^2, of the NMR line of a solid with a rigid structure equals the sum of the contributions from all the internuclear vectors r_i. These contributions are proportional to the values $r_i^{-6}(3\cos^2\theta_i - 1)^2$ (where r_i is the internuclear distance and θ_i is the angle between the vector $\overleftarrow{r_i}$ and the magnetic field vector). In an isotropic polymer all the angles θ_i are equally probable (if the origins of all the vectors $\overleftarrow{r_i}$ are placed at one point, then their ends are uniformly distributed over the surface of a sphere). In this case, in the expression for an isotropically rigid structure of the polymer $(\Delta H_2^2)_{ir}$ there appears the mean value $(3\cos^2\theta - 1)^2 = \frac{4}{5}$, which is independent of the position of the sample in the field.

If structural elements, namely macromolecules, bundles, and crystallites, in the polymer are oriented, then rotation of the sample in the field leads to a change in the angles θ_i and, consequently, in the value ΔH_2^2.

To establish the relation of the second moment of the line to the position of the sample in the field, it is necessary to define the character of the orientation of the structural elements in the polymer sample. Kitaigorodskii [63] considers two main types of orientation, namely, the axial texture and the planar texture. A more detailed classification of the possible types of orientation of the structural elements in a polymer is given by Heffelfinger and Burton [347], who distinguish six types of orientation: 1) random; 2) planar; 3) single-plane; 4) axial; 5) planar—axial; 6) single plane—axial. The commonest type of orientation in practice is type 4, which appears in fibers and which corresponds to the axial texture of Kitaigorodskii. In films there may be orientation of types 2, 3, and 6, depending on the method of production of the film and the structure of the polymer.

Let us examine the simplest case of axial orientation in a drawn fiber. The axes of all the molecules are parallel to the axis of the fiber and form an angle θ' with the magnetic field (Fig. 54). The molecule contains pairs of nuclei and the line connecting the nuclei is perpendicular to the axis of the molecule as, for example, in the CH_2 groups of a polyethylene chain. If we place the origins of the internuclear vectors $\overleftarrow{r}$ at one point, their ends do not lie on the surface of a sphere, but on a circle, whose plane is perpendicular to the direction of orientation of the axes of the molecules. It may be shown geometrically that the mean value

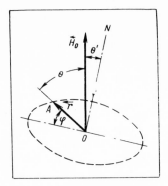

Fig. 54. Orientation of internuclear vectors perpendicular to the ON axis of the molecule ($\overleftarrow{r}$ is the internuclear vector; $\vec{H}_0$ is the magnetic field vector).

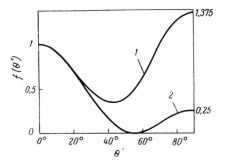

Fig. 55. Graphs of the functions: 1) $f(\theta') = {}^{27}/_8\sin^4\theta' - 3\sin^2\theta' + 1$; 2) $f(\theta') = {}^1/_4(3\cos^2\theta' - 1)^2$ (see page 183).

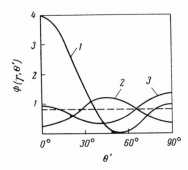

Fig. 56. Graph of the function $\Phi(\gamma, \theta')$. 1) $\gamma = 0°$; 2) $\gamma = 45°$; 3) $\gamma = 90°$ (the broken line shows the value of Φ for an isotropic sample [139]).

$$\overline{(3\cos^2\theta - 1)^2} = \frac{27}{8}\sin^4\theta' - 3\sin^2\theta' + 1 \tag{III-4}$$

The second moment of the NMR line of a structure oriented with respect to one axis equals

$$(\Delta H_2^2)_{\text{orient}} = \frac{5}{4}(\Delta H_2^2) \text{ ir}\left(\frac{27}{8}\sin^4\theta' - 3\sin^2\theta' + 1\right) \tag{III-5}$$

A graph of the function $\left(\frac{27}{8}\sin^4\theta' - 3\sin^2\theta' + 1\right)$ is given in Fig. 55. It is obvious that the maximum value of ΔH_2^2 is obtained when the fiber is perpendicular to the magnetic field ($\theta' = 90°$), and the minimum value when $\theta' \approx 45°$.

For a more accurate calculation of the relation $\Delta H_2^2 = F(\theta')$ it is necessary to take into account the contribution of all the internuclear vectors in the fiber, which form different angles with the axis of orientation. The corresponding formula was derived by Tsvankin and Fedin [139]. It contains the mean value $\overline{(3\cos^2\theta - 1)^2} = \Phi(\gamma, \theta')$, which depends on the angle γ formed by the internuclear vector with the axis of orientation, and the angle θ' between the axis of orientation and the vector of the magnetic field, as is shown in Fig. 56.

Complete orientation of the structural elements of the polymer along the fiber axis is never achieved in a fiber, and there is a certain distribution of the angles ω, formed by the axes of the macromolecules with the direction of drawing. In the expression for ΔH_2^2, instead of the angles γ and θ' we should have the distribution function $\psi(\omega)$. The formula for calculating ΔH_2^2 when there is a distribution of the axes of the molecules in the polymer was obtained by Yamagata and his co-workers [739] and Chujo and Sudzuki [239]:

$$\Delta H_2^2 = S_0 - S_2\sin^2\theta' + S_4\sin^4\theta' \tag{III-6}$$

The coefficients S_i represent the linear combination of the products of the values R_{2m} and C_{2m}.

The values R_{2m} completely determine the structure of the molecule (or crystallite):

$$R_{2m} = \frac{3}{4}(1 + I^{-1})\mu^2\Sigma r_i^{-6}\sin^{2m}\gamma_i \quad \text{where } m = 0,\ 1,\ 2$$

where I is the spin of the nucleus; μ is the magnetic moment of the nucleus; γ_i is the angle of the internuclear vector with the axis of the molecule.

The values C_{2m} depend on the character of the orientation of the structural elements of the macromolecules in the polymer and are the mean values:

$$C_2 = \overline{\cos^2 \omega} \qquad C_4 = \overline{\cos^4 \omega}$$

To calculate C_{2m}, Yamagata [739] used the distribution function $\psi(\omega)$ proposed by Kratky [422].

Chujo [239] obtained expressions for C_{2m} for a model of a linear macromolecule, which consisted of a chain of freely connected segments each of length l, stretched with a force f. The values of C_{2m} and, consequently, the form of the curve $\Delta H_2^2 = F(\theta')$ are determined by the parameter $\alpha = fl/kT$; the curves calculated for polyethylene for different values of α are given in Fig. 57.

A calculation for a Gaussian distribution of the angles of the internuclear vector with the axis of the fiber was carried out by Berendsen [173] (in an NMR study of the adsorption of water on a fiber).

The orientation of the structural elements in a film is a more complex case. Slonim and Urman [114] derived a general formula for the relation $\Delta H_2^2 = F(\theta')$, which is suitable for both fibers and films:

$$\Delta H_2^2 = \frac{3}{2} I (I+1) g^2 \mu_n^2 [E_1 \Gamma_1 \cos^4 \theta' +$$
$$+ (E_2 \Gamma_1 + E_4 \Gamma_2) \cos^2 \theta' + E_3 \Gamma_1 + E_5 \Gamma_2 + \Gamma_3] \qquad \text{(III-7)}$$

where g and μ_n are the nuclear g-factor and nuclear magneton.

The values Γ_i are linear combinations of the terms $\Sigma r_i^{-6} \cos^{2m} \gamma_i$ (where m = 0, 1, 2) and depend only on the structure of the molecule (structural element). The coefficients E_i are determined by the character of the orientation in the sample. Formulas are given in [114] for calculating the coefficients E_i for the distribution $\psi(\omega)$ of Kratky and also for an elliptical distribution in which the number of polymer molecules forming an angle ω with the axis of the fiber is proportional to the radius vector of the

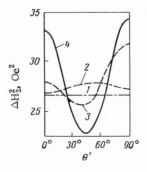

Fig. 57. Relation of the second moment of the NMR line (ΔH_2^2) of drawn polyethylene to the angle between the direction of drawing and the field vector (θ') with different values of the parameter α (see page 180) [239]. 1) $\alpha = 0$; 2) $\alpha = 1$; 3) $\alpha = 10$; 4) $\alpha = \infty$.

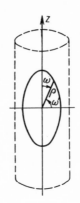

Fig. 58. Elliptical distribution of the axes of the structural elements in a polymer [114].

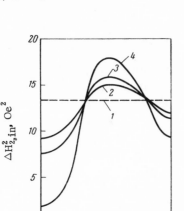

Fig. 59. Relation [see formula (III-7)] of the intramolecular contribution ($\Delta H_{2,\text{in}}^2$) to the second moment of the NMR line of polyformaldehyde to the angle between the direction of orientation of the fiber and the field vector (θ'). 1) Isotropic sample; 2,3,4) monoaxially drawn samples with elliptical distributions with $e = 5$, 20, and ∞, respectively.

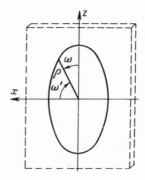

Fig. 60. Distribution of the axes of the structural elements in a film [114].

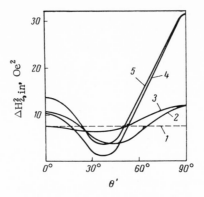

Fig. 61. Relation [see formula (III-7)] of the intramolecular contribution ($\Delta H_{2,in}^2$) to the second moment of the NMR line of a film of polyethylene terephthalate to the angle between the direction of greatest drawing and the field vector (θ'). 1) Isotropic sample; 2,3) axes of the molecules parallel to the plane of the film for an elliptical distribution with $e = \infty$ and $e = 1$, respectively; 4,5) planes of the molecules parallel to the plane of the film with $e = \infty$ and $e = 1$, respectively.

ellipse ρ (Fig. 58). The main axis of the ellipse lies along the axis of the fiber, the ratio of the semiaxes equals e, and the more complete the orientation, the greater the value of e. Figure 59 gives the results of calculation for polyformaldehyde; the values of Γ_i were determined from the known [674] structure. In contrast to polyethylene, for polyformaldehyde we obtain a maximum and not a minimum of ΔH_2^2 when $\theta' \approx 45°$.

It is assumed for a film that all the axes of the structural elements are parallel to the plane of the film, while all the angles of rotation of the elements about its axis are equally probable (planar orientation). Expressions have been derived for E_i for an elliptical distribution, when the value $\psi(\omega)$ is proportional to the radius vector of the ellipse ρ, whose major axis coincides with the direction of maximum drawing (Fig. 60). By using these expressions for E_i, it is possible to calculate from formula (III-7) the relation of ΔH_2^2 to the angle of rotation of the film about the three axes.

Figure 61 gives the theoretical curves of $\Delta H_2^2 = F(\theta')$ for a film of polyethylene terephthalate; the structural data for the calculation were taken from [263]. The same figure gives the curves of $\Delta H_2^2 = F(\theta')$ for the case where the planes of the macromolecules are parallel to the plane of the film, i.e., one-plane orientation according to Heffelfinger.

All the calculations presented above were carried out for samples with a rigid lattice. A great advantage of the NMR method in studying the orientation of molecules in polymers is the fact that this method makes it possible to study the structure of polymers with intensive molecular motion. Let us return to the simplest case of a pair of nuclei, for which the internuclear vector $\overleftarrow{r}$ is perpendicular to the axis of the molecule ON (see Fig. 54), and we will assume that the vector $\overleftarrow{r}$ is rotating rapidly (with a frequency greater than 10^4-10^5 Hz) about the axis ON. Then there is averaging of the local magnetic fields and

$$\Delta H_2^2 = \frac{5}{4}(\Delta H_2^2) \text{ ir} \cdot \frac{1}{4}(3\cos^2\theta' - 1)^2 \qquad \text{(III-8)}$$

The change in the function $f(\theta') = \frac{1}{4}(3\cos\theta' - 1)^2$ is given in Fig. 55; for all values of θ' apart from $\theta' = 0°$, the coefficient is less than the factor $(2\frac{7}{8}\sin^4\theta' - 3\sin^2\theta' + 1)$. With an increase in the molecular motion, ΔH_2^2 falls more, the closer the angle θ' is to 90°. A calculation of the relation of ΔH_2^2 to θ' for the case where the molecular chains rotate about their axes, taking into account the distribution of crystallites with monoaxial drawing, was carried out in the work of Yamagata [737].

The more general case was examined in the work of Slonim and Urman [117]. A formula was derived for the relation of ΔH_2^2 to θ' for an oriented structure, in which the internuclear vectors perform harmonic oscillations about the axis of the structural element. The formula has the same form as the expression for a rigid lattice [see equation (III-7)], but the values Γ_i appearing in it depend not only on the structure of the molecule, but also on the angular amplitude of the oscillations δ. When $\delta = 0$, the general formula changes into the expression (III-7) for a rigid lattice, while when $\delta = \infty$, it changes into the formula for free rotation of molecules.

Experimental Study of the Relation of the Second Moment of an NMR Line to the Position of the Sample in the Magnetic Field. The relation $\Delta H_2^2 = F(\theta')$ at low temperatures, when molecular motion is restrained,

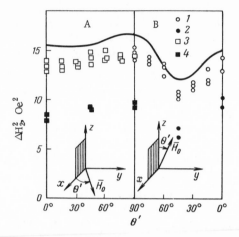

Fig. 62. Relation of the second moment of the NMR line (ΔH_2^2) to the disposition of a monoaxially drawn film in the magnetic field [133]. A) θ' is the angle with the x axis, rotation about the z axis; B) θ' is the angle with the z axis, rotation about the x axis. 1,2) Samples cut across the axis of drawing, at −196 and + 20°C; 3,4) samples cut along the axis of drawing, at −196 and + 20°C. Curve − relation calculated theoretically.

has been studied for a series of polymers. Yamagata and Hirota [739] determined this relation at −196°C for polyethylene, at −150°C for polyvinyl alcohol, and at −156°C for polytetrafluoroethylene.

The best agreement between the calculated and measured values was found for polytetrafluoroethylene; for polyvinyl alcohol the authors recommend an interproton distance in the methylene group of 1.72 Å and not 1.79 Å at −150°C.

With drawn polyformaldehyde [552, 736], a maximum of ΔH_2^2 is obtained at $\theta' \approx 45°C$ in accordance with theory. Figures 62 and 63 give the results of measurements of ΔH_2^2 for films of polyethylene terephthalate [133]. It is assumed that the sample of film lies in the plane. Monoaxially drawn film (Fig. 62) was rotated about the direction of drawing (z axis) and about the direction perpendicular to this axis (x axis). The theoretical relation $\Delta H_2^2 = F(\theta')$ was calculated from formula (III-7) for an elliptical distribution with a ratio of the semiaxes e = 100 on the assumption that part of the polymer (about $\frac{1}{3}$) remains unoriented, while the intermolecular contribution to the second moment of the NMR line equals approximately 6 Oe^2 (as in unoriented polymer).

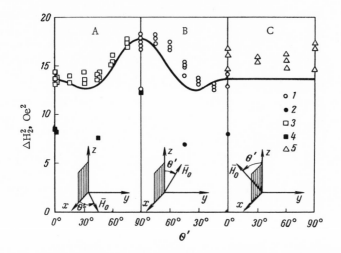

Fig. 63. Relation of the second moment of the NMR line (ΔH_2^2) to the position of a biaxially drawn film in the magnetic field [133]. A) θ' is the angle with the x axis, rotation about the z axis; B) θ' is the angle with the z axis, rotation about the x axis; C) θ' is the angle with the z axis, rotation about the y axis. 1,2) Samples cut along the first axis of drawing, at −196 and +20°C; 3,4) samples cut along the second axis of drawing, at −196 and +20°C; 5) sample rotated about the axis perpendicular to the plane of the film, at −196°C. Curve − relation calculated theoretically.

The theoretical curve closely follows the course of the change in ΔH_2^2 with a change in the angle of rotation of the film in the magnetic field, but it lies somewhat above the experimental values of ΔH_2^2, and this may be explained by errors in estimating the intermolecular contribution. The change in ΔH_2^2 with rotation of the film about the direction of drawing undoubtedly indicates the presence of planar orientation, since with purely axial orientation this effect would not be observed.

A biaxially drawn film (Fig. 63) was rotated about the directions of both the axes of drawing (z and x) and about a direction perpendicular to the plane of the film (y axis). The experimental data for a temperature of −196°C are described satisfactorily by the theoretical curve for an elliptical distribution with e = 1. The results obtained for rotation about the y axis, which is perpendicular to its plane, are less reliable due to the difficulties of preparing the corresponding sample. Nonetheless, it is evident that, in

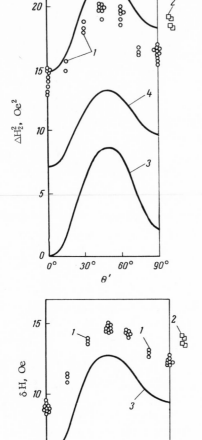

Fig. 64. Relation of the second moment of the NMR line (ΔH_2^2) of polyoxymethylene at −196°C to the angle of rotation of the sample in the magnetic field (θ') [118]. Points represent experimental data: 1) for single-crystal samples; 2) for a polycrystalline sample (independent of θ'). Curves show the calculated values: 3) contribution to ΔH_2^2 of interproton interaction of CH_2 groups of Z crystals; 4) contributions of CH_2 groups of Z and W crystals; 5) the overall second moment of the NMR lines. (The calculated second moment of the NMR line of isotropic polyoxymethylene equals approximately 20 Oe^2).

Fig. 65. Relation of the width of the NMR line (δH) at −196°C to the angle of rotation of a sample of polyoxymethylene in the magnetic field (θ') [118]. 1,2) Experimental data for single-crystal and polycrystalline (independent of θ') samples; 3) calculated values for the interaction of the protons in CH_2 groups of Z crystals of polyoxymethylene with a broadening of 8.4 Oe^2.

this case, the second moment remains constant within the limits of experimental error, as should be the case for a film drawn equally in both directions.

Orientation of the structural elements in a polymer may occur not only during drawing of an isotropic sample, but also during the synthesis process itself, namely, during polymerization in the solid phase. We determined [118, 499] the NMR spectra of poly-

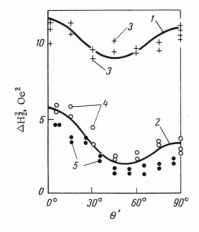

Fig. 66. Relation of the second moment of NMR line (ΔH_2^2) of a drawn sample of polytetrafluoroethylene to the angle between the direction of drawing and the field vector (θ') [737]. 1) Calculated curve for an immobile lattice; 2) calculated curve for a lattice with rotation of the macromolecules about their axes; 3,4,5) experimental data obtained at -156, $+13$, and $+27°C$, respectively.

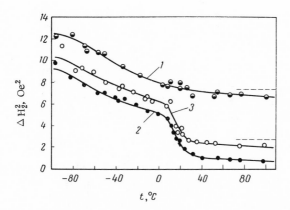

Fig. 67. Relation of the second moment of the NMR spectrum (ΔH_2^2) of polytetrafluoroethylene to the temperature with different angles between the direction of drawing of the fiber and the field vector [370]. 1) $\theta' = 0°$; 2) $45°$; 3) $90°$. The broken lines show the theoretical values of ΔH_2^2 calculated for a lattice with free rotation.

oxymethylene obtained by γ-irradiation of a crystal of trioxane. An x-ray study showed [221] that such a "single-crystal" sample actually consists of crystallites arranged in two ways: with the axis coinciding with the macroscopic axis of the crystal ("Z crystal") and inclined at an angle of $76°7'$ ("W crystal"). The relation of ΔH_2^2 to the angle coincides in the first approximation with that calculated from formula (III-7) for a rigid structure, taking into account intra- and intermolecular contributions. As Fig. 64 shows, the change in the overall value of ΔH_2^2 is caused mainly by the contribution of the interproton interaction in the CH_2 groups of the Z crystals, which are completely oriented along the axis of the sample. This is also confirmed by the nature of the relation of the line width (δH) to θ' (Fig. 65). The theoretical curve of $\delta H = F(\theta')$ in Fig. 65 was constructed by means of the $g(h)$ function derived in the same work for an isolated pair of nuclei with a uniform distribution of internuclear vectors over the surface of a cone.

Analysis of the relation of ΔH_2^2 to θ' for an oriented sample of a polymer over a wide range of temperatures makes it possible to obtain information on the character of the molecular motion. A series of studies has been made of oriented polytetrafluoroethylene (Teflon). Thus, Slichter and Yamagata and his co-workers [737, 739] showed that the nature of the relation $\Delta H_2^2 = F(\theta')$ at room temperature confirms the rotation of macromolecules about their axes (Fig. 66).

A comparison of the curves of the temperature dependence of ΔH_2^2 obtained for Teflon fibers at angles of 0, 45, and 90° to the field vector (Fig. 67) shows that at 12°C there begins rotation in the crystalline regions of the polymer [370]. If the axis of rotation coincides with the direction of the field ($\theta' = 0°$), the F—F internuclear vector always remains perpendicular to the field vector and there is no step on the curve of $\Delta H_2^2 = f(t)$ for $\theta' = 0°$ (curve 1), while at $\theta' = 45$ and 90°, a fall in ΔH_2^2 is observed (curves 2 and 3).

The experimental dependence of ΔH_2^2 found experimentally for linear polyethylene [472] is close to the theoretical dependence if we assume that harmonic oscillations of all chains occur with an angular amplitude of 15°.

The relation of $\Delta H_2^2 = f(t)$ for oriented samples of polyethylene lying at different angles to the field vector has also been studied by Peterlin and his co-workers [537, 551], Fischer [289], and Iwayanagi [382]. The samples consisted of drawn film or a layer of lamellar crystals precipitated from solution (in such a layer all the axes of the molecules are perpendicular to the plane of the lamellae). In a series of investigations it was shown that with a rise in temperature there is an increase in the rotational motion of the chains.

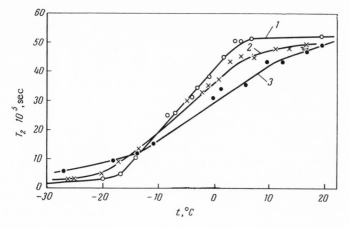

Fig. 68. Relation of the transverse relaxation time (T_2) to the temperature for rubber at different relative elongations of the sample [242]. 1) 10%; 2) 170%; 3) 300%.

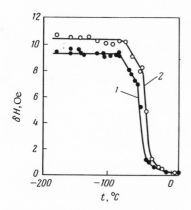

Fig. 69. Relation of the width of the line (δH) of the NMR spectrum of rubber to the temperature [539]. 1) Unstretched sample; 2) with a relative elongation of the sample of 500%.

In an oriented fiber of polyethylene terephthalate [372], a rise in temperature from —93 to 122°C produces a large fall in ΔH_2^2, when the fiber lies perpendicular to the field vector. This demonstrates the presence of restrained rotation of the CH_2 group about an axis parallel to the fiber axis.

Kazaryan and Urman [55] studied the relation of ΔH_2^2 to θ' for an amorphous monoaxially drawn film of polyethylene terephthalate. A comparison with the values of the second-order moment of the NMR line calculated for different models shows that at room temperature there is no rotation of the chains, but there is a statistical set of molecules with different angles of rotation of the chains about their axes.

The orientation of the molecules in fibers of polyvinyl alcohol has been studied by the NMR method in the work of Vol'kenshtein and his co-workers [12, 66]. The mean values of $\overline{\sin^2\omega}$ and $\overline{\sin^4\omega}$ were calculated for samples with different draw ratios.

Study of Deformed Polymers. An ordered arrangement of structural elements may also be observed in a deformed sample of polymer under stress. Natural rubber and synthetic fibers have been used as subjects for studying this phenomenon. The NMR spectrum of vulcanized natural rubber with a slight stretch (relative elongation of 70%) does not change substantially in comparison with the spectrum of an unstretched sample [338]. However, as Cohen-Hardia and Gabillar [242] showed, with strong stretching the form of the temperature dependence of the transverse relaxation time (T_2) for natural rubber changes (Fig. 68). At low temperatures (from −20 to −30°C), T_2 is greater for stressed samples, while at high temperatures (from +10 to +20°C), it is greater for unstressed samples. With elongations from 70 to 300%, and at temperatures from −10 to +10°C, the slope of the lines ($\Delta T_2 / \Delta t$) falls linearly with an increase in stretching. The authors give no theoretical explanation for the observed effect. Lösche [447] has suggested that it is connected with a decrease in the intermolecular distances in the rubber.

Oshima and Kusumoto [539] observed the NMR of vulcanized rubber from natural rubber with a high stretch (relative elongation of 500%) over a wide range of temperatures (Fig. 69). At a temperature from room temperature to −15°C the stretching does not affect the width of the line δH. With a further fall in temperature, an increase in δH occurs with an unstretched sample at a temperature from −38 to −52°C, while with a stretched sample it occurs at a temperature from −30 to −45°C; the orientation of the molecules evidently interferes with the motion of the segments. At a low temperature, ΔH_2^2 for a stretched sample is 20.3 Oe^2, and for an unstretched sample, 18.5 Oe^2.

It is considered [108, 561] that the increase in ΔH_2^2 with an increase in stretching is caused by the denser packing, which leads to a decrease in the distances between the protons. Vol'kenshtein [36] has pointed out that this conclusion is hardly valid, since the density of rubber hardly changes during stretching.

Very interesting results were obtained by Zhurkov and Egorov [46]. During the stretching of a bundle of fibers of polycaprolactam (Kapron) and polyethylene terephthalate (Lavsan) there is a fall in the intensity of the narrow component of the NMR line, which indicates a fall in the mobility of the segments in the polymer. When the load is removed, the form of the line is restored.

It is probable that more detailed information could have been obtained by studying the NMR of deformed samples with the axis of deformation of the polymer at different angles to the field vector.

4. Study of the Chain Structure

of a Polymer

Order of Addition of the Units in the Chain. The properties of a linear polymer depend substantially on the order of addition of the units. In the case of polymers obtained from unsymmetrical vinyl monomers, there is the possibility of two modes of addition of the units in the chain, namely, "head-to-tail" and "head-to-head." Sometimes these structures can be distinguished by means of the value of ΔH_2^2 of the broad NMR lines of a block of polymer. For a copolymer of trifluorochloroethylene with vinylidene fluoride, Lyubimov and his co-workers [80] calculated theoretically that the intramolecular contribution to ΔH_2^2 from the fluorine nuclei for structure I is 7.6 Oe^2 and, for structure II, 8.6 Oe^2.

$$
\begin{array}{cccc}
\text{F} & \text{F} & \text{H} & \text{F} \\
| & | & | & | \\
\cdots-\text{C}-\text{C}-\text{C}-\text{C}-\cdots \\
| & | & | & | \\
\text{F} & \text{Cl} & \text{H} & \text{F}
\end{array}
\qquad
\begin{array}{cccc}
\text{F} & \text{F} & \text{F} & \text{H} \\
| & | & | & | \\
\cdots-\text{C}-\text{C}-\text{C}-\text{C}-\cdots \\
| & | & | & | \\
\text{F} & \text{Cl} & \text{F} & \text{H}
\end{array}
$$

I $\qquad\qquad\qquad\qquad$ II

The experimental value of ΔH_2^2 at $-150°C$ equals 12.5 Oe^2; the estimated intermolecular contribution equals approximately 4.0 Oe^2; consequently, structure II is more probable.

More complete information is provided by studying high-resolution NMR spectra. In one of the first studies of the high-resolution NMR spectra of polymers, Borodin [23] determined the content of the terminal groups

$$-CF_3, -CF_2Cl \text{ and } -CCl_2F$$

in a series of samples of liquid polymers of the type

$$\cdots -CF_2 - CFCl - \cdots$$

In order to determine the structure of polyvinylidene fluoride, Naylor and Lasoski [498], using a working frequency of 30 MHz measured the chemical shift of fluorine ($\Delta\nu$ in frequency units) relative to trifluoroacetic acid for solutions of a series of polymers and obtained the following data (the atom whose resonance was observed is shown in heavy type):

		$\Delta\nu$, Hz
Polytrifluoroethylene	$\left(\begin{array}{cccc} F & \mathbf{F} & F & F \\ -C-C-C-C- \\ F & H & F & H \end{array}\right)_n$	. . . . −3950
	$\left(\begin{array}{cccc} F & F & \mathbf{F} & F \\ -C-C-C-C- \\ F & H & F & H \end{array}\right)_n$	. . . . −1150
Polytrifluorochloro-ethylene	$\left(\begin{array}{cccc} F & \mathbf{F} & F & F \\ -C-C-C-C \\ F & Cl & F & Cl \end{array}\right)_n$	. . . . . −1500
Perfluorodecanoic acid	$F_3C-\overset{\displaystyle \mathbf{F}}{\underset{\displaystyle F}{C}}-(CF_2)_7-COOH$	. . . . −1250
Polyvinyl fluoride	$\left(\begin{array}{cccc} H & \mathbf{F} & H & F \\ -C-C-C-C- \\ H & H & H & H \end{array}\right)_n$	. . . . −3100
Polyvinylidene fluoride	$\left(\begin{array}{cccc} H & \mathbf{F} & H & F \\ -C-C-C-C- \\ H & F & H & F \end{array}\right)_n$	. . . . . −400

Analysis of the table makes it possible to formulate general rules for the effect of a substituent on the resonance of ^{19}F.

1. When H is replaced by Cl and F at the carbon atom to which F is attached, the resonance is shifted to higher frequencies. This agrees with data obtained for low-molecular substances: with an increase in the electronegativity of the substituent the resonance is shifted to higher frequencies.

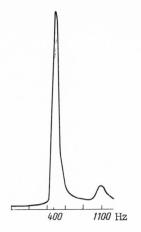

Fig. 70. NMR spectrum of fluorine (a solution of polyvinylidene fluoride in dimethylformamide was investigated [498]).

2. When H is replaced by F at a carbon atom adjacent to that to which the F is attached, the resonance is shifted to lower frequencies. This shift is apparently caused by the interaction of the F atom with the substituent at the adjacent C atom.

With a working frequency of 30 MHz, a solution of polyvinylidene fluoride in dimethylformamide gives a spectrum (Fig. 70) which consists of two lines lying at $\Delta \nu = 400$ Hz and 1100 Hz. They correspond to the two possible structures of the polymer, namely, "head-to-tail" and "head-to-head." The ratio of the intensities of the maxima shows that polyvinylidene fluoride obtained by polymerization with an initiator of the diazo type contains 8-10% of "head-to-head" structures.

Better resolution was obtained in the spectrum of a solution of polyvinylidene fluoride in N,N-dimethylacetamide (Fig. 71), plotted with a frequency of 56.4 MHz [718] (see also [722]) and 4 fluorine lines are found corresponding to the structures:

$$\cdots - CF_2 - CH_2 - CF_2^* - CH_2 - CF_2 - CH_2 - \cdots$$
$$I$$
$$\cdots - CH_2 - CH_2 - CF_2^* - CH_2 - CF_2 - CH_2 - \cdots$$
$$II$$
$$\cdots - CF_2 - CH_2 - CF_2^* - CF_2 - CH_2 - CF_2 - \cdots$$
$$III$$
$$\cdots - CH_2 - CH_2 - CF_2^* - CF_2 - CH_2 - CH_2 - \cdots$$
$$IV$$

From the relative intensities of the lines it is possible to calculate the content of "head-to-tail," "head-to-head," "tail-to-tail," and "tail-to-head" structures, and thus to determine the microstructure of the chain completely.

It was proposed [257] that polymers formed by polymerization of dimethylvinylsilane (I) have the structure (II)

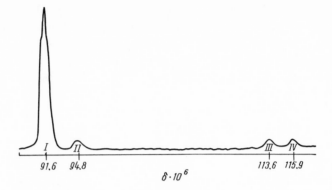

Fig. 71. High-resolution NMR spectrum of fluorine (a 25% solution of polyvinylidene fluoride in N,N-dimethylacetamide was investigated at room temperature with a working frequency of 56.4 MHz). The chemical shift δ is relative to an internal standard $CFCl_3$ [718].

$$
\begin{array}{ccc}
CH_3 & \overset{3}{C}H_3 & \overset{3}{C}H_3 \\
| & | & | \\
H-Si-CH=CH_2 & \cdots-Si-\overset{2}{C}H_2-\overset{2}{C}H_2-\cdots & \cdots-Si-CH-\cdots \\
| & | & | \quad | \\
CH_3 & \underset{3}{C}H_3 & \underset{3}{C}H_3 \ \underset{1}{C}H_3 \\
\text{I} & \text{II} & \text{III}
\end{array}
$$

However, the NMR spectrum of a solution of the polymer was found to contain three signals: the doublet of methyl protons in the group $> CH-CH_3$, which are denoted by (1) in formula III; the singlet of protons of the methylene groups (2), and the singlet of protons of methyl groups at a silicon atom (3). This demonstrates the presence of two polymer structures, namely, II and III [258]. The ratio of the areas of the signals was used to determine the composition of the polymer, which consisted of 75% of structure II and 25% of structure III.

An analogous analysis was carried out for polymers of diethylvinylsilane, diphenylvinylsilane, and methylphenylsilane.

The high-resolution spectra of solutions of polybutadiene and polydimethylbutadiene make it possible to determine the content of 1,2- and 1,4-units in the chain [104, 606].

The chain of polychloroprene may contain units of four types: 1,4-trans (I); 1,4-cis (II); 1,2- (III); and 3,4- (IV). In 1,4-polymerization there is the possibility of addition of the units "head-to-tail" (V), "head-to-head" (VI), and "tail-to-tail" (VII). In the NMR spectrum of solutions of polychloroprene in carbon disulfide obtained with a frequency of 100 MHz, the chemical shift for the olefinic protons in the cis-configuration equals $5.51 \cdot 10^{-6}$; in the trans-configuration it is $5.35 \cdot 10^{-6}$. The order of addition of the units in the chain is determined from the form of the NMR signal of the methylene groups. In the trans-polymer this signal consists of a doublet at $2.37 \cdot 10^{-6}$ and $2.33 \cdot 10^{-6}$, corresponding to addition of type V, a peak at $2.50 \cdot 10^{-6}$, corresponding to type VI, and a multiplet at $2.18 \cdot 10^{-6}$ corresponding to type VII. The signal of the methylene protons of cis-polychloroprene consists of a doublet at $2.56 \cdot 10^{-6}$ and $2.52 \cdot 10^{-6}$ (type VI) and a peak at $2.35 \cdot 10^{-6}$ (type V). By combining the data obtained from NMR and IR spectra, Ferguson [284] completely determined the structure of the chain of a series of samples of polychloroprene.

From the relative intensity of the peaks of methyl and methylene protons of a polymer of 3-methylbutene-1 it was established [59] that the chain of the polymer consists of α-dimethylpropane units

$$\cdots -CH_2-CH_2-C(CH_3)_2- \cdots$$

By treating germanium diiodide with acetylene and methylating the product with methylmagnesium iodide, Vol'pin and his coworkers [37] synthesized a polymer with alternating double bonds and germanium atoms in the chain. The NMR spectrum of a solution of the polymer contained only two peaks with a ratio of intensities of $1:3$ and chemical shifts relative to benzene of $0.3 \cdot 10^{-6}$ and $7.0 \cdot 10^{-6}$. These peaks correspond to methyne and methyl protons, proving the structure of the polymer

$$\cdots -CH=CH-Ge(CH_3)_2- \cdots$$

The structure of the chain of the organogermanium polymer

$$\cdots -CH_2-CH=CH-CH_2-Ge(CH_3)_2- \cdots$$

which was obtained by methylation of the reaction product of trichlorogermane and excess butadiene, was also established through the NMR spectrum [91].

The NMR method has been used in the study of the structure of polymers obtained from the nitriles of aromatic and aliphatic acids [69], and also from 9-vinylanthracene [476], cyclooctadiene-1,5 [584], 4-methylpentene [324, 703], methylcyclohexene [453], dimethylhexadiene [328], trichlorobutadiene [123], vinyl benzoate [172], acetone [487], norbornadiene [408], acetaldehyde [699], polymers obtained from ditolylethane and diisopropylbenzene [68], in the study of the structure of cellulose derivatives [310], phenol-formaldehyde resins [673, 726], polyurethane elastomers [669], and the so-called "phantom" polymers obtained by polymerization with the shift of a proton or a functional group [495].

Analyses of the high-resolution spectra of copolymers makes possible a detailed study of the microstructure of the chain. In the NMR spectrum of solutions of the copolymer of vinyl chloride and vinylidene chloride in o-dichlorobenzene (Fig. 72), determined at 100°C, the protons of the CH_2 groups give a triplet [238]. The two extremely large peaks correspond to segments which consist solely of units of vinyl chloride and vinylidene chloride, while the central, weakly expressed peaks correspond to the boundary structures

$$\cdots - CHCl - CH_2 - CCl_2 - CH_2 - \cdots$$

The ratio of the intensities of the peaks made it possible to calculate the content of vinyl—vinyl, vinyl—vinylidene, and vinylidene—vinylidene structures. It was shown [237] that in a sample of a copolymer obtained at a low degree of conversion from a mixture of monomers with a ratio [vinylidene chloride]/[vinyl chloride] = 25/75, contains 45% of "head-to-head" structures. With an increase in the ratio [vinylidene chloride]/[vinyl chloride] to 85/15, the fraction of these structures fell to 10% and the chain of the homopolymer of vinylidene chloride consisted solely of "head-to-tail" structures. The chain structure of copolymers of vinyl chloride and vinylidene chloride was also studied in the work of Okuda [536].

By means of semi-empirical rules relating the magnitude of the chemical shift and the structure, Ferguson [283] analyzed the high-resolution NMR spectra of a series of copolymers of vinylidene fluoride with hexafluoropropylene (Fig. 73). By using data on the magnitude of the chemical shift and the intensity of nine resonance lines of ^{19}F, the author showed that the copolymer had a

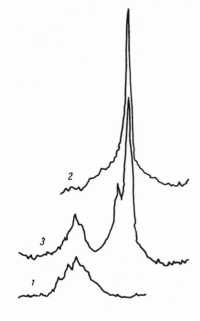

Fig. 72. Signal of the CH_2 group in NMR spectra of solutions of polymers in o-dichlorobenzene at 100°C [238]. 1) Polyvinyl chloride; 2) polyvinylidene chloride; 3) copolymer of vinyl chloride and vinylidene chloride.

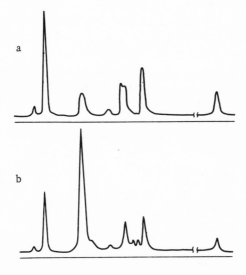

Fig. 73. NMR spectrum of fluorine (copolymers of vinylidene fluoride with hexafluoropropylene were investigated as a 50% solution in acetone) [283]. a) 61 mol.%; b) 85 mol.% vinylidene fluoride.

"random" linear structure and that it contained no sections, which consisted of adjacent hexafluoropropylene units, sections of block polymers, or branches. The structure of the copolymers is described by the formula

$$\left[\left(\underset{-CH_2CF_2CF_2CF-}{\overset{CF_3}{|}}\right)_{0.93}\left(\underset{-CH_2CF_2CFCF_2-}{\overset{CF_3}{|}}\right)_{0.07}\right]_{1-n}-$$
$$-[(-CH_2CF_2CH_2CF_2-)_{0.95}\,(-CH_2CF_2CF_2CH_2-)_{0.05}]_{n/2}$$

with $n = (x-50)/x$ and $x > 50$, where x is the amount of vinylidene fluoride in the copolymer in mol.%.

From the NMR spectrum of a copolymer of propylene with styrene it was concluded [156] that the chain contained only short sequences of styrene units. It was found that the phenyl protons gave only one peak, while from the work of Bovey and his co-workers [204] it is well known that polystyrene gives two phenyl peaks when the degree of polymerization is greater than ten (see page 257).

In the NMR spectra of solutions of copolymers of methyl methacrylate with styrene in carbon tetrachloride, the protons of the methoxy groups of methyl methacrylate give two peaks [505]. One of them has the same chemical shift as in the spectrum of polymethyl methacrylate. If a styrene unit is adjacent to a methyl methacrylate unit in the copolymer chain, then, because of the diamagnetic screening effect of the phenyl group the signal of the protons of the OCH_3 group of this unit is shifted to higher fields and a second peak is obtained in the spectrum. The same phenomenon has been observed in the spectra of chloroform solutions of copolymers of methyl methacrylate with p-xylylene [506]. The NMR spectrum, like the IR spectrum, confirms the formation of a copolymer structure. The chain structure of copolymers of methyl methacrylate has also been studied by Kato and his co-workers [399] and in the work of Bovey [194] and Harwood and Ritchey [345], which is examined below in connection with the study of the stereoregularity of polymers.

By using as standards natural rubber, cis- and trans-polybutadienes, and low-molecular substances, Chen Hung Yu [228] completely determined the microstructure of copolymers of iso-

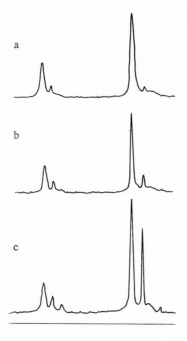

a

b

c

Fig. 74. High-resolution NMR spectra of copolymers of isoprene with butadiene (5% solution in carbon tetrachloride) at 25°C with different ratios of isoprene to butadiene [228]. a) 1:10; b) 1:4; c) 1:1.

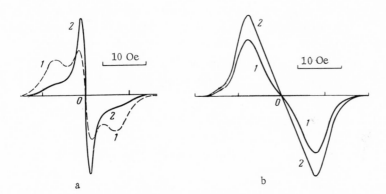

Fig. 75. NMR lines of two samples of polyethylene at 20°C (a) and −170°C (b) [549]. 1) Linear; 2) branched polyethylene.

prene and butadiene. From the spectra of 5% solutions of the co-polymers in carbon tetrachloride (Fig. 74) the concentrations of the following units were determined with an accuracy of 2-3%:

$$CH=CH_2$$
$$|$$
$$\cdots -CH_2-C-\cdots$$
$$|$$
$$CH_3$$

isoprene-1,2

$$CH_3-C=CH_2$$
$$|$$
$$\cdots -CH-CH_2-\cdots$$

isoprene-3,4

$$CH_3$$
$$|$$
$$\cdots -CH_2-C=CH-CH_2-\cdots$$

isoprene-1,4
cis and trans

$$CH=CH_2$$
$$|$$
$$\cdots -CH_2-CH-\cdots$$

butadiene-1,2

$$\cdots -CH_2-CH=CH-CH_2-\cdots$$

butadiene-1,4
cis and trans

Chain Branching. The degree of branching affects the packing density and the mobility of molecular groups in a polymer and, therefore, it appears in the NMR spectra of solid polymers. Thus, in a series of papers [306, 307, 472, 630] it was reported that the NMR line of polyethylene with branched chains is narrower and, consequently, the mobility of the protons higher than in linear polyethylene.

Peterlin and his co-workers [549, 550, 553] undertook a special study of the degree of branching and crystallinity of polyethylene by dilatometric and NMR methods. Figure 75 shows the difference in the form of NMR lines obtained both at 20°C and at −170°C. The residual intensity of the narrow component at −170°C is proportional to the $[CH_3]/[CH_2]$ ratio in the polymer.

The curves of the temperature dependence of δH and, to an even greater extent, ΔH_2^2 (Fig. 76) depend on the degree of chain branching of the polymer. The correlation frequency ν_c was calculated from the decrease in the line width with temperature (Fig. 76a). For linear polyethylene a straight line was obtained for $\log \nu_c = F(1/T)$, whose slope corresponded to an activation energy of 2.8 kcal/mole.

Two straight lines are obtained for branched polyethylene, and these intersect at −63°C. The activation energies determined from the slopes of the curves were 2.0 and 5.8 kcal/mole.

The form of an NMR line may be characterized by the "coefficient of the form of the line," which equals the ratio $(\Delta H_4^4)^{1/4}/(\Delta H_2^2)^{1/2}$. As Fig. 77 shows, the degree of branching of polyethylene [688] has a very considerable effect on the curve of the temperature dependence of the coefficient of the form of the line when t > −100°C.

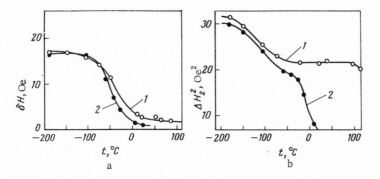

Fig. 76. Relation of the line width (δH, a) [549] and the second moment of the line (ΔH_2^2, b) [553] of the NMR spectrum of samples of polyethylene to temperature. 1) Linear; 2) branched polyethylene.

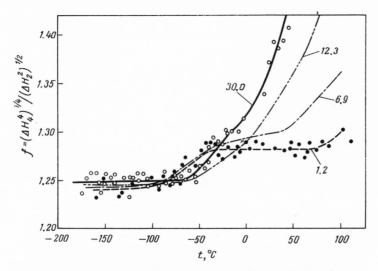

Fig. 77. Relation of the coefficient of the form of the NMR line $f = (\Delta H_4^4)^{1/4}/(\Delta H_2^2)^{1/2}$ to temperature for four samples of polyethylene with different degrees of branching (the figures on the curve give the number of CH_3 groups per 1000 C atoms in the chain [688]).

Glazkov [38] reported that for polysaccharides, the degree of branching of the chain also affects ΔH_2^2. If the number of terminal groups in the polymer is large enough, i.e., of the order of 1% or more, it is possible to determine their content from the high-resolution NMR spectrum of a solution. Thus, by measuring the ratio of the intensities of the signals of CH_2 and CH_3 groups [557] it was possible to estimate the degree of branching of low-density polyethylenes.

Stereochemical Structure.   One of the most urgent problems in the physical chemistry of polymers is the study of the effect of the stereochemical structure of the polymer on its properties. It is possible to use broad-line NMR spectroscopy, since the configuration and conformation of a chain substantially affect the form, width, and second moments of the line, and also the NMR relaxation time of the polymer in bulk.

High-resolution NMR is also used widely, since the form of the NMR line of polymer solutions also depends on the stereochemical structure of the chain.

The reflection of the stereochemical structure in NMR spectra and the effect of stereoregularity on the relaxation time of polymers in bulk have been examined in a series of studies, in which the subjects for the investigation were polyacrylates [13, 14, 507, 574], polypropylene [334, 407, 600, 602, 644], and other polymers.

In the work of Bazhenov, Vol'kenshtein, Kol'tsov, and Khachaturov [13, 14] the NMR absorption lines were plotted at temperatures from room temperature to 240°C for six samples of polymethyl methacrylate, of which two were technical Plexiglas (atactic), obtained by radical polymerization, and three samples were isotactic and one syndiotactic, obtained by catalytic polymerization.

Figures 78a and 78b show the temperature dependence of δH and ΔH_2^2 for three samples of polymethyl methacrylate. It is obvious that the break on the curves of $\Delta H_2^2 = f(t)$ in the region of the glass transition point t_g for stereoregular polymers is sharper than for atactic polymer. The figure shows that the isotactic sample has a glass transition point of about 50°C, while the syndiotactic sample has t_g above 200°C, which agrees with the lower softening point of the isotactic polymer and also with data on dielectric losses and birefringence. For atactic polymethyl methacrylate, $t_g = 100-120$°C. The authors consider that atactic polymethyl methacrylate is a particular form of copolymer or mixture of two stereoregular forms. Analogous results were obtained for stereoregular polymethyl methacrylates by Nishioka and his co-workers [507].

A comparison of curves of the temperature dependence of the spin—lattice relaxation time T_1 for samples of polymethyl methacrylate [574] shows that the high-temperature minimum of T_1 of the isotactic polymer is shifted by 40° in comparison with the minimum of T_1 for the syndiotactic polymer. For technical polymethyl methacrylate the relation $T_1 = f(t)$ is close to the curve for the syndiotactic polymer.

A detailed study has been made of the effect of stereoregularity on the NMR spectrum of polypropylene. Slichter and Mandell [644, 646] showed that the reduction in the width of the narrow component of the line on heating for atactic polypropylene occurs at a lower temperature than for isotactic polymer. This is explained by the fact that crystallites present in the isotactic polymer interfere with the molecular motion in the amorphous regions. The broad component of the line remains constant and, consequently, motion remains of low intensity in the crystalline regions (Fig. 79).

The low-temperature decrease in ΔH_2^2 for both polymers lies in the same temperature range (Fig. 80). As an investigation of the NMR spectrum of deuterated polypropylene showed [334] the low-temperature decrease in ΔH_2^2 is explained by motion of the side methyl groups of the chain. It is obvious that the rotation of the CH_3 groups is independent of the order of the structure.

Sauer et al. [600] found that the fall in ΔH_2^2, corresponding to the β transition (glass transition in amorphous regions) in isotactic polypropylene occurs at temperatures which are approximately 30° higher than in atactic polymer.

The effect of stereoregularity on the NMR spin—lattice relaxation time of the protons in polypropylene is evident from Fig. 81, which is taken from the work of Kawai and his co-workers [407]. Over a wide range of temperatures, T_1 of isotactic polypropylene is twice as high as for atactic polymer. The low-temperature minimum in T_1 at −130°C is the same for the two samples, which confirms that stereoregularity has little effect on the motion of the CH_3 groups (see Chapter IV). The high-temperature minimum of T_1 of the isotactic polymer is observed at 100°C and that of the atactic polymer at 50°C. This is in qualitative agreement with data obtained on the temperature dependence of the NMR line width.

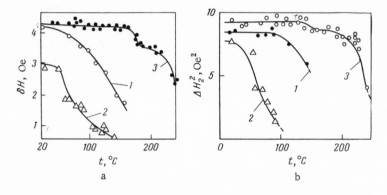

Fig. 78. Relation of the width of the line (δH, a) and the second moment of the line (ΔH_2^2, b) of the NMR spectrum of polymethyl methacrylate to the temperature [14]. 1) Atactic polymer from radical polymerization; 2) isotactic polymer from catalytic polymerization; 3) syndiotactic polymer from catalytic polymerization.

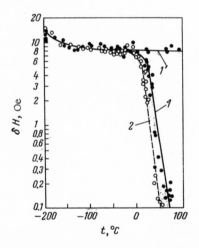

Fig. 79. Temperature dependence of the width of the NMR line (δH) of isotactic (curves 1 and 1') and atactic (curve 2) polypropylene [644].

    The NMR method has also been used to study other stereoregular polymers, namely, isotactic polybutene and polystyrene [628] and polymethylpentene-4 [223]. Slichter [628] found that ΔH_2^2 of atactic polybutene is less than ΔH_2^2 of the isotactic polymer (Fig. 82a), while ΔH_2^2 of polystyrene (Fig. 82b) is independent of the stereoregularity at low temperatures. However, with a rise in

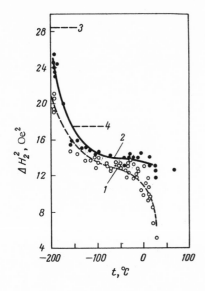

Fig. 80. Temperature dependence of the second moment of the NMR line (ΔH_2^2) of polypropylene [644]. 1) Isotactic; 2) atactic polypropylene; 2) theoretical value of ΔH_2^2 for a structure with stationary CH_3 groups; 4) the same for a structure with rotating CH_3 groups.

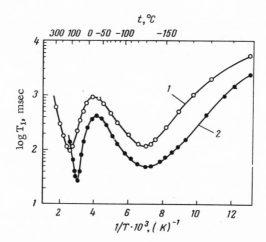

Fig. 81. Relation of the spin–lattice relaxtion time to temperature for isotactic (curve 1) and atactic (curve 2) polypropylene [407].

temperature the fall in ΔH_2^2 for atactic polystyrene is greater than for the isotactic polymer, since the molecular motion in the amor-

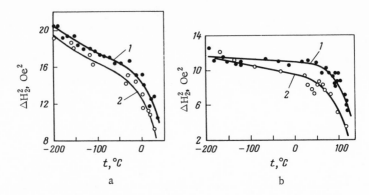

Fig. 82. Temperature dependence of the second moment of the NMR line (ΔH_2^2) of polybutene (a) and polystyrene (b) [628]. 1) Isotactic; 2) atactic polymer.

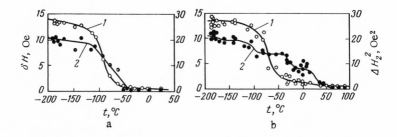

Fig. 83. Temperature dependence of the line width (δH, curve 1) and the second moment of the line (ΔH_2^2, curve 2) of the NMR spectrum of cis-polybutadiene (a) and trans-polybutadiene (b) [676].

phous polymer is stronger than in the crystalline polymer due to the lower packing density.

Polymers of regular structure with double bonds in the chain may exist in two geometrically isomeric forms, namely, cis and trans. Broad-line NMR spectroscopy makes it possible to observe the substantial difference in the character of the molecular motion of the two forms.

Takeda and his co-workers [676] and Gupta [331] studied the synthetic rubbers cis- and trans-polyisoprenes and cis- and trans-polybutadienes. The decrease in the line width corresponding to the glass transition point for the trans-isomers was observed at a

higher temperature than for the cis-isomers. Thus, for polybutadiene the temperatures were −46 and −90°C, respectively (Fig.
83a,b). For trans-polybutadiene (Fig. 83b) the decrease in ΔH_2^2
with temperature occurred in two stages, namely, at −120 and
−40°C. The authors suggested that the two transitions correspond
to the beginning of motion of the molecular chains in the amorphous and crystalline regions of the polymers (see Chapter IV). For
cis-polybutadiene the temperatures of these transitions almost coincide and ΔH_2^2 decreases smoothly (Fig. 83a).

The properties of polymers with a system of conjugated
bonds depend substantially on whether the macromolecule has a
planar structure or adjacent units are noncoplanar. In our work
[116] an attempt was made to use the NMR method to elucidate the
structure of polyphenylene and its derivatives. We plotted the
NMR spectra of polyphenylene (I), polyazophenylene (II), the methyl
derivative of polyazophenylene (III), and the model substance
quaterphenyl (IV) at temperatures of −196, +20, and +175°C.

The value 12.6 Oe^2 found for ΔH_2^2 for quaterphenyl at −196°C agrees
satisfactorily with the value 12.0 Oe^2 calculated from the known
structure [555]. The value determined experimentally for polyphenylene $\Delta H_2^2 = 10.4$ Oe is considerably less than that calculated
for a planar structure (14.8 Oe^2). The same discrepancy was found
for polyazophenylene and the methyl derivative of polyazophenylene. This indicates that neighboring phenyl nuclei are noncoplanar
in molecules of these polymers; the calculated and experimental

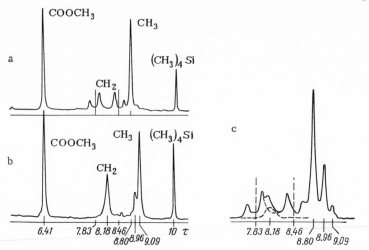

Fig. 84. NMR spectra of 2% solutions of polymethyl methacrylate in chloroform at 120°C [210]. a) 95% isotactic; b) 85% syndiotactic; c) 81% isotactic.

values of ΔH_2^2 agree if we assume that the planes of neighboring nuclei are turned 35° about the bond.

The development of a method of determining the stereochemical structure of polymers from the high-resolution NMR spectra
of their solutions is probably most important.

As a rule, the NMR spectra of polymers are complex and
different methods are used for interpreting them. Comparisons
are made of the spectra of the polymer samples investigated and
the spectra of polymers of known stereoregularity and mixtures of
them and also the spectra of model compounds (dimers, trimers,
and oligomers). Appropriate solvents are selected and the spectra
are plotted at high temperature to improve the resolution. Deuteration of the polymers and the double resonance method (elimination
of spin—spin interaction) are used to simplify the spectra.

As a result of the work of a series of investigators in 1959-
1964, it became possible to develop a method of determining the
stereoregularity of polymethacrylates, polyacrylates, polystyrene,
polymethylstyrene, polyvinyl chloride, polypropylene, polyvinyl alcohol, polyalkyl vinyl ethers, polytrifluorochloroethylene, and other
polymers. It is obviously difficult at present to formulate general
rules for the relation of the form of the NMR spectrum to the configuration and conformation of a macromolecule in solution, and we
will limit ourselves here to merely giving the main results obtained.

Polymethyl methacrylate

$$\cdots - CH_2 - \overset{\overset{\displaystyle CH_3}{|}}{\underset{\underset{\displaystyle COOCH_3}{|}}{C}} - \cdots$$

has been studied in most detail and a series of papers has been devoted to the analysis of its spectra.* The results of these studies agree with each other well.

Figure 84 gives typical spectra of two samples of polymethyl methacrylate, namely isotactic (a) and syndiotactic (b). The spectra were determined for 2% solutions of the polymers in chloroform at 120°C (the resolution was inadequate at a lower temperature). The extreme right-hand line belongs to the internal standard, tetramethylsilane, while the positions of the other peaks are given in τ units, i.e., ppm, with respect to the signal of tetramethylsilane, for which the value 10τ is taken.† The signal of the ester group $COOCH_3$ appears at 6.41τ in both spectra. The two protons of the methylene group are magnetically equivalent and give one line at 8.18τ if the unit adjacent to the CH_2 group in the chain has a configuration opposite to the configuration of the given unit, i.e., is attached syndiotactically. An isotactic link, when the unit adjacent to the CH_2 group has the same configuration as the given unit, eliminates the magnetic symmetry of the two methylene protons and as a result of spin—spin interaction, a quadruplet is obtained (Fig. 84a). The centers of its component doublets lie at 7.83 and 8.46τ. The spectrum of a polymer containing 81% of isotactic links (Fig. 84c) contains simultaneously a singlet at 8.18τ and two doublets at 7.83 and 8.46τ.

The protons of the α-methyl groups give three lines (at 8.80, 8.96, and 9.09τ), depending on how the two neighboring units are connected:

1) with isotactic links on both sides of the given unit (isotactic unit), the line is at 8.80τ;

---

*See the literature [15, 16, 193, 202, 210, 391, 392, 400, 477, 479, 510, 511, 707] and also the section on polymerization in Chapter V.

†Sometimes the chemical shift of tetramethylsilane as taken as 0 and the chemical shift is determined as millionths of the field strength from the signal of tetramethylsilane in the low field direction (δ scale) and then $\delta = 10-\tau$.

2) with an isotactic link on one side and a syndiotactic link on the other (heterotactic unit), the line is at $8.96\,\tau$;

3) with syndiotactic links at both sides (syndiotactic unit), the line is at $9.09\,\tau$.

In the spectrum of the isotactic polymer (Fig. 84a) the strongest line is at $8.80\,\tau$, while the strongest in the spectrum of the syndiotactic polymer is the peak at $9.09\,\tau$. Consequently, from the ratio of the intensities of the lines in the high-resolution NMR spectrum of polymethyl methacrylate it is possible to determine the fraction of isotactic and syndiotactic links and the fraction of isotactic, heterotactic, and syndiotactic units. *

These data are used widely in studying the polymerization of methyl methacrylate (see Chapter V).

The general rules observed in the NMR spectra of polybutyl methacrylate are similar to those established for polymethyl methacrylate, but are less clearly expressed. Only a qualitative estimate of the type of stereoregularity is possible from the form of the spectrum [15].

For studying the structure of polymethacrylic acid [195, 441] it was methylated with diazomethane and the NMR spectrum of a solution of the polymethyl methacrylate obtained was determined. Analogously, samples of polymers of methacrylic anhydride [368, 480, 689], a polymer of methacrylonitrile [658, 659], a polymer of the menthyl ester of methacrylic acid [657], and also polymers of the dihydrate of barium methacrylate [74] were converted into polymethyl methacrylate by hydrolysis and methylation since the method of determining stereoregularity from the NMR spectrum has been developed best for polymethyl methacrylate.

The NMR spectrum of polymethyl acrylate

$$\cdots - CH_2 - \underset{\underset{\displaystyle COOCH_3}{|}}{CH} - \cdots$$

---

*The following terms are also used: isotactic and syndiotactic diads; isotactic, heterotactic, and syndiotactic triads.

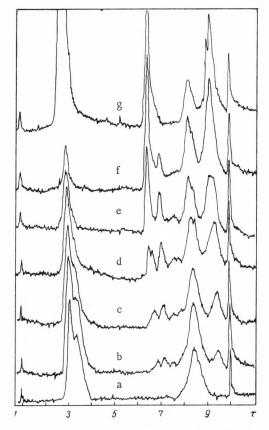

Fig. 85. NMR spectra of 20% solutions of copolymers of
methyl methacrylate with styrene in carbon tetrachloride
(a-f) and chloroform (g) at 90°C with different initial
ratios of the monomers in the reaction mixture [194].
a) 0:100; b) 10:90; c) 25:75; d) 50:50; e) 75:25; f)
90:10; g) 100:0.

is also affected by the stereoregularity, and this appears as a
change in the form of the signal of the methylene protons [457]. In
the syndiotactic polymer the CH_2 protons give a triplet, while in
the isotactic polymer they give a multiplet, corresponding to the
AB part of a spectrum of the ABKL type. Therefore, for determining the stereoregularity of polyacrylonitrile it is convenient to convert it into polymethyl acrylate.

It was possible [194, 227] to determine completely the stereo-chemical configuration and order of the units in a copolymer of methyl methacrylate (MMA) with styrene (ST) and methylstyrene

$$\cdots -CH_2-\underset{\underset{COOCH_3}{|}}{\overset{\overset{CH_3}{|}}{C}}-\cdots$$

MMA unit

$$\cdots -CH_2-CH-\cdots$$

ST unit

In the spectra of solutions of copolymers of methyl methacryl-ate and styrene (MMA + ST) in carbon tetrachloride or in chloro-form (Fig. 85) there are the peaks of the phenyl protons of ST in the region of 3.0-$3.5\,\tau$, the protons of the OCH_3 groups of MMA at 6.4-$7.9\,\tau$, the CH_2 groups of MMA and the CH and the CH_2 groups of ST at 8.1-$8.5\,\tau$, and the α-CH_3 groups of MMA at 8.8-$9.5\,\tau$.

The structure of the copolymer may be determined from the form of the signal of the protons of the OCH_3 groups. The follow-ing forms are possible for the signal of the methoxyl group of a MMA$_{(2)}$ unit:

1. With methyl methacrylate units on both sides of the unit MMA$_{(2)}$

$$\cdots -MMA_{(1)}-MMA_{(2)}-MMA_{(3)}-\cdots$$

there is one peak at $6.40\,\tau$, regardless of the stereochemical con-figuration of MMA$_{(1)}$ and MMA$_{(3)}$.

2. With a styrene unit on one side the unit MMA$_{(2)}$

$$\cdots -MMA_{(1)}-MMA_{(2)}-ST_{(3)}-\cdots$$

there are two peaks at 7.0-$7.2\,\tau$ and 6.4-$6.8\,\tau$, corresponding to structures in which ST$_{(3)}$ has the same configuration as MMA$_{(2)}$ or the opposite.

3. With styrene units on both sides of the unit MMA$_{(2)}$

$$\cdots -ST_{(1)}-MMA_{(2)}-ST_{(3)}-\cdots$$

there are three peaks at 6.9, 7.2, and $7.5\,\tau$, depending on whether the two ST units have the same configuration as MMA$_{(2)}$ ("co-isotactic structure"), one ST unit has the same configuration, while the second has the opposite ("coheterotactic"), or both ST units are opposite in configuration to the MMA$_{(2)}$ unit ("cosyndiotactic structure").

In the spectra of copolymers of MMA + ST in aromatic solvents the ratio of the intensities of the peaks of the signal from the OCH$_3$ group depends not only on the configuration of the two nearest neighbors of the given MMA unit, but also on the configuration of more remote units, so that it is possible to determine the order of the monomers in a section of five units [345].

Clark [240] studied the effect of the stereochemical structure on the NMR spectra of model compounds, namely, dimers of acrylonitrile. The signal of the methylene protons was found to be most sensitive to the configuration; the meso (isotactic) dimer gives a septet and the racemic (syndiotactic) dimer, a quadruplet.

The high-resolution NMR spectra of solutions of polystyrene have been studied (Fig. 86) [412, 746]. The lines were assigned by means of the spectrum of a deuterated sample [746]. The spectrum contains peaks corresponding (starting from low magnetic fields) to meta- and para-protons (1) of the phenyl group, ortho-protons (2) of the phenyl group, and methyne (3) and methylene (4) protons of the chain.

The position of the peaks depends on the stereoregularity of the polymer and the nature of the solvent. In spectra of isotactic polystyrene in solutions in thionyl chloride or carbon disulfide, the signals (3) and (4) are well resolved, while for the technical atactic polymer, and also for polystyrene obtained with an alkyllithium or alkylsodium catalyst, the peaks (3) and (4) merge together and there is only one line of the protons of the chain.

In the NMR spectrum of solutions of poly-β-dideuterostyrene obtained at 100°C it is possible to distinguish three peaks of the α-protons, corresponding to the isotactic, heterotactic, and syndiotactic sequences in the polymer chain [214]. In all the samples studied, apart from those prepared with a Ziegler catalyst, the syndiotactic sequences predominated.

As Brownstein and his co-workers [215] showed, in the NMR spectrum of poly-α-methylstyrene the signal of the α-methyl group is sensitive to the stereochemical configuration. This group lies

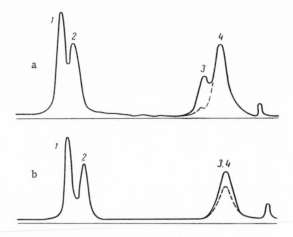

Fig. 86. NMR spectra of 10% solutions of polystyrene (solid line) and poly-α-deuterostyrene (broken line) in carbon disulfide. a) Isotactic; b) atactic polymer. (See text for assignment of peaks.)

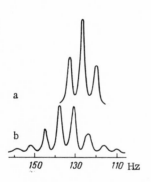

Fig. 87. Theoretically calculated NMR spectra of the methylene protons of polyvinyl chlorides [691]. a) Syndiotactic polymer (part A of a spectrum of type A_2B_2); b) isotactic polymer (part AB of a spectrum of type ABC_2).

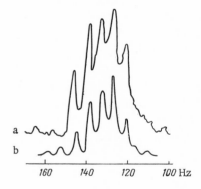

Fig. 88. NMR spectrum of methylene protons [691]. a) Experimental spectrum for a 10% solution of technical polyvinyl chloride in o-dichlorobenzene at 170°C with a working frequency of 60 MHz; b) theoretically calculated spectrum for a structure containing 55% isotactic and 45% syndiotactic links.

between two phenyl nuclei and its diamagnetic screening depends on the structure of the polymer. In the spectrum of a 20% chloroform solution of poly-α-methylstyrene determined at 80°C with a frequency of 56.4 MHz, the protons of the α-methyl group give a triplet with lines at 7.7, 20.0, and 46.2 Hz relative to tetramethylsilane. The ratio of the areas of the components of the triplet is proportional to the fraction of the syndiotactic, heterotactic, and isotactic units. This interpretation of the peaks in the triplet of the α-methyl group was disputed in [594], but was confirmed by Braun [212].

The spectrum of polyvinyl chloride solutions is very complex and its form depends markedly on the frequency of the measurement, the solvent, and the temperature and the data of different investigators do not agree completely [197, 199, 238, 269, 270, 271, 380, 597, 691, 744]. Different methods have been used to interpret the spectra of polyvinyl chloride. These include comparison of the spectrum of the polymer with the spectra of model compounds (d,l-2,4-dichloropentane and meso-2,4-dichloropentane [269-271, 597]), plotting the spectra of deuterated polymers [199, 744]

and eliminating the spin—spin interaction between the methyne and methylene protons by the double resonance method [197, 597, 744].

Convincing results were obtained in [691] in which the spectra were plotted at 170°C at a frequency of 60 MHz and a theoretical calculation was carried out on an electronic computer. The author considers that the methyne protons (1) of polyvinyl chloride give a quartet, i.e., part A of a spectrum of type AB_4 (see page 83) as a result of interaction with protons of the two adjacent CH_2 groups, regardless of the stereochemical configuration:

The methylene protons (2) and (3) in the syndiotactic polymer are magnetically equivalent and give in the NMR spectrum a triplet, i.e., part A of a spectrum of type A_2B_2 (Fig. 87a). In isotactic polyvinyl chloride the protons of the CH_2 group are nonequivalent (one of them is closer to the chlorine atoms than the other) and give a complex signal, i.e., part AB of a spectrum of type ABC_2 (Fig. 87b). The spectrum of technical polyvinyl chloride of type "Opalon 300" (Fig. 88) coincides with that calculated for a polymer structure containing 55% isotactic and 45% syndiotactic links.

Yoshino and Komiyama [744] consider that to explain the spectrum of polyvinyl chloride it is necessary to take into account the effect of the configuration of remote asymmetric carbon atoms. The assignment of the peaks in the spectrum proposed by these authors is based on the difference in configuration of segments of four units, i.e., tetrads. Fresh experimental data are evidently required.

In polypropylene, as in polyvinyl chloride

$$
\begin{array}{cc}
\overset{2)}{CH_3}\,H \quad CH_3\,H & CH_3\,\overset{(2)}{H}\,\overset{(1)}{H} \quad H \\
|\quad| \quad\; |\quad| & |\quad|\quad| \quad\; | \\
\cdots -C-C-C-C-\cdots & \cdots -C-C-C-C-\cdots \\
|\quad| \quad\; |\quad| & |\quad|\quad| \quad\; | \\
H\;\;H \;\; H\;\;\;H & H\;\;H\;\;CH_3\,H \\
{\scriptstyle (3)\;\;(1)} & {\scriptstyle (3)} \\
\text{isotactic} & \text{syndiotactic} \\
\text{structure} & \text{structure}
\end{array}
$$

the protons of the CH_2 groups $H_{(2)}$ and $H_{(3)}$ are equivalent if the neighboring unit has the opposite configuration and nonequivalent for an isotactic structure. However, due to the effect of the α-protons and the methyl protons the spectrum is very complicated. To interpret the spectrum and determine the local stereoregularity of polypropylene, investigators have made use of comparison of experimental spectra with calculated spectra [598], comparison with the spectra of a model substance (2,4-dimethylpentane) and deuterated derivatives [662], and plotting spectra at high temperature (up to 150°C) at two frequencies, namely, 60 and 100 MHz [725]. It was possible to establish that the signal of the methylene protons consists of three doublets, which correspond to syndiotactic, heterotactic, and isotactic triads in the polypropylene chain.

The chemical shifts and spin—spin interaction constants for the spectra of isotactic and syndiotactic polypropylene are also given in the work of Kato and Nishioka [401] and Ohnishi and Nukada [534].

Danno and Hayakawa [261] and Ramey and Field [579, 580] studied the effect of stereoregularity on the NMR spectrum of solutions of polyvinyl alcohol (I), polyvinyl acetate (II), and polyvinyl trifluoroacetate (III).

$$\cdots -CH_2-\overset{\overset{\displaystyle H}{|}}{\underset{\underset{\displaystyle OH}{|}}{C}}-\cdots \qquad \cdots -CH_2-\overset{\overset{\displaystyle H}{|}}{\underset{\underset{\displaystyle OCOCH_3}{|}}{C}}-\cdots \qquad \cdots -CH_2-\overset{\overset{\displaystyle H}{|}}{\underset{\underset{\displaystyle OCOCF_3}{|}}{C}}-\cdots$$

I   II   III

The structure of polyvinyl acetate may be determined from the signal of the CH_3 protons. A solution of polyvinyl acetate in methylene chloride at 100 MHz gives a three-component signal with peaks at 7.98, 8.00, and 8.02, which correspond to syndiotactic, heterotactic, and isotactic units in the chain. The stereoregularity of polyvinyl alcohol may be determined from the form of the signal of the CH_2 groups of a solution of polymer in heavy water at 70°C and above [261]. The signal of the methyne protons in polyvinyl alcohol and polyvinyl trifluoroacetate is split into three components, corresponding to iso-, hetero-, and syndiotactic triads, but only after illumination of a spin—spin interaction with the methyl groups. A study of the stereoregularity of polyvinyl alcohol was also reported in [16]: the polyvinyl alcohol was esterified with formic acid and the structure determined from the spectra of a solution of polyvinyl formate in heavy water.

In the first study of the very complex NMR spectrum of solutions of polymethyl vinyl ether [411]

$$\cdots -CH_2-\overset{\overset{\displaystyle H}{|}}{\underset{\underset{\displaystyle OCH_3}{|}}{C}}-\cdots$$

the authors were unable to assign the poorly resolved peaks. Only by an empirical method was a correlation found between the form of the line of the methylene group of the ether and the degree of stereoregularity. This correlation was used in assessing polymerization catalysts. To improve the resolution in later work [216,

582], solvents were selected (the most suitable solvent was found to be nitromethane at high temperature and a mixture of chlorobenzene and methylene chloride at room temperature) and double resonance was used. The spectrum of a solution of polymethyl vinyl ether in a mixture of chlorobenzene and methylene chloride determined at 100 MHz [582] consists of three signals at 6.49, 6.73, and 8.25 τ, corresponding to the α-protons, the methoxyl protons, and the β-protons. The signal of the α-protons consists of a triplet (after elimination of spin—spin interaction with the methylene protons). The components at 6.43, 6.48, and 6.54 τ correspond to syndiotactic, heterotactic, and isotactic triads in the polymer chain. The signal of the protons of the OCH_3 groups is also split into three components at 6.70, 6.72, and 6.74 τ, corresponding to syndiotactic, heterotactic, and isotactic triads.

The signal of methylene protons, after elimination of their spin—spin interaction with the α-protons, consists of a quartet of type AB superimposed on a singlet. The AB quartet is from isotactic diads and the singlet from syndiotactic diads.

The spectrum of a solution of poly-α-methylvinyl methyl ether

$$\cdots -CH_2-\overset{\overset{\displaystyle CH_3}{|}}{\underset{\underset{\displaystyle OCH_3}{|}}{C}}- \cdots$$

in chlorobenzene, determined by Goodman and You-Ling Fan [323], was found to be unexpectedly simple: three separate peaks were obtained, one from the protons of the OCH_3 groups at 6.86 τ, one from the CH_2 group at 8.04 τ, and one from the α-CH_3 group at 8.57 τ. The sample was evidently completely syndiotactic.

Polytrifluorochloroethylene is difficultly soluble, and to determine the structure of this polymer, Tiers and Bovey [690] synthesized special samples with a low molecular weight. The spectrum of a solution of such a sample is given in Fig. 89. From the ratio of the intensities of the components of the doublets from CF_2 and CFCl groups it is possible to determine the fraction of isotactic and syndiotactic links between units in the chain.

The determination of the chain structure of polyacetaldehyde was found to be a difficult problem [207-209, 321].

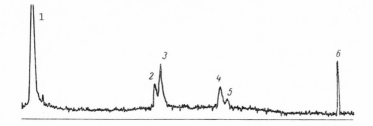

Fig. 89. Fluorine lines in the NMR spectrum of a 20% solution of polytrifluorochloroethylene in 3,3'-bis-(trifluoromethyl)-biphenyl at 150°C with a
working frequency of 40 MHz [690]. 1) Solvent; 2) CF_2 groups in isotactic
diads; 3) CF_2 groups in syndiotactic diads; 4) CFCl groups in syndiotactic
diads; 5) CFCl groups in isotactic diads; 6) calibration line. The distance
between lines 1 and 6 equals 4000 Hz.

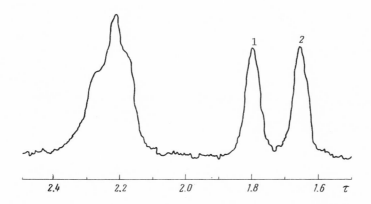

Fig. 90. Proton lines in NMR spectrum of a solution of a mixture of cis- and
trans-1,4-polyisoprenes in benzene at 25°C with a working frequency of 60
MHz [229]. 1) Cis-methyl group; 2) trans-methyl group.

$$\cdots -O-\underset{\underset{\displaystyle CH_3}{|}}{\overset{\overset{\displaystyle H}{|}}{C}}-\cdots$$

The form of the spectrum of this polymer depends on the type of
solvent. By using the double resonance method it was possible to
establish that the signal of the protons of the methyl group consists
of two doublets superimposed. The doublet in the low-field region
is due to heterotactic triads and that in the high-field region, iso-

tactic triads. In the chain of amorphous rubbery polyacetaldehyde, obtained with cationic catalysts, there are no or very few syndiotactic triads. The chain consists of *lll* and *ddd* blocks. The mean length of the blocks equals three units, i.e., considerably less than is required for crystallization.

Analysis of the high-resolution NMR spectra of solutions also makes it possible to distinguish trans- and cis-isomers. Chien and Walker [231] showed that the spectra of solutions of trans- and cis-isomers of the dimethyl ester of 1,4-cyclohexanedicarboxylic acid are different, while the spectra of the polyesters obtained from their reaction with ethylene glycol are identical. Comparison of the intensities of the maxima in the spectrum of the polyester and the dimethyl esters shows that isomerization occurs during the transesterification reaction and, regardless of the structure of the starting ester, we obtain a polymer with a ratio of units with trans- and cis-configurations of 3 : 2.

From the spectrum a solution of 1,4-polyisoprene [229, 608] (Fig. 90) it is possible to determine the content of cis- and trans-units with an error of less than 0.5%. The cis- and trans-forms of polybutadiene were determined in the work of Lombardi and Segre [442] and also in the work of Chen Hung Yu [228], which was examined above.

By comparing the NMR spectra of alicyclic epoxide resins with the spectra of model compounds, it was possible to demonstrate that the resins have a trans-ether structure [159].

It has been pointed out [608] that the NMR method has advantages over the determination of cis- and trans-isomers from infrared spectra, namely, accuracy, rapidity, and simplicity of interpretation of the NMR spectra. Particularly high accuracy and sensitivity in the determination of the structure of natural and synthetic polyisoprenes was achieved in [316], in which the spectra were plotted at a frequency of 100 MHz and an electronic data storage unit, "Mnemotron," was used to increase the signal-to-noise ratio.

Chapter IV

Study of Molecular Motion in Polymers by the NMR Method

A considerable number of the problems in the physics and chemistry of polymers which may be solved by the NMR method are directly or indirectly related to the effect of molecular motion in the polymer on the NMR spectrum. Thus, the effect of the degree of crystallinity and the chain structure of the polymer on the form of the NMR line, examined in Chapter III, is explained ultimately by molecular motion. In this chapter we discuss only certain trends in the study of molecular motion in polymers.

1. Effect of Molecular Motion on the Form, Width, and Second Moment of Lines in the NMR Spectrum of a Polymer in Bulk

The mechanical and physicochemical properties of polymers in bulk in a vitreous and particularly highly elastic state are determined to a large extent by the character of the molecular motions of individual groups and segments of the chains. The NMR method is particularly suitable for studying molecular motions: it is precisely this method which gave one of the first direct demonstrations of rotation about bonds in the chains of polymers [359].

Just about the greatest number of all papers on NMR in polymers are concerned with the study of molecular motions in polymers through the temperature dependence of the form, width, and second moment of the NMR line and therefore it is not profitable to examine them in detail in this monograph and to generalize all the re-

sults obtained, since this would be far beyond the framework of
the book. The reader will find a detailed discussion in special reviews of this problem [541, 601, 628, 636, 639, 729, 730]. Here we
will examine only some aspects of the problem.

As a rule, the complex form of the NMR line of polymers is
observed only at definite temperatures. Thus, the form of the line
due to the presence in the polymer chain of weakly interacting
groups appears only at low temperatures when the molecular motions are restricted. The form of the line caused by the presence
in the polymer of regions with different mobilities appears only at
high temperatures (see Chapter III).

The form of the simple NMR line also changes as a result of
the effect of molecular motion. For example,* for 6,6 polyamide
the ratio $\delta H_{ms}/(\Delta H_2^2)^{1/2}$ falls from 3.0 to 2.5 over the temperature
range from −253°C to room temperature, while the ratio
$(\delta H_{1/2})^2/\Delta H_2^2$ falls from 6.9 at 47°C to 5.3 at 61°C and 4.2 at 97°C
[394]. Figure 77 (page 201) shows the change in the coefficient of
the form of the line $(\Delta H_4^4)^{1/4}/(\Delta H_2^2)^{1/2}$ with temperature for a series
of polyethylene samples.

Sinnot [623] constructed the NMR line of polymethyl methacrylate for a rigid structure and also for a structure with rotation
of one CH_3 group and both CH_3 groups. For the construction he
used the NMR line of a stationary CH_3 group from the spectrum of
trichloroethane $C_2H_3Cl_3$ at −196°C, the line of a rotating CH_3 group
from the spectrum of methyl chloroformate $ClCOOCH_3$ at −196°C,
and the NMR line of a CH_2 group from the spectrum of malonic
nitrile $CH_2(CN)_2$ at −196°C. The spectrum constructed agrees well
with the experimental spectrum (Fig. 91).

The decrease in the width of the NMR line on heating is
caused by averaging of the local magnetic fields (see Chapter I).
For a series of polymers the narrowing of the lines on heating (or
broadening on cooling) occurs in stages with each change in δH
corresponding to a definite change in the character of the molecular motion in the polymers.

---

*Here, δH_{ms} represents the width of the line between the points of maximum slope (be-
tween the extrema of the derived function) and $\delta H_{1/2}$ is the width of the line at the
half height.

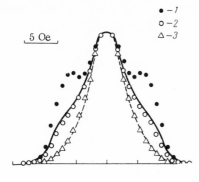

● −1
○ −2
△ −3

Fig. 91. Comparison of the form of the experimental line in the NMR spectrum of polymethyl methacrylate at −196°C (solid line) and + 27°C (broken line) with the form of the line calculated for a rigid structure (1), for a structure with rotation of one (2) and two (3) methyl groups [623].

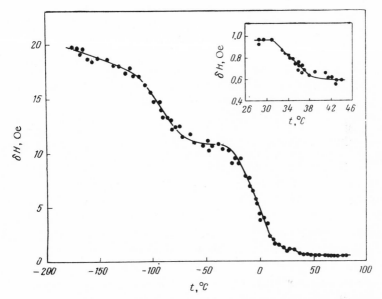

Fig. 92. Relation of the width of the line (δH) in the NMR spectrum of polyisobutene to the temperature [686].

   Thurn [685, 686] measured the width of the lines of high-molecular polyisobutene and polymethyl methacrylate. Three regions were found in which the width of the line of polyisobutene (Fig. 92) fell, namely at ~100°C, over the range from −30 to +10°C, and at 30-40°C. By comparing the NMR data with results obtained by other methods, the author came to the conclusion that the narrowing of the line was caused by transitions in the polymer. The first transition is connected with rotation of the CH_3 groups about their axes of symmetry C_3 and the second, with the beginning of motion of the main chain. The transition at 30-40°C was observed only for samples of high molecular weight and is apparently explained by motions of large sections of the chain.

The spectral line of polymethyl methacrylate showed four regions of narrowing (Fig. 93), corresponding to transitions in the polymer which are also observed by the dynamic mechanical method. The transition at 75° is apparently connected with the presence of 0.2% of moisture in the sample studied by the author, while the other changes in the width of the spectral line are caused by the beginning of motion of the CH_3 groups, individual units, and large sections of the chains.

Transitions producing changes in the width of the lines are largely interpreted as the result of the freeing of vibrational or rotational motions in amorphous and crystalline regions of the polymer. However, if a reduction in δH is observed down to a very small value (for example, to 0.019 Oe at $-73°C$ and above for polydimethylsiloxane) [365], then it must be assumed that longitudinal motions also appear in the crystallites.

This paradoxical conclusion has also been confirmed by other methods.

For some polymers a change in the molecular motion has no effect on the width of the NMR line. For example [313], with high-density polyethylene a considerable decrease in the width of the line δH_{ms} occurs only in the region of $-73°C$ (Fig. 94). On the curve of the change in ΔH_2^2 with temperature there are two transitions, namely, at a temperature from -73 to $-23°C$, which is caused by the presence of cooperative motions in a small fraction of the segments of the chain in amorphous regions of the polymer, and at a temperature of about 77°C, which is connected with disordering and fusion of crystalline regions. In this case the second transition has no effect on the value of δH_{ms} since the maximum on the the curve corresponding to the derived function is determined mainly by the magnitude of the narrow component of the complex NMR line, while melting in the crystalline regions affects the broad component. In general, for a line of complex form the concept of "line width" becomes quite indefinite. For a line whose complex form is due, for example, to resonance of a methyl group, the separation of the line into broad and narrow components is arbitrary. Graphs

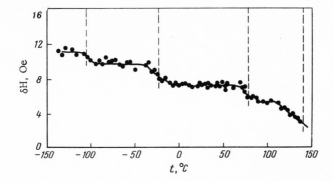

Fig. 93. Relation of the width of the line (δH) in the NMR spectrum of polymethyl methacrylate containing 0.2% water to the temperature [686].

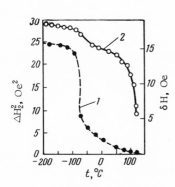

Fig. 94. Temperature dependence of the width of the line (δH, curve 1) and the second moment of the line (ΔH$_2^2$, curve 2) in the NMR spectrum of high-density polyethylene [313].

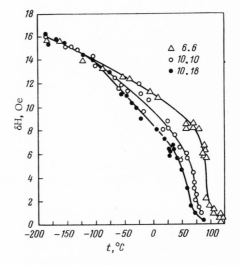

Fig. 95. Temperature dependence of the width of the line (δH) in the spectrum of 6,6-10, 10- and 10,18-polyamides [631].

of the relation δH = f (t) in the low-temperature region with two curves (for broad and narrow components) presented for such polymers as polymethyl methacrylate should be regarded as a description of the change in the form of the line.

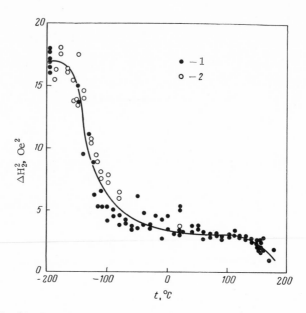

Fig. 96. Temperature dependence of the second moment of the line (ΔH_2^2) in the NMR spectrum of polycarbonate [120]. 1) "Diflon" polycarbonate; 2) "Makrolon" polycarbonate.

The study of the temperature dependence of the form of the NMR line makes it possible to assess the character of the intermolecular forces which determine molecular mobility in a polymer. The role of intermolecular forces appears clearly, for example, in the change in the form of the curves $\delta H = f(t)$ with an increase in the number of CH_2 groups between amide groups in a series of polyamides [628, 631] (Fig. 95). In polyethylene, narrowing of the NMR line is observed at −73°C and in polyamides at temperatures from 52 to 102°C, since the replacement of a CH_2 group by an amide group leads to the appearance of hydrogen bonds between the chains. The shorter the paraffin segments between the amide groups, the higher is the frequency of hydrogen bonds in the chain, the lower is the mobility, and the higher the temperature of narrowing of the NMR line.

The second moment is the most important characteristic of the NMR line of a polymer since the second moment may be calculated for a rigid structure and for a structure with individual mobile molecular groups, and the experimental and theoretical results may be compared. The literature contains theoretically calculated values of ΔH_2^2 for almost all the polymers used widely and a summary is given in the review [112]. The same review gives experimentally determined values of the second moment at low temperatures and at room temperature.

The change in the second moment with temperature makes it possible to estimate the motions of the main chain and the side groups of the polymer (particularly the motion of methyl groups). Figure 96 shows the relation $\Delta H_2^2 = f(t)$ for two samples of polycarbonate [79, 120]. The theoretical value of ΔH_2^2 calculated for polycarbonate of known structure [97] equals 20-25 Oe^2 (rigid structure, different conformations of the CH_3 groups). The experimental value of ΔH_2^2 at −196°C equals 16-17 Oe^2; the discrepancy is evidently explained by the fact that at this temperature some mobility of the methyl groups is still preserved. The fall in ΔH_2^2 with temperature in the region from −150 to −80°C is caused by rotation of the CH_3 group. At a temperature of 150-170°C, ΔH_2^2 decreases because there is the beginning of motion of segments of the chain molecules: the polymer changes from the glassy to a highly elastic state. The glass transition point of polycarbonate determined by other methods [97] equals 150°C.

If the NMR line has a complex form it is possible to calculate the second moment separately for the narrow and broad components of the line and to compare these values with the theoretical values. Figure 97 shows the results of such a calculation for linear and branched polyethylene [472].

If the monomer unit contains several molecular groups, then it is possible to calculate the value of ΔH_2^2 for a rigid molecule and for a molecule in which some or all of the molecular groups rotate. Thus, Powles [560] showed that over the range from −73 to +67°C the experimental value of ΔH_2^2 of polymethyl methacrylate coincides with the theoretical value, calculated for a structure with rotation of one CH_3 groups. Both methyl groups begin to rotate at a higher temperature. Analogous calculations were carried out by Sinnot for polyethyl methacrylate [623].

Odajima and his co-workers [528, 531] made a detailed study of the motion of methyl groups in a series of polymers: polymethacrylic acid, sodium polymethacrylate, polymethyl methacrylate, poly-α-methylstyrene, poly-o-, m-, and p-methylstyrenes, and

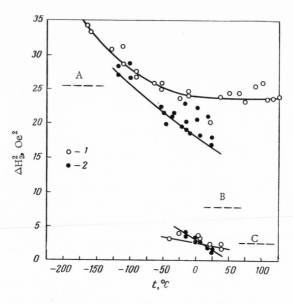

Fig. 97. Temperature dependence of the second moment of the line (ΔH_2^2) of the NMR spectrum of polyethylene. 1) Linear polymer; 2) polymer with branched chains. A,B,C) Calculated values of ΔH_2^2 for a rigid structure and for structures with synchronous and nonsynchronous rotation of neighboring CH_2 groups, respectively.

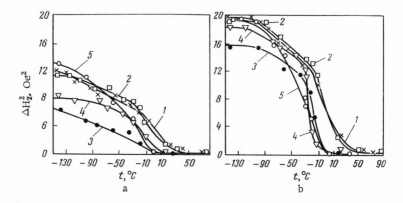

Fig. 98. Temperature dependence of the second moment of the fluorine (a) and hydrogen (b) lines in the NMR spectrum of polymers [18]. 1) Copolymer of vinylidene fluoride and trifluorochloroethylene; 2) copolymer of vinylidene fluoride and hexafluoropropylene; 3) polymer of perfluoromethoxyperfluoropropyl acrylate; 4) copolymer of vinyledene fluoride, trifluorochloroethylene, and perfluoromethoxyperfluoropropylacrylate; 5) polymer of hexafluoropentamethylene adipate.

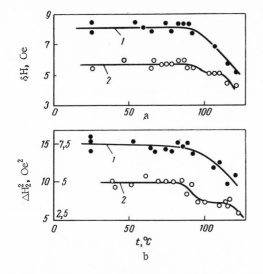

Fig. 99. Temperature dependence of the line width (δH, a) and second moment of the line (ΔH_2^2, b) in the NMR spectrum of poly-2-fluoro-5-methylstyrene [1]. 1) For protons; 2) for fluorine.

polydimethylstyrenes. A comparison of the experimental values of ΔH_2^2 with those calculated for a rigid structure showed that methyl groups attached to benzene rings rotate freely even at −196°C, while reorientation of α-methyl groups begins in the range from −173 to −23°C.

The information on the character of the molecular motions in a polymer may be increased substantially if the polymer contains two forms of magnetic nuclei (usually hydrogen and fluorine) and it is possible to compare the temperature dependences of δH and ΔH_2^2 for both nuclei.

Lyubimov and his co-workers [80] plotted the NMR spectra of protons and fluorine over a wide range of temperatures for a series of fluorine-containing rubbery polymers. The curves of $\Delta H_2^2 = f(t)$ for protons (Fig. 98b) may be divided into three sections: 1) constant values of ΔH_2^2; 2) a slow fall in ΔH_2^2; and, 3) a relatively rapid fall in ΔH_2^2. The end of the first and the beginning of the second section for all the polymers lay at about −110°C and the decrease in ΔH_2^2 over the second section was due to reorientation of CH_2 groups. The temperatures of the transitions from the second to the third sections correspond to the glass transition points, determined on a Kargin balance. Over the third section, ΔH_2^2 falls as a result of motion of the segments of the molecular chains. It is obvious that the chains of polyhexafluoropentamethylene adipate

are more mobile than for the other polymers and this is explained
by the presence of the ester groups OCO in them, which act as
"hinges." For fluorine (Fig. 98a) the character of the curves of
$\Delta H_2^2 = f(t)$ is the same as for protons, but in general, the fluorine-
containing groups of the polymers investigated were found to be
more mobile than the methylene groups.

A study was made [1, 3, 147] of the structure and the molecular motion in poly-2-fluoro-5-methylstyrene:

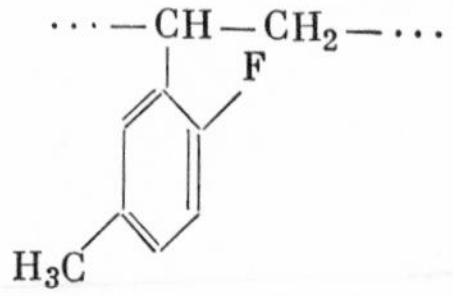

Values were calculated theoretically for the second moment
of the NMR line for fluorine $(\Delta H_2^2)_F$ and hydrogen $(\Delta H_2^2)_H$ for two
possible conformations of the chain (isotactic spiral and syndiotactic planar) and two possible positions of the radical. The best
agreement with experiment was obtained for a planar syndiotactic
chain with the orientation of the disubstituted phenyl group where
the plane of the radical is perpendicular to the plane of the chain
and the distance between the F atom and the H atom of the methyne
group is minimal. It was found that $(\delta H)_H$ and $(\Delta H_2^2)_H$ for polyfluoromethylstyrene fall monotonically beginning at 80°C (Fig. 99).
The values of $(\delta H)_F$ and $(\Delta H_2^2)_F$ fall in two clearly expressed steps.
The transition at 115°C is trivial since the polymer softens at this
temperature. To elucidate the nature of the transition at 85°C, the
effect of vibrations of the radicals on $(\Delta H_2^2)_F$ was calculated. It
was shown that the decrease in $(\Delta H_2^2)_F$ found experimentally may
be produced only by cophasal torsional oscillations of the radicals
of sufficiently great amplitude.

The conclusions drawn in this work were confirmed by a subsequent study of the NMR spectra of poly-para-fluorostyrene (I)
[322] and poly-2,5-difluorostyrene (II) [35]:

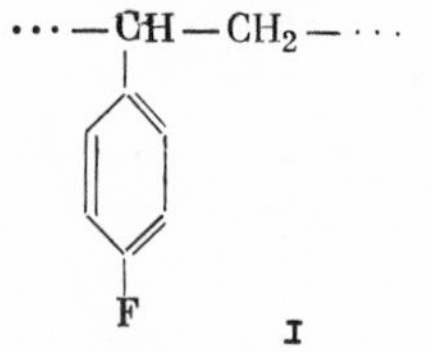

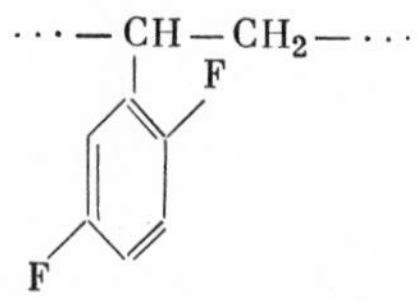

On the curves of the temperature dependence of δH and ΔH_2^2 of fluorine and hydrogen for para-fluorostyrene there is only one transition at 125°C, which corresponds to softening. The steric conditions for motion of the radicals in poly-para-fluorostyrene are evidently more favorable than in polyfluoromethylstyrene. However, the fluorine NMR spectrum is not affected by these oscillations, since the fluorine atoms lie on the axis of the oscillations. For poly-2,5-difluorostyrene the curve of $(\Delta H_2^2)_F = f(t)$ shows a fall at the softening point (110°C) and also a transition at 72°C, produced by torsional oscillations of the radicals. The magnetic field at fluorine atoms in ortho- and meta-positions relative to the chain is averaged out in these oscillations.

In the study of molecular motions in polymers by the NMR method the interpretation of the results is facilitated by the use of isotopic replacement. Deuteration, i.e., replacement of ^{1}H by D, is used most widely. The gyromagnetic ratios of the isotopes ^{1}H and D differ markedly from each other $(\gamma/2\pi = 4257$ and $654\ Hz/Oe$, respectively), so that deuterium does not interfere with the observation of hydrogen resonance.

Isotopic replacement has been used, in particular, in studying molecular action in polyethylene terephthalate [282, 433, 706], polypropylene [334, 531, 733], polystyrene [528], polyamide [312], and polyvinyl alcohol [743]. The NMR spectrum of polyethylene terephthalate and deuterated polyethylene terephthalates was studied by Ward and his co-workers [282, 433, 706]:

From the change in δH and ΔH_2^2 it was possible to assess the mobility of the CH_2 groups and the phenyl rings. For example, it was shown that the ratio of the intensities of the broad and narrow components of the signal does not change when the hydrogen in the CH_2 groups is replaced by deuterium. This is evidently explained by the fact that with a rise in temperature equal fractions of the benzene rings and methylene groups acquire mobility.

An analogous investigation for polypropylene and its deutero derivatives:

$$\cdots - CH_2 - \underset{\underset{\textstyle CH_3}{|}}{CH} - \cdots \qquad \cdots - CH_2 - \underset{\underset{\textstyle CD_3}{|}}{CH} - \cdots \qquad \cdots - CD_2 - \underset{\underset{\textstyle CH_3}{|}}{CD} - \cdots$$

was carried out by Woodward and his co-workers [531, 733] and Gupta [334]. As Fig. 100 shows, the value of ΔH_2^2 falls in the low-temperature region as a result of reorientation of the CH_3 groups of polypropylene. After deuteration, the value ΔH_2^2 falls only insignificantly in the region from -196 to $-73°C$.

The study of the motion of individual molecular groups is facilitated considerably if the mobility of the chains of the polymer in bulk is so great that the NMR signal has a complex form as a result of the appearance of a chemical shift as, for example, in fluoro rubbers (see Chapter III, page 153) [11].

Some information on molecular motion in polymers may be obtained by measuring the temperature dependence of the width of the NMR line at different pressures. Noller and Billings [518] showed (Fig. 101) that an increase in pressure from 1 to 1000 atm leads to an increase in the width of the line, indicating a decrease in the intensity of the molecular motion. The authors explain the observed effect by the decrease in free volume in the polymer.

In the study of molecular motion in irradiated polymers, results obtained by the NMR method successfully augmented information obtained by the EPR method [44, 678].

From the change in δH or ΔH_2^2 with temperature it is possible to calculate the correlation frequency ν_c and the activation energy of molecular motions E. If the theory of nuclear magnetic relaxation of Bloembergen, Purcell, and Pound applies and the change in

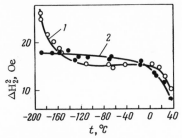

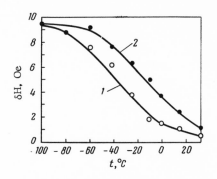

Fig. 100. Temperature dependence of the second moment of the line (ΔH_2^2) in the NMR spectrum of polymers [733]:

$1 - \cdots -CH_2-CH-\cdots;$
     $\quad\quad\quad\;\;|$
     $\quad\quad\quad CH_3$
$2 - \cdots -CH_2-CH-\cdots$
     $\quad\quad\quad\;\;|$
     $\quad\quad\quad CD_3$

Fig. 101. Temperature dependence of the line width (δH) in the NMR spectrum of polyisobutene [518]. 1) At atmospheric pressure; 2) at a pressure of 1060 atm.

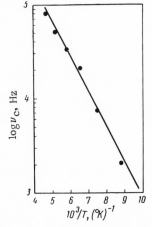

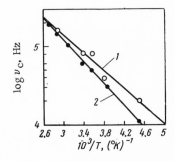

Fig. 102. Relation of the correlation frequency (ν_c) to the reciprocal temperature (1/T) for polycarbonate [120].

Fig. 103. Relation of the correlation frequency (ν_c) to the reciprocal temperature for polytrifluoroethylene [634]. 1) For hydrogen; 2) for fluorine.

the correlation frequency with temperature follows the Arrhenius law

$$\nu_c = \nu_{c,\,0}\,e^{-E/RT} \tag{IV-1}$$

then a graph in the coordinates $\log \nu_c = F(1/T)$ gives a straight line, from whose slope it is possible to determine the activation energy

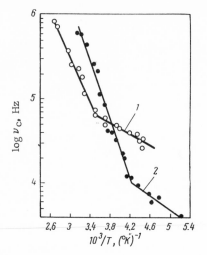

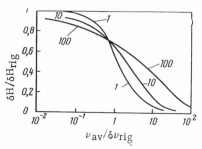

Fig. 105. Relation of the line width (δH) in the NMR spectrum to the correlation frequency for different values of the parameter b (given on the curves) [561].

Fig. 104. Relation of the correlation frequency (ν_c) to the reciprocal temperature (1/T) [472]. 1) For linear polyethylene; 2) for polyethylene with branched chains.

$$E = -2.3R\frac{\Delta(\log v_c)}{\Delta(1/T)} \qquad \text{(IV-2)}$$

and from whose point of intersection with the ordinate axis, it is possible to determine the preexponential factor $\nu_{c,0}$.

A graph was constructed of the relation $\Delta H_2^2 = f(t)$ for polycarbonate [120] (see page 233) (Fig. 102). The relation $\log \nu_c = F(1/T)$ was satisfactorily linear; the activation energy equaled 1.7 kcal/mole.

If the polymer contains hydrogen and fluorine, it is possible to calculate ν_c from the narrowing of the NMR lines of both types of nuclei. The activation energies 2.6 and 3.2 kcal/mole and similar values of ν_c were obtained for polytrifluoroethylene (Fig. 103) [634].

A graph in the coordinates $\log \nu_c - 1/T$ for polyethylene [472] consists of two straight-line sections (Fig. 104). For linear polyethylene, the activation energies of molecular motions corresponding to the low-temperature and high-temperature narrowing of the lines equal 1.5 and 6.1 kcal/mole and for branched polyethylene, 2.0 and 8.8 kcal/mole.

Uo and Fedin [126] derived a simplified formula for determining the potential barrier (E) to restrained rotation in solids. If the

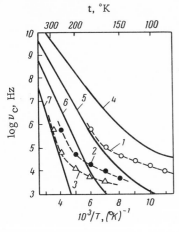

Fig. 106. Relation of the correlation frequency (ν_c) to the reciprocal temperature $(1/T)$ [531]. Broken curves – experimental values: 1) for α-methylstyrene; 2) for polymethylmethacrylate; 3) for poly-α-methylstyrene. Solid curves – curves calculated theoretically from the frequencies of tunnel transitions through the potential barrier for CH_3 groups; 4) with E = 4.4 kcal/mole; 5) with 5.9 kcal/mole; 6) with 7.4 kcal/mole; 7) with 9.9 kcal/mole.

relation of E to the angle of rotation of the molecule (φ) is described by the equation

$$E = \frac{E_0}{2}\,(1 - \cos n\varphi) \qquad\qquad \text{(IV-3)}$$

then

$$E_0 = 2.5kT_{nar}\log\frac{n}{\delta H_{low}}\sqrt{\frac{kT_{nar}}{2I_{in}}} \qquad\qquad \text{(IV-4)}$$

where T_{nar} is the temperature of narrowing of the line; δH_{low} is the width of the line at low temperature; I_{in} is the moment of inertia of the molecule.

For the calculation of E_0 from formula (IV-4) there is no need to carry out a large number of accurate measurements of the width of the NMR line. It is sufficient to estimate only the order of the moment of inertia of the rotating group. The potential barrier may be estimated with an accuracy of ±10% from the temperature at which line narrowing occurs: E_0 (kcal/mole) = $37T_{nar}$.

The activation energy of molecular motion has been calculated from NMR data for many polymers. The values of E obtained are summarized in [108, 472].

A comparison of the values for the activation energy of molecular motion calculated from the narrowing of NMR lines with values obtained by other methods shows that the calculation of E from the correlation frequency should be approached with caution.

Powles [561, 563] gives examples of low values of E due to the effect of a broad distribution of correlation frequencies of molecular motion in a polymer.

In a number of cases the molecular motion in a polymer cannot be described by means of one value of the correlation frequency and it must be assumed that there is a spectrum of correlation frequencies. This spectrum may be characterized by the distribution of the correlation frequencies $I(\nu_c)$. Powles [452, 561, 563] examined the case of a rectangular logarithmic distribution:

$$I\,(\ln \nu_c) = \frac{1}{2\ln b} \quad \text{when} \quad \frac{\nu_{av}}{b} < \nu_c < b\nu_{av}$$

$$I\,(\ln \nu_c) = 0 \quad \text{when} \quad \nu_c < \frac{\nu}{b} \quad \text{and} \quad \nu_c > b\nu_{av} \tag{IV-5}$$

The value ν_{av} characterizes the mean value of the correlation frequency, while b is the width of the distribution of correlation frequencies. If b = 1, then there is only one correlation frequency, which equals ν_{av}. When b = 10, there are frequencies from $\frac{1}{10}\nu_{av}$ to $10\nu_{av}$, etc. The change in the line width with frequency ν_{av} for b = 1, 10, and 100 is shown in Fig. 105.* It is obvious that with b = 10 and b = 100, a greater change in ν_{av} corresponds to the same change in the line width than with b = 1. If we ignore the presence of a distribution of correlation frequencies, then the values of E_{app} obtained are less than the true values. Connor examined cases of rectangular, Gaussian, and more complex asymmetric functions for the distribution [250] of the correlation frequencies both from NMR data and from the results of dielectric measurements.

For some polymers containing methyl groups the temperature dependence of the correlation frequencies found experimentally does not agree with that calculated theoretically. The broken curves in Fig. 106 show the relation of $\log \nu_c$ to 1/T for α-methylstyrene (1), polymethyl methacrylate (2), and poly-α-methylstyrene (3). The solid curves were constructed from a quantum mechanical calculation, using data on the frequency of tunnel transitions through the potential barrier for a CH_3 group. A comparison of the curves

---

* In Fig. 105, δH_{rig} and $\delta \nu_{rig}$ denote the width of the line for a rigid lattice in units of field strength (Oe) and frequency (Hz).

shows that the theoretical curve does not coincide with the experimental curves for any values of E. The authors consider that, in addition to the presence of a spectrum of correlation frequencies, the reason for the discrepancy may also be the nonsinusoidal form of the potential barrier.

2. NMR Relaxation Time and Molecular Motion in Polymers

Detailed information on molecular motions in a polymer may be obtained by measuring the relaxation times T_1 and T_2 over a wide range of temperatures. Unfortunately, the interpretation of the results obtained is hampered by the fact that as yet there is no complete theory of NMR relaxation for such complex systems as polymers. The theory of relaxation developed by Bloembergen, Purcell, and Pound [179] for simple liquids is usually used.* The scheme for the relation of T_1 and T_2 to the correlation time and reciprocal temperature is presented in Fig. 107. The spin—lattice relaxation time T_1 is minimal at the value of τ_c when the correlation frequency (ν_c) is close to the resonance frequency. From the form of the curve of $\log T_1 = F(1/T)$, it is possible to calculate the activation energy E and the pre-exponential factor in the Arrhenius equation (see page 233) and the effective distance r between the protons, whose dipole—dipole interaction is responsible for the relaxation.

The temperature dependence of T_1 and T_2 found experimentally for polymers differs in many cases from the relation predicted by the BPP theory. Thus, for natural rubber [339] we obtain a sharper minimum in T_1 than according to theory (Fig. 108). The authors explain this deviation by the cooperative character of molecular motions in the sample. On the other hand, for a series of polymers there is a flatter minimum on the curve of $\log T_1 = F(1/T)$ than according to the BPP theory. The discrepancy may be explained by the presence of a broad spectrum of correlation frequencies. Figure 109 shows curves of $\log T_1 = F(1/T)$ calculated by Powles [561] for the rectangular logarithmic distribution of correlation frequencies with a width b, which was examined above (see

---

*The abbreviation "BPP theory" is often used in NMR literature.

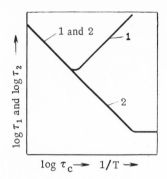

Fig. 107. Relation of the spin–lattice relaxation time (T_1, curve 1) and spin–spin relaxation time (T_2, curve 2) to the correlation time (τ_c) and the reciprocal temperature (1/T) according to the theory of Bloembergen, Purcell, and Pound [179].

page 236); it is obvious that with an increase in b the minimum in T_1 is blunted. Particularly marked discrepancies between experiment and theory are found for the ratio T_1/T_2 at the minimum of T_1: the theory predicts that the ratio T_1/T_2 should equal approximately 2, while experiment gives 73 for polyisobutene [571], 340 for polyethylene [349], and even 1300 for polydimethylsiloxane [565]. In these cases there are apparently two relaxation mechanisms operating simultaneously, for example, fast and slow mechanisms, which may be characterized by two correlation times τ_1 and τ_2. Thus, intra- and intermolecular dipole–dipole interaction may produce fluctuations in the local magnetic field of different frequencies and strengths. In the first approximation it may be assumed that the relaxation mechanisms are additive:

$$\frac{1}{T_1} = \frac{1}{T_{11}} + \frac{1}{T_{12}} \quad \text{and} \quad \frac{1}{T_2} = \frac{1}{T_{21}} + \frac{1}{T_{22}} \tag{IV-6}$$

A scheme for the temperature dependence of T_1 and T_2 for this case is given in Fig. 110 [469]. For the whole system there is a considerable range of temperatures over which T_1 is determined by the fast process, while T_2 is determined by the slow process. As a result, at the minimum of T_1 we find that $T_1 \gg T_2$.

A theory of relaxation times for liquid-phase polymers has been developed in the work of Odajima [520] and Khazanovich [134]. Odajima calculated the relaxation times in polymer solutions on the basis of a model of segmental motion which does not take into account the linking of the segments. Khazanovich used the Kargin–Slonimski [56] model in this theory. In this model the chain is divided into segments connecting "beads" to which only frictional forces apply, and it is assumed that a Gaussian distribution holds for the lengths of the segments. Moreover, it is assumed that there is intrinsic rotation of the separate groups. Account is taken

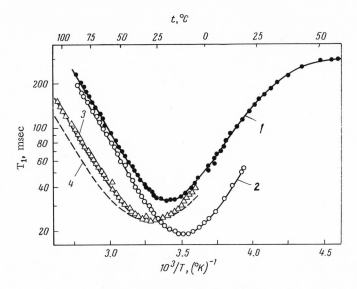

Fig. 108. Relation of the spin–lattice relaxation time (T_1) to the reciprocal temperature (1/ T) [339]. 1,2) Natural rubber with working frequencies of 28 and 20 MHz; 3)
vulcanized rubber at 20 MHz; 4) theoretical curve for vulcanized rubber.

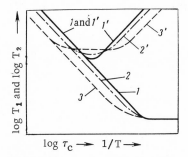

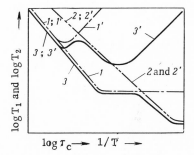

Fig. 109. Relation of the spin–lattice
relaxation time (T_1, curves 1',2',3')
and spin–spin relaxation time (T_2,
curves 1, 2, 3) to the correlation time
(τ_c) and the reciprocal temperature
(1/ T) with different widths for the distribution of correlation frequencies (b)
[561]. 1,1') b = 1; 2,2') b = 10; 3,3')
b = 100 (see page 236).

Fig. 110. Relation of the spin–lattice
relaxation time (T_1, curves 1',2',3')
and spin–spin relaxation time (T_2,
curves 1,2,3) to the correlation time
(τ_c) and the reciprocal temperature
(1/ T) in a system with two relaxation
mechanisms [469]. 1,1') Slow process;
2,2') fast process; 3,3') overall curve
for system.

only of the magnetic dipole—dipole interaction inside the groups, in which the distances between the magnetic nuclei are identical and constant (for example, in CH_2 or CH_3 groups). For this model the spectrum of correlation frequencies is described approximately by the function

$$I(\nu_c) = \frac{1}{\nu_{c,\,up}} \left(1 - \sqrt{\frac{\nu_{c,\,low}}{\nu_c}} \right) \qquad \text{(IV-7)}$$

where $\nu_{c,up}$ and $\nu_{c,low}$ are the upper and lower limits of the spectrum:

$$\nu_{c,\,low} \ll \omega_0 \ll \nu_{c,\,up} \qquad \text{(IV-8)}$$

where ω_0 is the resonance frequency.

The condition (IV-8) means that the temperature at which the spectrum is plotted is higher than the temperature corresponding to the minimum in T_1 (see Fig. 107). With the condition

$$\nu_{c,\,low} > \frac{1}{T_2} \qquad \text{(IV-9)}$$

which always holds for polymer solutions, but only holds for polymer melts at a sufficiently high temperature (for example, above 100°C for polyethylene with a molecular weight $M < 10^4$ and above 200°C for $M < 10^5$), the relaxation times are determined from the formulas:

$$\frac{1}{T_1} = \frac{2\gamma^2 \Delta H_{2,\,gr}^2}{\nu_{c,\,up}} \ln \frac{\nu_{c,\,up}}{\omega_0} \qquad \text{(IV-10)}$$

$$\frac{1}{T_2} = \frac{0,7}{T_1} + \frac{1,2}{\nu_{c,\,up}} \gamma^2 \Delta H_{2,\,gr}^2 \ln n \qquad \text{(IV-11)}$$

where γ is the gyromagnetic ratio of the resonating nucleus; $\Delta H_{2,gr}^2$ is the contribution of the group to the second moment of the line with an isotropic distribution of internuclear vectors relative to the field; if the group has no intrinsic rotation, the value of $\Delta H_{2,gr}^2$ is taken for a rigid lattice, while with rapid rotation $\frac{1}{4}$ of this value is taken; n is the number of segments in the chain.

The parameter $\nu_{c,low}$ is the reciprocal of the maximum time time of mechanical or dielectric relaxation. It may be determined, for example, from the viscosity η and the density ρ of a melt of the polymer:

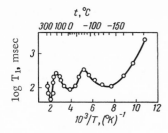

Fig. 111. Relation of the spin–lattice relaxation time (T_1) to the reciprocal temperature (1/T) for polyethyl methacrylate [440].

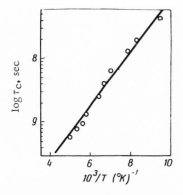

Fig. 112. Relation of the correlation time (τ_c) of the molecular motion of $COOC_2H_5$ groups in polyethyl methacrylate to the reciprocal temperature (1/T) [404].

$$\nu_{c,\,low} = \frac{\pi^2 RT\varrho}{6M\eta} \qquad (IV-12)$$

The parameter $\nu_{c,up}$ is related to the mean square length of a segment

$$\nu_{c,\,up} = \frac{4kT}{\eta\,(\overline{l^2})^{3/2}} \qquad (IV-13)$$

From Khazanovich's theory it follows that

1) for liquid polymers, $T_1 \gg T_2$;

2) for dilute solutions of polymers T_1 is independent of M, while T_2 depends on it very little;

3) T_1 and T_2 are independent of the solution concentration until chain entanglement begins;

4) for polymers in the melt, T_1/T_2 increases with M.

Formulas (IV-10) and (IV-13) make it possible to calculate the length of a segment from measured values of the relaxation time. Analysis of literature data shows that Khazanovich's theory agrees with experiment at least qualitatively. The mean size of the segment of a polymer in a melt is found to be of the same order as the size of a monomer unit, but in solution, it is considerably greater.

The fact that T_1 is independent of molecular weight for dilute polymer solutions also follows from the theory developed by Ullman [696].

The relaxation times in polymers may be measured with broad-line NMR spectrometers. For example, T_1 may be determined by the saturation method and T_2 from the width of the line

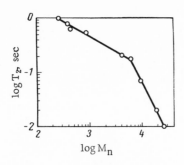

Fig. 113. Relation of the spin–spin relaxation time (T_2) to the number average molecular weight (M_n) for n-paraffins and polyethylene at 150°C [469].

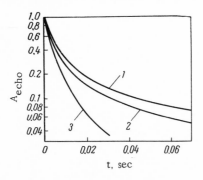

Fig. 114. Relation of the amplitude of the spin echo (A_{echo} in relative units) to the time between the first pulse and the appearance of the spin echo for melts of three samples of polyethylene at 150°C [469]. 1) Sample with a broad molecular weight distribution; 2) sample with an average distribution; 3) sample with a narrow distribution.

(see Chapter II). However, the most accurate data on T_1 and T_2 for polymers in bulk and in melts have been obtained by the spin-echo method. The relaxation times for a series of polymers have been measured over a wide range of temperatures.

NMR relaxation has been studied in detail for polyacrylates and polymethacrylates [404, 406, 568, 572]. Curves in the co-ordinates $\log T_1 - 1/T$ for acrylic polymers (polymethyl acrylate, polyethyl acrylate, and polybutyl acrylate) have two minima, one at 50-70°C, which corresponds to motion of the segments of the polymer chain, and one in the low-temperature region (from −125 to −200°C), which is due to internal rotation of the alkyl ester groups. With methacrylic polymers (polymethyl methacrylate, polyethyl methacrylate, and polybutyl methacrylate) there are three minima, one at 200-250°C, which is connected with motion of the segments, one close to 0°C, which corresponds to coopera-tive motion of the side methyl groups and ester groups, and one at −150°C and below, which is caused solely by the motion of the ester alkyl groups. Figure 111 gives as an example the rela-tion $\log T_1 = F(1/T)$ for polyethyl methacrylate, and Fig. 112 gives the relation $\log \tau_c = F(1/T)$ for the rotation of the ethoxyl groups.

With amorphous polyacetaldehyde there is a minimum in T_1 at −100°C, which is due to rotation of the methyl groups, and at 44°C, which is connected with the motions of the main chain [249].

For partly crystalline polymers such as polyethylene [405], on the curve of $\log T_1 = F(1/T)$ there are two minima, corresponding to motion in the crystalline and amorphous regions. McCall and Douglass [466], who measured T_1 and T_2 for a series of samples of polyethylene, which differed in the degree of chain branching and crystallinity, consider that the low-temperature minimum in T_1 is connected with the rotation of CH_3 groups and the high-temperature minimum, with liquid-like motions in the amorphous regions (see also [367, 695]). From the temperature dependence of T_1 and the form of the signal of free precession of the nuclear spins after a 90° pulse, it was possible [341, 342] to elucidate in detail the picture of the molecular motion in linear polyethylene. In the solid polymer over the temperature range from −120 to +50°C, NMR relaxation is connected with statistical oscillations of the C_2H_4 groups through an angle of ±10°; in the melt the molecular motion may be represented clearly as a combination of "vortical rotations" and "serpentine" motions of the molecular chains.

For a melt of polyethylene at 150°C the transverse relaxation time depends on the molecular weight (Fig. 113). The authors of [469] consider that measurement of T_2 by the spin-echo method makes it possible to determine the molecular weight of polyethylene conveniently and rapidly. Moreover, the form of the curve of the relation of the spin-echo amplitude to the time between the 90° pulse and the appearance of the echo signal depends on the molecular weight distribution (Fig. 114), so that it is also possible to estimate the width of the molecular weight distribution. The relation of the relaxation time to the chain length (molecular weight) has also been investigated [149, 236] for a series of polyethylene glycols

$$\cdots - CH_2 - CH_2 - O - \cdots$$

It would be desirable to study this problem in more detail.

For polypropylene [354, 407, 573], the curve of $\log T_1 = F(1/T)$ has two minima in T_1, one at low temperatures (from −120 to −145°C), which is connected with rotation of the CH_3 groups, and one at high temperatures (from 75-80 to 100°C), which is due to motion of segments of the main chain in the amorphous regions of the polymer (see page 205). This was demonstrated by comparison of the NMR spectra of polypropylene with normal [573] and deuterated methyl groups. Because of the smaller magnetic moment of

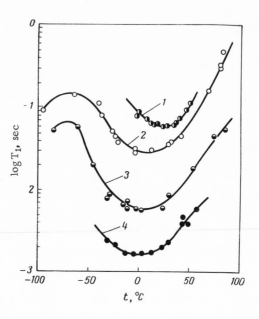

Fig. 115. Relation of the spin–lattice relaxation time (T_1) to temperature for natural rubber at different working frequencies [640]. a) 60 MHz; 2) 30 MHz; 3) 7.5 MHz; 4) 2 MHz.

D nuclei, the rotation of CD_3 groups should have less effect on the relaxation of the other protons of the molecule than the rotation of CH_3 groups and, consequently, for the deuterated polypropylene the low-temperature minimum in T_1 should be reduced. This was actually observed in the experiment. The deuteration method has also been used to interpret data on the temperature dependence of T_1 and T_2 for polyethylene terephthalate. It was established that over the range of temperatures from −73 to +127°C there is restricted rotation of the CH_2 groups about the C—C bond and the activation energy for rotation equals approximately 2 kcal/mole [704]. In polyamides [463] the two transitions observed on the temperature dependences of T_1 and T_2 are caused by reorientation about the axes of the macromolecules and the beginning of liquid-like motions in the noncrystalline regions.

NMR relaxation for polyisobutene has been studied in [518, 571] and for low-molecular samples in [675]. A minimum in T_1 was observed at 50°C and two values of T_2 at a temperature above

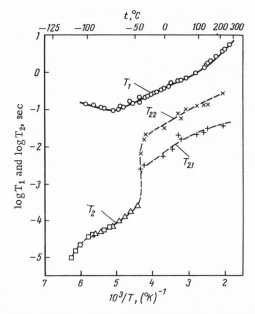

Fig. 116. Relation of the relaxation times (T_1 and T_2) to the reciprocal temperature ($1/T$) for polydimethylsiloxane ($n \approx 2200$) [565].

$170°C$. The results may be explained if we assume that in the polymer there are two types of molecular motions, which differ in frequency by a factor of 10^3. For each of these motions there is a distribution of correlation times. The two relaxation processes are evidently caused by motions of chain sections of different sizes. There is a discrepancy between the values of the activation energy of molecular motion in a melt of polyisobutene calculated from NMR data and from the viscosity of the melt [518]. The authors explained this by the fact that the width of the NMR line is affected by cooperative motion of a large number of segments. The spin–lattice relaxation time of polyisobutene increases with a rise in pressure and with an increase in the frequency at which the NMR is observed, but the change in T_1 with frequency does not obey the equations of the BPP theory of relaxation, also indicating a distribution of correlation times.

The temperature dependence of the relaxation time has also been studied for polyethylene glycols [155], for polymers of propylene oxide [713], and polyoxymethylene [695].

Above we have already given the relation $\log T_1 = F(1/T)$ for natural rubber and the vulcanizate [339] (see page 238). More detailed information on the character of the molecular motion may be obtained by measuring the temperature dependence of T_1 over a wide range of working frequencies. Unfortunately, the difficulties associated with obtaining a sufficiently homogeneous and stable magnetic field of high strength and the fall in the sensitivity of an NMR spectrometer at low working frequencies limit the frequency range of the measurements as yet. In the work of Slichter and Davis, measurements of T_1 were carried out for polyisobutene and polybutadiene at 20, 30, and 50 MHz [642], for polyolefins [641] at frequencies from 10 to 50 MHz, and for rubber [640] at frequencies from 2 to 60 MHz (Fig. 115).

With an increase in frequency the minimum in T_1 is shifted to higher temperatures.

The temperature dependence of the viscoelastic properties of elastomers is described well by the semi-empirical relation of Williams, Landel, and Ferry [714]. It was found that the experimental dependence of the temperature of the minimum in T_1 of natural rubber to the working frequency may be described satisfactorily by means of the same equation.

Data on the activation energy of molecular motion in natural rubber confirm that there is a distribution of the correlation times of Brownian motion of the segments of the polymer chains [521].

NMR relaxation in polydimethylsiloxanes [431, 469, 565] $[-\,Si\,(CH_3)_2 - O -]_n$ has a very complex character.

For polymers with n > 1000, the minimum in T_1 on the curve of $\log T_1 = F(1/T)$ lies at $-77°C$ and there is a break at $+90°C$. The spin—spin relaxation time increases with a jump at $-42°C$; above this temperature there are two values of T_2, which are denoted by T_{21} and T_{22} (Fig. 116).

Powles and Hartland [565] consider that one mechanism of spin—spin relaxation is connected with the interaction inside the methyl groups, while the second is connected with the interaction between the chains. Kusumoto et al. [431] consider that in polydimethylsiloxanes there are spiral motions of the $Si\,(CH_3)_2$ groups so that the protons of neighboring units of the same chain may interact.

The spin-echo method makes it possible to determine the self-diffusion coefficient D simultaneously with the relaxation times. The limits of applicability of the method ($D \geq 10^{-7}$ cm^2/sec) make it possible to measure D only for melts and polymer solutions.

McCall et al. [467] measured D of polyethylene melts at 130-200°C and calculated the activation energy of self-diffusion E. For chains with more than 20-30 carbon atoms the value of E is no longer dependent on the chain length. The relation of D to the molecular weight is given by the equation

$$D = \text{const} \cdot n^{-5/3}$$

where n is the number of C atoms in the molecule.

Odajima [520] measured the self-diffusion coefficient for oligomeric dimethylsiloxanes. The values of the coefficient D for water in a gel of carboxymethylcellulose, measured by the spin-echo method and by the isotope method, agree well with each other [350].

3. Comparison of Data Obtained
by the NMR Method and Other Methods

As we have already pointed out, the NMR method should not be used in isolation, but together with other physical and chemical methods. In the study of molecular motion in polymers, particularly valuable results are obtained by the dynamic mechanical method. The molecular motion in polymers containing polar groups may also be studied by dielectric measurements.

A comparison of the data of the NMR and dynamic mechanical methods was made even in the first work on NMR in polymers (1947) [151]. Data from the NMR method has been compared with the results of mechanical and dielectric measurements in a series of reviews [504, 601, 685, 700, 730, 734] and here we will limit ourselves to only a few characteristic examples.

In many cases the transition in polymers which is observed through the narrowing of the NMR line also appears in mechanical measurements as a sharp maximum of the logarithmic decrement (Δ) and a fall in the dynamic shear modulus (G). This transition is observed, for example, for polyvinyl chloride [601] at 77-87°C

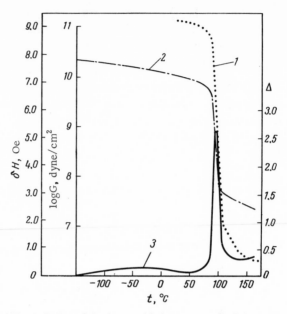

Fig. 117. Relation of the line width (δH, curve 1) of the NMR spectrum, the dynamic shear modulus (G, curve 2), and the logarithmic decrement (Δ, curve 3) to temperature for polyvinyl chloride [601].

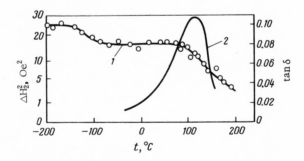

Fig. 118. Relation of the second moment of the NMR line (ΔH_2^2, curve 1) and the dielectric loss tangent ($\tan \delta$, curve 2) to temperature for polymethyl methacrylate [560].

and corresponds to the glass transition point (Fig. 117). The second flat maximum on the logarithmic decrement at about −33°C evidently corresponds only to small oscillations of the methylene groups and has no appreciable effect on the width of the NMR line.

For polymethyl methacrylate [560] the narrowing of the line above +70°C, which is caused by motion of the main chains, coincides with the maximum of the dielectric loss tangent at a frequency of 10^4 Hz (Fig. 118).

McCall and Anderson [463] observed good agreement between the temperatures for the minimum of the spin—lattice relaxation time found experimentally and predicted from dielectric measurements (Fig. 119).

Satisfactory agreement between the temperatures of the maximum dielectric loss tangent, the maximum mechanical loss, and the minimum of T_1 is also observed for high-molecular polyoxyethylenes [254].

Sauer and Woodward [601] compared the results of the NMR method and dynamic mechanical measurements for a series of both amorphous (polyisobutene and polymethyl methacrylate) and also partly crystalline polymers (polyvinyl chloride, polyethylene, irradiated polyethylene, polypropylene, polybutene, polyamides, and polytetrafluoroethylene). The authors came to the conclusion that the melting of the crystallites appears more strongly in the change in the dynamic mechanical properties of a polymer than in NMR spectra, since local rotation and reorientation in the solid state usually narrow the NMR line before melting of the crystallites and when the spectrum is plotted on a broad-line NMR spectrometer the subsequent narrowing is no longer appreciable.

The dynamic mechanical method is also preferable in the observation of secondary transitions in amorphous regions, which are connected with the motion of segments of the main chains, for example, the low-temperature transitions in polyamides. The transition of a polymer from the glassy to the highly elastic state is readily observed by both methods; there is only some difference in the temperature of the transition and this is explained by the difference in the ranges of frequencies used. The NMR method is much more sensitive than the dynamic mechanical method to the rotation of side chains such as the CH_3 groups in polyisobutene, polymethyl methacrylate, and polypropylene. Thus, the two methods successfully complement each other.

It is convenient to compare data on molecular motion in polymers obtained by the NMR and other methods by means of diagrams

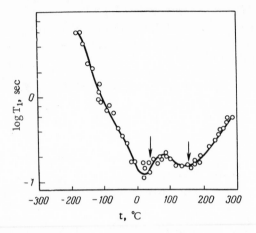

Fig. 119. Relation [463] of the spin–lattice relaxation time (T_1) to temperature for 6,6 polyamide. (The arrows show the minima of T_1, which were predicted from dielectric measurements.)

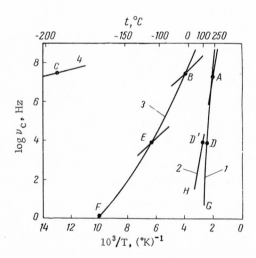

Fig. 120. Relation of the correlation frequency (ν_c) to temperature for polymethyl methacrylate according to NMR data (A,B,C,D,E), mechanical (F,G), and dielectric (H) measurements [572]. 1) Motion of the main chains; 2) side chains; 3) α-CH_3 groups; 4) CH_3 groups of the side chains.

of the correlation frequency against the reciprocal temperature. Figure 120 shows such a diagram for polymethyl methacrylate [572]. The points A, B, and C at $\nu_c = 10^7$ Hz correspond to minima on the curve of $\log T_1 = F(1/T)$. The points D, D', and E at $\nu_c = 10^4$ Hz correspond to the temperatures of narrowing of the NMR line. The points F and G were obtained from dynamic mechanical measurements and the point H from measurements of the dielectric loss tangent. The lines 1, 2, 3, and 4 in Fig. 120 show the change with temperature of the correlation frequency of the motion of the main chains, side chains, α-CH_3 groups, and CH_3 groups of the side chains of the polymer. Short lines are drawn through points A, B, C, and E, and their slopes correspond to the activation energies of molecular motion calculated from NMR data. For points B and E it is substantially less than the slope of curve 3, and this is explained by the presence of a distribution of correlation frequencies in the polymer.

4. Study of Molecular Motion in the

System Polymer—Low-Molecular Substance

Swollen Polymers. By studying the NMR spectrum of a swollen polymer it is possible to follow how the mobility of the chain changes as the macromolecules surrounding it are replaced by small solvent molecules. The number of papers in this field is still small but, obviously, by using a solvent and a polymer containing different nuclei (for example, fluorinated solvents or deuterated compounds for organic polymers) and by studying the NMR over a wide range of temperatures and component ratios in the system it is possible to obtain valuable results. At the same time it is possible to obtain information on the character of molecular motion of the solvent in the polymer.

As a measure of the swelling of the polymer it is possible to observe the change in the form, width, and second moment of the NMR line and the change in the relaxation time.

As Kosfeld and Vosskötter [420] have shown, when polymethyl methacrylate is swelled in chlorobenzene and other solvents, instead of a simple NMR line in the temperature region from −70 to +10°C there is a line of complex form, which consists of two components. The broad component ($\delta H \approx 8$ Oe) corresponds to protons of the methylene groups, while the narrow component ($\delta H = 1$-4 Oe) corresponds to protons of the methyl groups. The splitting of the line into components is explained by the fact that over a certain temperature range the plasticizer increases the mobility of the CH_3 groups, but has relatively little effect on the mobility of the CH_2 groups in the main chain of the polymer.

As early as 1951 it was observed [357] that when a 1 : 1 co-polymer of butadiene with styrene (Hycar) is swelled in benzene there is a decrease in the width of the NMR line and the tempera-ture at which the line narrows is lowered (Fig. 121). The authors considered that the decrease in the width of the line was caused by an increase in the intermolecular distances on swelling. Slichter [108] reported that this explanation is untrue, since the intermole-cular component of the line width is considerably less than the in-tramolecular component and a small change in the distance be-tween the polymer chains cannot produce a substantial narrowing of the line. The effect of the solvent is explained by the fact that it increases the motion of the chain segments of the polymer.

Powles [559, 562] studied the NMR spectra of the systems polyisobutene—benzene and polyisobutene—carbon tetrachloride. It was found that there is a critical concentration of the solvent at which "freeing" of the polymer chains is observed. This concentra-tion corresponds to a ratio of six monomer units of polyisobutene to one solvent molecule. With a molecular ratio of polymer to benz-ene of less than 6 : 1, the fall in the second moment on heating oc-curs at a temperature 30° lower than for the pure polymer (Fig. 122).

The activation energy of molecular motion was calculated from the temperature dependence of the second moment of the line. The values were almost identical for pure and swollen polymer, while the entropy of activation increased markedly on swelling, namely, from 6.1 to 22.3 entropy units.

Narrowing of the NMR line on swelling was also reported by Odajima and Nagai [524], who studied the swelling of polystyrene in carbon disulfide and the swelling of other polymers in solvents which contained no hydrogen.

An interesting subject for the study of swelling is a polymer swollen in its own monomer. One such system (polystyrene—styrene) has been studied by two methods, namely, by following the change in the width of the NMR line [191] and the spin—lattice re-

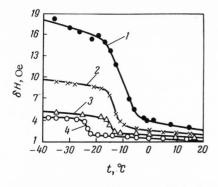

Fig. 121. Relation of the line width (δH) in the NMR spectrum of a copolymer of butadiene and styrene (Nycar OS-10) to temperature with different degrees of swelling in benzene [357]: 1) 5%; 2) 10%; 3) 25%; 4) 150%.

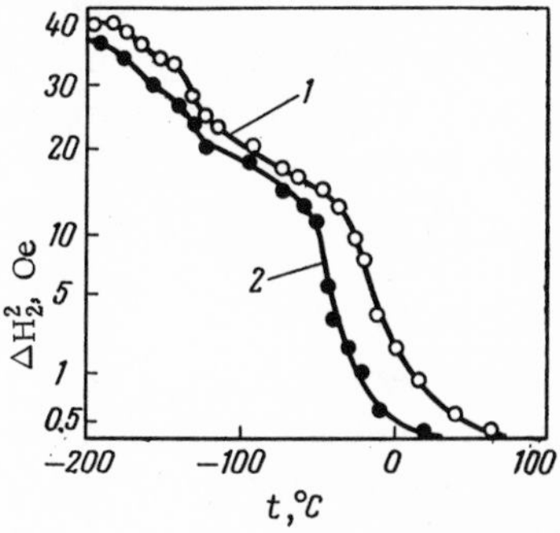

Fig. 122. Relation of the second moment of the line (ΔH_2^2) in the NMR spectrum of polyisobutene to temperature [562]. 1) Pure polymer; 2) swollen in benzene to a ratio of 4.88 monomer units per benzene molecule.

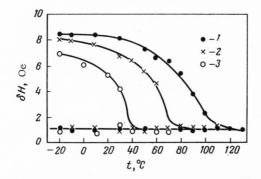

Fig. 123. Relation of the width of the narrow component of the line (a) and the broad component of the line (b) of the NMR spectrum of polystyrene to the temperature with different degrees of swelling of the polymer in dichloromethane [418]: 1) 5.5%; 2) 11%; 3) 15.7% CH_2Cl_2.

laxation time [182] in relation to the composition of the system and the temperature. Both methods confirm that the mobility of chain segments of the polymer increases sharply when monomer molecules are introduced between the chains.

The change in the mobility of a low-molecular-weight substance in a polymer was studied by Kosfeld [418, 419] through the

NMR spectra of plasticized polystyrene. Plasticizers containing hydrogen (dichloromethane and cyclohexane) and containing no hydrogen (carbon tetrachloride) were used. The NMR line of the system polystyrene—dichloromethane has a complex form and consists of broad and narrow components; with a rise in temperature the width of the narrow component remains constant, while the broad component narrows (Fig. 123). Evidently, even in this case, when the system is in a glassy state, the plasticizer molecules retain great freedom of motion. With cooling below the glass transition point the intensity of the narrow component falls and consequently an increasing number of plasticizer molecules become immobile. The authors measured the diffusion coefficient for the same system. Below the glass transition point the diffusion coefficient falls by several orders. It is probable that the motion of plasticizer molecules then becomes predominantly rotational rather than translational.

The narrow line did not appear with the system polystyrene—carbon tetrachloride and only a decrease in the line width was observed on swelling.

In the study of the mobility of a low-molecular substance in a swollen polymer such as benzene in rubber [724], valuable results are obtained by measuring the self-diffusion coefficient D by the spin-echo method. The advantage of the method lies in the fact that no concentration gradient is required for measuring D. It has also been reported [470] that the results of measuring the self-diffusion coefficient are interpreted comparatively readily, since it depends only on translational motions, while the NMR relaxation time also depends on rotational motion in the system.

In the NMR spectra of the water—starch system there is broadening of the water line from $0.4 \cdot 10^{-6}$ to $1.1 \cdot 10^{-6}$ Oe when a gel is formed [248], and this is due to the formation of hydrogen bonds between H_2O molecules and OH groups of the starch.

The swelling of partly crystalline polymers generally involves only the amorphous regions. Thus, when polyethylene is swelled in carbon tetrachloride the narrow component of the NMR line becomes still narrower, while the broad component does not change [462]. Swelling occurs only in regions which give the narrow component of the NMR signal. By using the formula for the correlation time τ_c from the BPP theory and the expression for τ_c

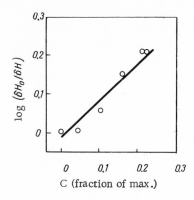

Fig. 124. Change in the width of the narrow component of the NMR line of linear polyethylene when it is swollen in carbon tetrachloride [462].

from Eyring's theory, the authors showed that the change in the width of the narrow component of the NMR line with the concentration of the solvent C in the polymer must be described by the equation

$$\delta H = \delta H_0 e^{-\rho C} \qquad \text{(IV-14)}$$

where δH_0 is the width of the line of the initial polymer; ρ is a coefficient which is independent of C; C is the concentration of CCl_4 as a fraction of the maximum (at saturation) concentration. Experiment confirms the linearity of the relation $\log(\delta H_0 / \delta H) = F(C)$ (Fig. 124).

If the solvent contains no protons, then the ratio of the integral intensities of the narrow and broad components of the NMR line of a partly crystalline polymer are independent of its degree of swelling. At the same time, the width of the narrow component of the line decreases with an increase in swelling and, consequently, the resolution of the complex line becomes clearer and the determination of the degree of crystallinity is facilitated. This was used in the work of Woodward and his co-workers [733]. In the determination of the degree of crystallinity of polybutene and polymethylpentene by the NMR method, tetrachloroethylene was used as the solvent. However, some part of the solvent may still penetrate into the surface layer of the crystals. By studying NMR in single crystals of polyethylene swollen in tetrachloroethylene, Fischer and Peterlin [290] showed that chains lying on the side surfaces of the crystals become mobile on swelling.

The penetration of a low-molecular liquid (acetic acid, ethanol, etc.) into amorphous regions of 6,6 polyamide produces a paradoxical effect, namely, an increase in the width and second moment of the NMR line at low temperatures [330]. There is probably rupture of hydrogen bonds between amide groups and regular packing of the chains is facilitated.

The change in the relaxation time on swelling of the polymer has been studied in detail in the system polyisobutene—carbon tetrachloride [559] (see also [470]). In the region of the critical concentration (one CCl_4 molecule to six monomeric polymer units) there is a rapid rise in the transverse relaxation time T_2 and a decrease in the correlation time τ_c.

A formula derived by Nolle is given in the work of Odajima [520] for the critical concentration

$$C_{cr} = \frac{7.1 M_{un}^{3/2}}{a N_A l^3 M^{1/2}} \qquad \text{(IV-15)}$$

where C_{cr} is the critical concentration, g/cm^3; M_{un} is the molecular weight of a monomer unit; α is the "excluded volume"; N_A is Avogadro's number; l is the length of the segment; M is the molecular weight (the formula holds when $M > 10^5$).

Chernitsyn and his co-workers [140] made use of measurements of the relaxation time (spin-echo method) in the system polyvinyl chloride—plasticizer to assess plasticizers. With a plasticizer content of 30–40 wt.%, its molecules are freed from the blocking action of the polymer macromolecules and T_2 increases. The higher T_2, the more effective is the plasticizer. NMR data are confirmed by other methods.

With high degrees of swelling it is possible to plot the high-resolution NMR spectrum of the solvent. In this case, the interaction between the solvent and the polymer may be assessed through the change in the chemical shift for different functional groups of the solvent [438].

Polymer Solutions. Data on molecular motion in polymer solutions may be obtained from high-resolution NMR spectra and also by the spin-echo method.

It is only possible to obtain a high-resolution NMR spectrum of a polymer solution when the molecular motion is sufficiently intense to average out the local magnetic fields created by the magnetic moments of adjacent nuclei. For a series of polymers, good resolution of multiplets is achieved only at a high temperature, for example, above 70°C for solutions of polyvinyl alcohol in heavy water [261] and only at 170°C for polyvinyl chloride in o-dichlorobenzene [691].

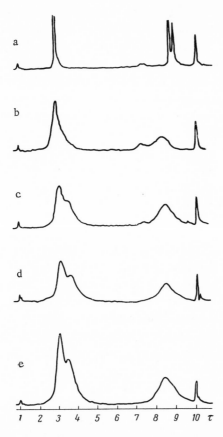

Fig. 125. High-resolution NMR spectra of solutions in carbon tetrachloride [204]. a) Cumene; b,c,d,e) polystyrene
of different degrees of polymerization (P_n = 5, 10, 20, 500).

By using the high-resolution NMR spectrum of a polymer solution it is possible to study the molecular motion of each molecular group individually as in this case there is the possibility of determining the line width and the relaxation times T_1 and T_2 for each
peak in the spectrum. For example, T_1 may be determined by the
direct method, i.e., by the restoration of the signal after saturation, while T_2 may be calculated from the line width. Sergeev [103]
used this method to compare the frequency of motion of solvent
molecules and methyl and methoxyl groups in the polymer molecule for solutions of polymethyl methacrylate in chloroform.

Analogous measurements were made by Liu [439] for solutions of poly-α-methylbenzyl methacrylate.

Bovey et al. [204] plotted the NMR spectra of solutions of a series of polystyrenes with different mean degrees of polymerization. In Fig. 125 we compare the spectra of solutions of polystyrene (d) with the number average degree of polymerization P_n = 500 and a low-molecular hydrocarbon of similar structure, namely, cumene (a).

The lines were assigned on the basis of the spectra of deuterated polystyrenes. The peaks in the spectrum belong to the ortho-protons, meta- and para-protons, protons of CH_2 and CH groups in the chain, and protons of the internal standard, tetramethylsilane. The authors also compared the spectrum of a solution of a high-molecular polystyrene with the NMR spectra of a series of low-molecular polystyrenes with P_n = 5-20 (see Fig. 125b, c, d). The width of the lines for polystyrene is considerably greater than for cumene and is independent of the molecular weight, and also the solution concentration if the concentration is no higher than 50-60%. This shows that there is only motion of the segments and no motion of the molecule as a whole.

The viscosity of the solution in the samples varied from the viscosity of the pure solvent to thousands of centipoises (a viscous liquid). In normal low-molecular liquids the value of $1/T_2$ and, consequently, the width of the line is proportional to the viscosity. It is obvious that in polymer solutions the line width is determined by the local viscosity in the immediate neighborhood of the chain segment. The local viscosity is independent of the molecular weight; the concentration affects it only when neighboring segments begin to interfere with motion. The value of T_1 for a benzene nucleus in polystyrene equals 0.4 sec, while for benzene at 25°C the calculated value T_1 = 18.9 sec. Consequently, benzene nuclei in a polystyrene chain rotate slower by a factor of 40-50 than a benzene molecule.

It should be noted that calculations based solely on measurements of the line width in high-resolution NMR spectra of polymers are not always reliable. As a subsequent study of polystyrene solutions showed [464] a great contribution to the line width is made by the spin—spin interaction of the ortho- and meta-protons. However, the conclusion on the small effect of the macroscopic viscosity on the motion of the molecular groups in a polymer solution is still apparently valid. For example, the same conclusion was reached by Bresler and his co-workers [28] on the basis of measurements of the dielectric loss tangent. The fact that the width of the line of the NMR spectrum of polymer solutions is not determined by the macroscopic viscosity was also confirmed by a study of the relation of δH to the concentration, which was carried out by Sergeev and Karpov [104]. As a subject for the study they selected solutions of polyisobutene in benzene and carbon tetrachloride. The value of δH extrapolated to zero concentration for methyl protons in both solvents was the same and equal to approximately 5 Hz, while the intrinsic viscosities were quite different.

By means of the high-resolution NMR method it was possible to obtain information on the mechanism of the transition observed in a polystyrene solution. The existence of this transition was observed as early as 1949. Chmutov and Slonim [121, 141] found that the diffusion coefficient D of solutions of polystyrene in ethylbenzene does not change monotonically on heating: there is a maximum and a minimum on the curve of $D = F(T)$ in the temperature region of 60-80°C. This effect was not confirmed by Bresler and his co-workers [29] or Varoqui [697], but in a series of studies it was shown that at the temperature corresponding to the glass transition point of the polymer in bulk or somewhat below it there is a change in a series of the properties of a polystyrene solution. In particular, anomalies were found in the curves of the temperature dependence of the surface tension [285, 286], the density [488], the dielectric loss tangent [128, 129], the radius of inertia and the second virial coefficient calculated for changes in light scattering [586], depolarization of scattered light [708], and the optical density in the ultraviolet region [437]. It was assumed [586] that the anomaly in the temperature dependence of the properties of the solution was due to a change in the relative disposition of the benzene rings. Lui [440] measured the temperature dependence of the width of the NMR line of aromatic protons in polystyrene in solution. It was found that over the temperature range of 40-80°C for atactic and 50-60°C for isotactic polymer the line width falls more rapidly than at other temperatures. Apparently, below 50°C comparatively stable ordered sections are formed as a result of van der Waals forces between neighboring phenyl rings, while over the range of 50-60°C they "melt."

The intensity of molecular motion in solution may also be assessed through the merging of the multiplets in high-resolution NMR spectra.

Amide protons in polyacrylamide [203] are magnetically non-equivalent at room temperature and a doublet is obtained in the spectrum. As the temperature is raised to 70°C, the rate of rotation of the amide groups increases and the two peaks merge into one; the activation energy of rotation of the amide group, calculated from the temperature dependence of the doublet splitting, equals 10.5 kcal/mole.

The spin–echo method was used to measure the relaxation time in solutions of polyisobutene and polystyrene in carbon tetrachloride in the work of Nolle [516, 517]. Using the data of Nolle and his own measurements, Odajima [520] calculated that the kinetic unit in a solution of polyisobutene is a segment of ten monomeric units. Calculation according to Khazanovich's theory [134], in which a more refined model was used than in Odajima's theory, gives for polyisobutene in CCl_4 at 50°C a mean-square segment length $(l^2)^{1/2} \approx 10$ Å, on the assumption that the methyl group rotates rapidly about the axis of symmetry.

The activation energy of molecular motion, calculated from the values of T_1 measured by the spin–echo method for the ortho-, meta-, and para-protons of the benzene ring in solutions of polystyrene in tetrachloroethylene [464] equals 3 kcal/mole.

Chapter V

Study of Chemical Processes in Polymers by the NMR Method

Chemical reactions of high-molecular substances play a great part in technology and biology and their study is one of the most important problems of polymer chemistry. The well-known chemical reactions of macromolecules may be divided into four main groups:

1) intramolecular reactions;

2) polymer analog conversions;

3) reactions involving a decrease in the mean degree of polymerization, which include breakdown processes under the action of various physical, chemical, and mechanical factors, including the action of high-energy radiation;

4) reactions involving an increase in the mean degree of polymerization, in particular, "cross linking" with the formation of three-dimensional structures (see, for example, the reviews of Berlin [19, 20] and reports at the International Symposium on Macromolecular Chemistry in Moscow [87]).

The NMR method may be used for solving many problems concerned with the study of reactivity of macromolecules and also with the study of the structure and properties of the products obtained. However, up to now the possibilities of this method have not been used at all fully. We will examine studies of polymerization, cross-linking processes, the effect of radiation on polymers, and the breakdown of polymers.

1. Study of Polymerization

There are two main trends in the study of polymerization by the NMR method. First, it is possible to follow the change in the

261

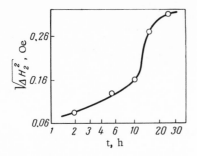

Fig. 126. Change in the second moment of the line (ΔH_2^2) in the NMR spectrum during the polymerization of ethyl acrylate with γ radiation (t is the irradiation time) [619].

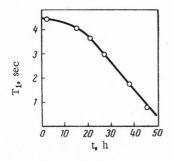

Fig. 127. Relation of the spin–lattice relaxation time (T_1) to the polymerization time (t) for styrene. (The polymerization was carried out at 90°C and T_1 was measured at 25°C [189].)

mobility of the units in the reaction medium through the change in the width of the NMR line or the relaxation time. This will give information on the kinetics of the process. Secondly, it is possible to determine the structure and particularly the stereoregularity of the polymers synthesized and to obtain information on the effect of the polymerization conditions on the mechanism of chain growth, etc.

The first field includes, for example, an experimental study of the polymerization of ethyl acrylate, acrylonitrile, and methyl methacrylate in the work of Shibata [619]. Figure 126 shows the increase in the second moment of the NMR line in the polymerization of ethyl acrylate under the action of γ rays from a ^{60}Co source. However, to explain the results obtained theoretically it is necessary to take into account the effect of the change in viscosity on the relaxation time and for such a complex system it is probably difficult. The possibility of studying polymerization through the change in the width and the second moment of the NMR line was pointed out by Lösche [444].

Valuable data for studying the polymerization process on a molecular scale are provided by measurements of the NMR relaxation time.

Bonera and his co-workers [185, 187-189] (see also [70, 311]) studied the polymerization of the vinyl compounds: styrene, methyl methacrylate, and vinyl acetate by heat and ultraviolet radiation. It was shown that during the polymerization there appear two spin—lattice relaxation times, namely, a high T_{1M}, corresponding to protons of monomer molecules, and a low T_{1p} corresponding to polymer molecules. It is very interesting that the monomer molecules retain a high mobility even when the macroscopic viscosity increases markedly. Thus, in the polymerization of styrene, during the first 20 h the viscosity increases from $0.73 \cdot 10^{-2}$ to 200 poise, while T_{1M} falls only insignificantly (Fig. 127). The values of T_{1p} hardly change during the polymerization. From the change in T_{1M} it is possible to determine the activation energy E for motion of the polymer molecules in the system. Values of E calculated by two methods, namely, from the temperature dependence of T_{1M} and from the degree of polymerization, are different [188]. The discrepancy is explained by the increase in the width of the spectrum of correlation times during polymerization.

The second field, namely the study of stereospecific polymerization, has developed rapidly in recent years due to the refinement of methods of determining and interpreting high-resolution spectra of polymers (see Chapter III).

There has been a series of studies of the polymerization of vinyl compounds and in particular detail, the polymerization of methyl methacrylate. *

During the polymerization of vinyl compounds there is the possibility of isotactic and syndiotactic addition of the new monomer unit to the growing chain. The stereochemical configuration of the macromolecule depends on the type of addition. Due to steric effects, dipole interactions, etc., these two possible types of addition are energetically nonequivalent. The process of chain growth during polymerization may be described by different mathematical models. The simplest of these is that in which it is assumed that the probability of isotactic or syndiotactic addition is independent of the configuration of the last unit of the growing chain. This assumption is justified, for example, in radical polymerization of methyl methacrylate. For this case there are two rate constants k_i and k_s for the two chain-growth reactions, and

---

*See literature [16, 192, 193, 202, 210, 244, 256, 293, 327, 391, 392, 403, 436, 477, 479, 510, 511, 701, 707, 711].

$$k = k_0 e^{-E/RT} \qquad\qquad (V-1)$$

where E is the activation energy of the addition of a unit. The probability p_i that during the polymerization there will be isotactic addition is determined from the expression

$$p_i = \frac{k_i}{k_i + k_s} \qquad\qquad (V-2)$$

and the probability of syndiotactic addition

$$p_s = 1 - p_i = \frac{k_s}{k_i + k_s} \qquad\qquad (V-3)$$

To describe the polymerization process it is necessary to know the values of p_i and p_s; their ratio is determined by the difference in activation energies:

$$\frac{p_i}{p_s} = \frac{k_i}{k_s} = e^{-\frac{E_i - E_s}{RT}} \qquad\qquad (V-4)$$

The values of p_i and p_s are proportional to the numbers of isotactic and syndiotactic bonds and may be determined by analysis of the NMR spectrum (see Chapter III, page 209) through the form of the signal from the methylene protons:

$$p_i = \frac{F_i}{F_i + F_s} \qquad\qquad p_s = \frac{F_s}{F_i + F_s} \qquad\qquad (V-5)$$

where F_i and F_s are the areas of the quadruplet and singlet at 8.18τ in the signal of the methylene protons.

As Bovey [192] showed, in the radical polymerization of methyl methacrylate

$$p_i = 0.25; \quad p_s = 0.75; \quad E_i - E_s = 775 \pm 75 \text{ cal/mole}$$

The activation energy of isotactic addition is somewhat higher than that of syndiotactic addition and therefore normal technical poly-methyl methacrylate contains 75% of syndiotactic bonds between the units and only 25% of the isotactic bonds.

In the anionic polymerization of methyl methacrylate the very simple assumption made above is not justified. In this case the probabilities p depend on both the conditions of the process and on how the last unit of the chain was added, i.e., on the type of the previous addition. The probability that isotactic addition follows isotactic addition may be denoted by p_{ii} and that it follows syndiotactic addition, by p_{si}; the probabilities p_{ss} and p_{is} are defined analogously. These values may also be obtained by analysis of the NMR spectrum through the form of the signal from the α-methyl protons (see Chapter III):

$$p_{ii} = \frac{F_{ii}}{F_{ii} + \frac{1}{2} F_{is,si}}$$

$$p_{is} = \frac{\frac{1}{2} F_{is,si}}{F_{ii} + \frac{1}{2} F_{is,si}}$$

$$p_{si} = \frac{\frac{1}{2} F_{is,si}}{F_{ss} + \frac{1}{2} F_{is,si}} \tag{V-6}$$

$$p_{ss} = \frac{F_{ss}}{F_{ss} + \frac{1}{2} F_{is,si}}$$

where F_{ii}, $F_{is,si}$, and F_{ss} are the areas of the peaks at $8.80\,\tau$, $8.96\,\tau$, and $9.09\,\tau$ in the signal of the protons of the α-methyl group.

The values p_{ii}, p_{is}, p_{si}, and p_{ss} depend on the activation energies:

$$\frac{p_{ii}}{p_{is}} = \frac{k_{ii}}{k_{is}} = e^{-\frac{E_{ii} - E_{is}}{RT}} \tag{V-7}$$

For polymethyl methacrylate obtained by polymerization with n-butyllithium in toluene at −70°C, the following values were found from the NMR spectrum, which was presented in Chapter III (see page 208) [210]:

$$p_i = 0.81 \qquad p_{ii} = 0.88 \qquad p_{si} = 0.44$$
$$p_s = 0.19 \qquad p_{is} = 0.12 \qquad p_{ss} = 0.56$$

The differences in activation energies equal

$$E_i - E_s = -620 \pm 35 \text{ cal/mole}$$
$$E_{ii} - E_{is} = -900 \pm 100 \text{ cal/mole}$$
$$E_{si} - E_{ss} = -100 \pm 10 \text{ cal/mole}$$

The polymer obtained by radical polymerization contains three syndiotactic links between units for each isotactic link and they are distributed randomly along the chain. In polymethyl methacrylate obtained with n-butyllithium, for each syndiotactic link there is an average of more than four isotactic links and, moreover, the distribution of the links of each type is not independent. If the last link in the growing chain is isotactic, then for the addition of the next unit the isotactic configuration is seven times as probable as the syndiotactic, since $p_{ii}/p_{is} = k_{ii}/k_{is} \approx 7$.

For the quantitative description of the microstructure of the chain we use the concepts of "isotacticity" I and "syndiotacticity" S, which denote the fractions of b o n d s (diads) connecting units with the same and with different configurations. In addition, we introduce the concepts of isotactic, syndiotactic, and heterotactic units. An isotactic unit has units of the same configuration on either side; a syndiotactic unit has two neighboring units with the opposite configuration; a heterotactic unit is attached to a unit with the same configuration on one side and a unit with the opposite configuration on the other side. The fractions of isotactic, heterotactic, and syndiotactic u n i t s (triads) are denoted by I*, H*, and S*, respectively. All these values may be obtained by analysis of high-resolution NMR spectra:

$$I = \frac{F_i}{F_i + F_s} \qquad S = \frac{F_s}{F_i + F_s}$$

$$I^* = \frac{F_{ii}}{F_{ii} + F_{is,si} + F_{ss}}$$

$$H^* = \frac{F_{is,si}}{F_{ii} + F_{is,si} + F_{ss}} \qquad \text{(V-8)}$$

$$S^* = \frac{F_{ss}}{F_{ii} + F_{is,si} + F_{ss}}$$

$$I^* = I - \frac{1}{2} H^* \qquad S^* = S - \frac{1}{2} H^*$$

It is also possible to calculate the mean length of isotactic and syndiotactic sequences (expressed in numbers of units):

$$\bar{l}_i = \frac{2I + H^*}{H^*} \qquad \bar{l}_s = \frac{2S + H^*}{H^*}$$

For example, for polymethyl methacrylate obtained by radical polymerization: I = 75%; S = 75%; I* = 6%; H* = 38%; S* = 56%; l = 2 units; l_s = 5 units. For isotactic polymethyl methacrylate:

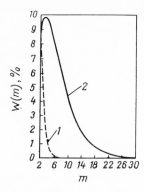

Fig. 128. Determination of the weight fraction of sequences consisting of m units of the same configuration as a function of m on the example of technical polymethyl methacrylate with a syndiotacticity of 75% [210]. 1) Isotactic; 2) syndiotactic sequences.

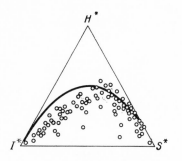

Fig. 129. Triangular diagram of the stereoregularity of the samples of polymethyl methacrylate obtained by different methods [479]. (The curve is drawn for a heterotacticity coefficient $\eta = 1$.)

$I = 81\%$; $S = 19\%$; $I^* = 72\%$; $H^* = 17.3\%$; $S^* = 10.7\%$; $l_i = 9$ units; $l_s = 3$ units.

Moreover, with the given mathematical model for the polymerization process [391], it is possible to calculate the distribution of the lengths of sequences of units with a definite configuration

$$w_i(m) = m p_i \frac{p_{is}^2\, p_{ii}^{m-1}}{1 - p_{is}\, p_{si}}$$

$$w_s(m) = m p_s \frac{p_{si}^2\, p_{ss}^{m-1}}{1 - p_{is}\, p_{si}} \qquad \text{(V-9)}$$

where $w_i(m)$ is the weight fraction of isotactic sequences consisting of m units; $w_c(m)$ is the same for syndiotactic sequences.

Figure 128 gives the distribution of sequences for polymethyl methacrylate calculated from NMR data.*

Miller [479] has proposed the description of the stereoregularity of a polymer by means of a triangular diagram. Figure 129 gives such a diagram, constructed from NMR data, for samples of polymethyl methacrylate obtained by different methods. The degree of stereoregularity may be characterized by the "coefficient of heterotacticity"

---

* The general theory for calculating the lengths of sequences of units with the same configuration in the polymer chain from NMR spectra was developed in [243].

$$\eta = \frac{H^*}{2IS}$$

For a completely regular polymer $\eta = 0$; for a polymer with randomly alternating units, $\eta = 1$; for a completely heterotactic polymer, $\eta = 2$. Figure 129 gives a curve corresponding to $\eta = 1$. It is obvious that most experimental points lie below this curve, and that $\eta < 1$ for them, indicating the formation of stereoblocks.

Thus, analysis of the high-resolution NMR spectrum of polymethyl methacrylate gives very complete information on the course of the polymerization process (values of p), the activation energies (values of E) and the microstructure of the polymer obtained (values of I, S, I*, H*, and S*).

By means of the NMR method a detailed study was made of the effect of temperature, solvent, the nature of the initiator, the degree of conversion, various additives, and other factors on the polymerization of methyl methacrylate. Thus, it was shown [192] that in radical polymerization of methyl methacrylate the probability of syndiotactic addition increases with a fall in the polymerization temperature. In a study of the anionic polymerization of methyl methacrylate, it was found [210] that the fraction of isotactic links is greatest in polymers obtained with organolithium initiators and falls with a change to sodium and potassium compounds; polymers of higher isotacticity are formed in nonpolar solvents, while the effect of temperature is insignificant. By controlling the synthesis through NMR measurements it was possible to select polymerization conditions which yield a purely atactic polymethyl methacrylate [327] with $I^* \approx 0$; $H^* = 0.51$; $S^* = 0.49$.

In a study of the polymerization of methacrylic anhydride [480, 689] and methacrylic acid [195, 487] the polymer was converted into polymethyl methacrylate for determination of the stereoregularity (see Chapter III). It was shown [480] that the polymerization of methacrylic anhydride with azoisobutyronitrile at high temperature yields predominantly the isotactic polymer and at low temperature, heterotactic polymer. It is evident that the formation of syndiotactic sequences is kinetically more probable, while the isotactic structure is thermodynamically more stable.

The pH of the solution affects the stereochemical configuration of polymethacrylic acid obtained by polymerization in an aqueous solution with a peroxide initiator [195]. With a rise in the pH there is a decrease in isotacticity, which is explained by the increase in the electrostatic repulsion between the monomer and the growing polymer radical.

In the polymerization of vinyl compounds the opening of the double bond of the monomer may be either of the cis- or trans-type. If one of the hydrogen atoms in the CH_2 group is replaced by deuterium, then from the NMR spectrum of the polymer it is possible to determine the type of opening of the double bond. In the polymerization of isopropyl α-cis-β-D_2-acrylate with a Grignard reagent, the polymer obtained is isotactic with respect to the configuration of the isopropyl groups, but has a random distribution of the configurations of the CHD groups [610]:

$$
\begin{array}{c}
\overset{(1)}{} \\
\cdots -\underset{\underset{\underset{OC_3H_7}{|}}{\underset{CO}{|}}{C}} - \underset{\underset{D}{|}}{C} - \underset{\underset{\underset{OC_3H_7}{|}}{\underset{CO}{|}}{C}} - \underset{\underset{D}{|}}{C} - \underset{\underset{\underset{\underset{(2)}{OC_3H_7}}{|}}{\underset{CO}{|}}{C}} - \underset{\underset{H}{|}}{C} - \underset{\underset{\underset{OC_3H_7}{|}}{\underset{CO}{|}}{C}} - \underset{\underset{D}{|}}{C} - \cdots
\end{array}
$$

This is demonstrated by the fact that the signals of the chemically nonequivalent protons $H_{(1)}$ and $H_{(2)}$ (after elimination of spin—spin interaction with deuterium) have the same intensity. Consequently, cis- and trans-openings of the double bond are equally probable. It was shown analogously [745, 747] that in the free-radical polymerization of methyl acrylate cis- and trans-opening of the double bond occur with the same frequency, while in polymerization with lithium aluminum hydride there is only trans-opening.

From the NMR spectrum of polypropylene it is possible (see Chapter III) to determine the content of diads I and S and triads I*, H*, and S*. For fractions of polypropylene obtained [725] with the catalyst $TiCl_4$—$Al(C_2H_5)_3$, the relation between these values shows that the polymer has a stereoblock structure and that during the polymerization of polypropylene the probability of the addition of the next unit to the growing chain isotactically or syndiotactically does not obey simple Bernoulli statistics.

In the polymerization of trifluorochloroethylene (for the NMR spectrum see Chapter III, page 219) there is predominantly syndiotactic chain growth; the fact that the ratio p_i/p_s is independent of the polymerization temperature shows that the character of the addition is due mainly to steric hindrance [690].

A study of the high-resolution NMR spectrum of polyvinylidene fluoride made it possible to calculate the probabilities of different types of addition of the monomer unit to the growing chain [718]. The growing radical may have at the end of the chain $-CF_2\cdot$ ("head") or $-CH_2\cdot$ ("tail"). Four types of addition of the monomer unit CF_2-CH_2 are possible: "head-to-tail" (probability p), "head-to-head" $(1-p)$, "tail-to-tail" (q), "tail-to-head" $(1-q)$. The values of the probabilities p and q are determined from the formulas

$$ p = \frac{1}{2} + \sqrt{\left(\frac{1-w}{2}\right)^2 - D} + \frac{w}{2} \; ; \quad q = p - w \qquad (V-10) $$

where w = $(C - D)/(C + D)$ and A + B + C + D = 1.

A, B, C, and D are the normalized intensities of the lines in the NMR spectrum of the polymer (see Chapter II, page 193).

For a series of technical and laboratory samples of polyvinylidene fluoride the following values were found: p = 0.93-0.95 and q = 0.90-0.96.

The NMR method has been used to study the formation of dimers, trimers, etc., in a solution of formaldehyde and to determine the corresponding formation constants [624].

The NMR method has also been used to study the kinetics of thermal decomposition of a polymerization initiator, namely, azo-isobutyronitrile [656].

High-resolution NMR spectra are used to elucidate polymerization mechanism [244, 277, 321, 323, 380] and to study the effect of catalysts on the structure of the polymers obtained, for example, polyisobutene [160] and poly-n-vinylcarbazole [348].

Analysis of the NMR spectra of samples of poly-α-methylstyrene (see Chapter III) showed [212, 215] that a Ziegler catalyst and butyllithium do not lead to the formation of a stereoregular polymer; syndiotactic poly-α-methylstyrene may be obtained with a cationic catalyst. According to data obtained by Sakurada and his co-workers [594], the degree of isotacticity of poly-α-methylstyrene changes with the catalyst in the series

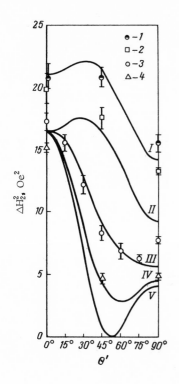

Fig. 130. Relation of the second moment of the line (ΔH_2^2) of the NMR spectrum of a single crystal of trioxane to the angle (θ') between the C axis of the sample and the field vector at different temperatures [117]: 1) at $-26°C$; 2) at $0°C$; 3) at $25°C$. Theoretical curves: I) ΔH_2^2 for a rigid lattice; II, III, IV, V) intramolecular contribution to ΔH_2^2 for a structure with harmonic oscillations of the molecules about the c axis with amplitudes δ equal to 0, 60, 90°, and ∞, respectively.

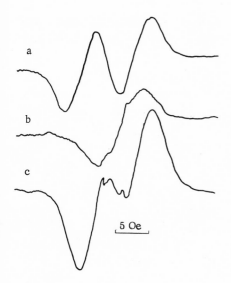

Fig. 131. Form of the NMR lines in spectra obtained at room temperature with the C axis of the samples along the field vector [130]. a) Trioxane single crystal; b) polyoxymethylene washed free from monomer; c) polyoxymethylene in a trioxane lattice.

$$AlCl_2C_2H_5 \gg BF_3(C_2H_5)_2O > TiCl_4 > Al(C_2H_5)_3 - TiCl_4 \gg$$
$$\gg AlCl(C_2H_5)_2 > KC_4H_9 \approx NaC_4H_9 \approx LiC_4H_9$$

Data which are important for understanding the mechanism of polymerization may be obtained by observing the NMR spectra of the catalyst. A study was made [41] of the change in the NMR spectrum during the polymerization of phenylacetylene with a Ziegler catalyst in octadeuterotoluene. In the first minutes there was a sharp fall in the intensity of the line of the proton of $Ti-CH_3$. This is explained by the insertion of the monomer at the $Ti-CH_3$ bond, since for the new group formed $Ti - CH = C \Big\langle {{}^{CH_3} \atop {}_{C_6H_5}}$ all the chemical shifts of the protons are different from the chemical shifts of the protons $Ti-CH_3$.

The structure of a complex catalyst used in the stereoregular polymerization of dienes and containing cobalt and aluminum was established [538] by means of the NMR spectrum of the ^{27}Al nuclei.

The NMR method has been used successfully to study the products from radiation telomerization [374] and particularly radiation polymerization in the solid phase [264, 441, 489, 683]. It was established that in the polymerization of methacrylic acid in the crystalline state under the action of γ radiation [489] twice as many isotactic bonds are obtained as in polymerization in solution.

A certain degree of mobility of the monomer molecules in the crystal lattice is necessary for polymerization to occur in the solid phase. It is in precisely the temperature range in which intensive molecular motion begins that there is a sharp increase in the polymerization rate and the polymer yield for a series of monomers. The "defreezing" of the molecular motions in the solid state as the temperature rises can be followed very well by means of the NMR method. The polymerization of trioxane by γ radiation, leading to the formation of an oriented, highly ordered polyoxymethylene (see Chapter III) with a high strength, is of particular interest. Japanese authors [136, 414] studied the NMR spectrum of polycrystalline trioxane. It was shown that above 45°C there is a change in the form of the NMR line because of a change in the character of the molecular motion and an increase in the polymer yield.

More detailed data on molecular motion in solid trioxane were obtained by us [117] in a study of the NMR spectrum of single-crystal samples. Figure 130 gives the relation of the second moment of NMR line of trioxane to the angle θ' in the magnetic field at different temperatures. Comparison with theoretical curves of $\Delta H_2^2 = F(\theta')$ shows that at −26°C and below the intensity of the molecular motion is low. At room temperature the molecules evidently execute oscillations about the C axis with an amplitude of about 60° and with a rise in temperature to 55°C, the amplitude of the oscillations increases* to 90°.

By observing the change in the NMR line it is possible to follow the formation of a polymer chain directly in a monomer lattice [130, 131]. We plotted the NMR spectra of a trioxane single crystal (a), polyoxymethylene (b) obtained by radiation polymerization of a single crystal of trioxane with subsequent washing out of excess monomer, and the polymer in the monomer lattice (c) (Fig. 131). As the figure shows, the form of the NMR line changes markedly during polymerization. During polymerization there is also the accumulation of low-molecular products and defects in the crystal lattice, which give a narrow component of the NMR line. With the formation of the polymer there is also a considerable change in the second moment of the line and the character of the relation of the form and the second moment of the line to the angle between the sample axis and the magnetic field. Thus, it is possible to follow the course of polymerization in the solid phase through the change in the form and the second moment of the NMR line in relation to the degree of polymerization, the temperature, and the position of the sample in the magnetic field.

2. Study of Cross-Linking Processes
in Polymers

The NMR method makes it possible to observe the change in structure and molecular mobility in polymers produced by cross-linking processes, i.e., vulcanization, hardening of resins, etc.

---

*For simplicity the calculations of ΔH_2^2 in [117] were made for harmonic oscillations of the molecules. Jumps between equilibrium positions are obviously more probable physically than oscillations.

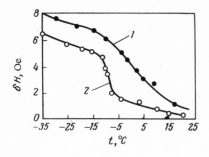

Fig. 132. Relation of the line width (δH) in the NMR spectra of a 50:50 copolymer of butadiene with styrene (Hycar, OS-10) to temperature [357]. 1) Vulcanized sample; 2) unvulcanized sample.

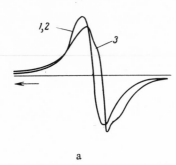 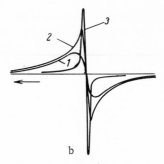

a   b

Fig. 133. Lines of ^{19}F (a) and ^{1}H (b) nuclei in the NMR spectrum of the fluoro copolymer Viton, vulcanized with hexamethylenediamine (the spectrum was determined at 30°C) [81]. 1) Starting polymer; 2) mixture of polymer with 10 parts by weight of hexamethylenediamine; 3) the same mixture after vulcanization at 120°C for 10 min.

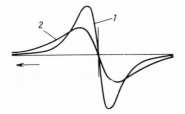

Fig. 134. Line of ^{19}F nuclei in the NMR spectrum of the fluoro copolymer Viton, vulcanized with hexamethylenediamine (10 parts by weight) with the addition of 10 parts by weight of magnesium oxide [81]. 1) Starting polymer; 2) after vulcanization at 120°C for 10 min.

The change in the NMR during vulcanization of rubbers has been studied by a series of authors. As early as 1950, Holroyd, Mrowca, and Guth [359] observed that during the vulcanization of a 70:30 butadiene—styrene rubber (GR-S grade), 50:50 butadiene—

styrene rubber (Hycar OS-10), and butyl rubber there is an increase in the width of the NMR line at room temperature. The temperature dependence of the line width for Hycar was measured by Holroyd, Mrowca, et al. [164, 357, 358, 490, 492] (Fig. 132), and this also shows a considerable difference in the character of the molecular motions of unvulcanized and vulcanized rubbers particularly in the temperature range from −15 to −5°C. The authors consider that vulcanization interferes with the formation of regions with free rotation. The effect of vulcanization on the width of the NMR lines in natural rubber was studied by Holroyd et al. [360]. They showed that with the introduction of 8% sulfur, the width of the NMR line of natural rubber at room temperature increases to a lesser extent than with a butadiene−styrene copolymer and put forward the hypothesis that the cross links formed during vulcanization in natural rubber are closer in strength to "physical" bonds, while those in synthetic rubber are closer to "chemical" bonds.

Great attention was paid to the effect of vulcanization on the NMR spectrum in the work of Gutowsky and his co-workers [338, 339]. It was found that with an increase in the vulcanization time of natural rubber there is a substantial rise in the temperature at which there is narrowing of the line, which is observed in the region from −133 to −103°C, and is due to motion of the CH_3 groups. The temperature at which there is narrowing of the line observed in the region of −48°C and which is connected with the movement of segments, is independent of the vulcanization time. The activation energy of the motion which is responsible for narrowing of the line in the low-temperature region equals 2.7 kcal/mole for a sample vulcanized for 30 min and increases to 4.5 kcal/mole with an increase in the vulcanization time to 90 min. The width of the line is affected both by the amount of bound sulfur in the vulcanizate and also by the sulfur content of the starting mixture. It was also observed that with an increase in the content of bound sulfur there is a rise in the temperature of the minimum of the spin−lattice relaxation time T_1. The authors proposed that the first process in the vulcanization of rubber is not the formation of cross links, but the formation of cyclic structures.

$$\cdots -CH_2-\underset{\underset{H_2C}{|}}{C}-\underset{\underset{CH_2}{|}}{CH}\quad\underset{}{C}-CH-CH_2-\cdots$$

Slichter [108] pointed out that this hypothesis is not adequately substantiated and requires a more complete investigation. Oshima and Kusumoto [540], who studied the change in the NMR spectrum of rubber containing 2-30% sulfur over a wide range of temperatures, suggest the practical use of the NMR method to check the degree of vulcanization. The use of the NMR method to study the kinetics of vulcanization of tire rubbers was recommended in [31].

The NMR method has been used [81] to study the vulcanization of rubbery fluoro copolymers such as Viton and Kel F-3700. It was observed that the vulcanization of fluoro rubbers by hexamethylenediamine produces a substantial change in the form of the resonance line of fluorine and hydrogen (Fig. 133). The fluorine line loses its symmetry. After vulcanization of the mixture, the resonance line of hydrogen consists of two components, namely, a broader line (of the order of tenths of an Oe) and a very narrow line superimposed on it. The change in the form of the NMR lines is explained by the abstraction of one hydrogen atom from CH_2 groups and one fluorine or chlorine atom from CF_2 or $CFCl$ groups of the polymer chains with the liberation of HF or HCl, the formation of CH and CF groups, and the formation of double bonds. In the proton NMR spectrum of the polymer after vulcanization the broader line corresponds to the protons of CH_2 groups. The narrow line is given by protons of CH groups, whose magnetic interaction with the other protons is comparatively weak. The disruption of the symmetry of the NMR line of ^{19}F during vulcanization is explained by the presence of a chemical shift of the fluorine nuclei in CF groups. Magnesium oxide is added to some mixtures as an acceptor of hydrogen fluoride. During the vulcanization of such mixtures, a large amount of MgF_2 is formed and this gives a broad NMR line; therefore, the overall resonance line broadens and its asymmetry is suppressed (Fig. 134).

The effect of the formation of cross links in the polymer on the NMR spectrum has also been studied in the hardening of resins. Jain [386] followed the change in the form of the NMR line during the hardening of a polyester-amide, modified with epoxide resin. The partly hardened resin gave a complex line, which indicated the presence of regions with different degrees of hardening. From the ratio of the areas of the components of the line it was possible to calculate the percentage of unhardened material.

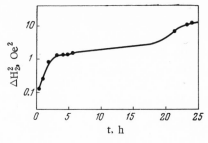

Fig. 135. Relation of the second moment of the line (ΔH_2^2) of the NMR spectrum of an epoxide resin to the hardening time [447].

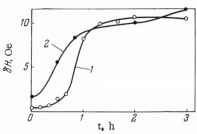

Fig. 136. Relation of the width of the line (δH) in the NMR spectrum of the resin ED-6, containing 15% maleic anhydride, to the heating time [651]. 1) Spectra were plotted immediately after heating the samples at 100°C; 2) after heating the samples were kept for 8 days at room temperature.

Fig. 137. NMR line of ED-6 resin containing 15% maleic anhydride after heating for 60 min at 100°C [651].

Matsushita [456] studied the hardening of resins (casting grades) based on phthalic anhydride and containing styrene and also observed a two-component NMR line with partial hardening. The simultaneous use of NMR and other methods made it possible to select the optimal hardening conditions.

The temperature dependence of the second moment of the NMR line of polyester resins, obtained by condensation of maleic anhydride with diethylene glycol and hardened by heating with styrene, indicates the presence of two transitions in the polymer [679]. The first transition at a temperature from −3 to −6°C is caused by the presence of the motion of protons in the methylene groups of diethylene glycol, remote from the site of cross linking. The form of the NMR line at room temperature shows that about 30% of all the protons are in a state of intensive molecular motion. The second transition is connected with the presence of motion of the polyester molecules. The temperature of the transition rises from 36 to 60°C with an increase in the styrene content of the mixture. Above the temperature of the second transition, which occurs in the polymer, the NMR spectrum of the resin contains a narrow line, but retains the "wings" corresponding to protons which lie close to bridges between the chains where the mobility is limited.

Chuvaev, Ivanova, and Zubov [142] studied the kinetics of hardening with styrene of an unsaturated polyester resin containing ethylene glycol, diethylene glycol, and maleic acid and observed simultaneously a change in the form of the NMR line of the system and a change in the spin—spin relaxation time T_2. It was shown that during hardening there is a reduction in the number of protons in the unhardened phase and also a reduction in their mobility. In the first stage of hardening T_2 falls from $30 \cdot 10^{-3}$ to $1 \cdot 10^{-3}$ sec, and the correlation frequency of molecular motion ν_c from 10^{10} to 10^8 Hz, and this is explained by the formation of a cross-linked network.

The hardening of epoxide resins was studied by Lösche [75, 444, 446, 447] and Slonim, Lyubimov, and Kovarskaya [651]. Figure 135 shows the change in the second moment of the NMR line during the hardening of an epoxide resin by an aliphatic polyamine at room temperature. It is obvious that the process occurs in two stages. First there is a decrease in the mobility of the chains and then suppression of the motion of the CH_3 groups begins.

In [651], an epoxide resin of the type ED-6 was hardened with maleic anhydride at an elevated temperature; the NMR spectra were plotted at room temperature. At the same time, thermomechanical measurements were made on the samples investigated. Figure 136 shows the change in the width of the NMR line in relation to the heating time of the resin with the hardener at 100°C. The unheated mixture gives a narrow NMR line, while the most rapid increase in the width of the line, corresponding to a rapid decrease in the mobility of the chains, occurs at the end of the first hour of hardening at 100°C. In the region corresponding to the rapid increase in the line width, for samples heated at 100°C for 40-60 min the NMR line has the complex form which is characteristic of two-phase systems. Figure 137 shows that the absorption curve in this case may be regarded as the superimposition of two lines, one broad and one narrow. It may be surmised that the two components

of the line belong to hardened and unhardened resin. If a completely hardened sample is stored at room temperature, processes occur in it which are accompanied by a further increase in the width
of the resonance line. In Fig. 136, curve 2, which corresponds to
the width of the lines of the sample after storage, lies above
curve 1 in the region of heating times which are inadequate for
complete hardening. Apparently, in the corresponding samples at
room temperature there is the gradual formation of a cross-linked
network and an increase in the viscosity of the resin, leading to a
decrease in the intensity of molecular motions and an increase in
the width of the NMR line.

It is interesting to note that although preliminary heating is
not an essential condition for this process (it also occurs in the
raw resin), there is a heating time (about 30 min at 100°C) with
which the hardening rate during subsequent storage is maximal.

The width of the NMR line is sensitive to changes which occur during the first stages of the hardening process, but does not
change during subsequent thermal treatment.

The properties of an epoxide resin change substantially after
heating to 250°C and subsequent cooling to room temperature.
These changes are readily observed by a thermomechanical investigation, but do not affect the width of the NMR line plotted at room
temperature. These changes in properties may evidently affect the
width of the NMR line if the spectrum is plotted at an elevated temperature and the broader line would be observed with resin samples
which had been subjected to thermal treatment.

3. Study of the Effect of
Radiation on Polymers

Ionizing radiation (γ rays, neutron beams, etc.) produces substantial changes in polymers: in large doses, radiation spoils the
properties of polymers, producing rapid aging of parts. On the
other hand, strictly controlled doses of radiation are used in polymer technology. Therefore, the study of the changes in polymers
under the action of ionizing radiation is an important problem.

The NMR method has been used for this purpose largely in
the investigation of the effect of radiation on polyethylene and to a
lesser extent, on polytetrafluoroethylene and other polymers.

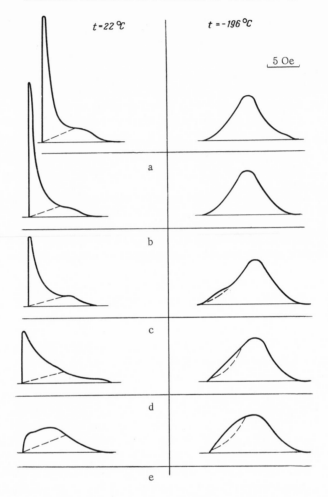

Fig. 138. Change in the form of the NMR line of polyethylene as a result of irradiation with a neutron beam (only the right-hand half of the
curve of the derived absorption function is shown) [308]. a) Unirradi-
ated polymer; b) irradiated with $0.3 \cdot 10^{18}$ neutrons/cm^2; c) $0.6 \cdot 10^{18}$;
d) $1.2 \cdot 10^{18}$; e) $8.0 \cdot 10^{18}$.

In a series of studies [308, 309, 599], Fuscillo and Sauer examined the effect of irradiation with neutrons in a nuclear reactor
and γ rays from a ^{60}Co source on polyethylene. The dynamic mechanical method was also used in addition to the NMR method. After
irradiation with an integral dose up to $8 \cdot 10^{18}$ neutrons/cm^2, the
form of the absorption curve plotted at room temperature is changed

markedly (Fig. 138). The NMR line of polyethylene has a complex form; after irradiation there is a change mainly in the form of the narrow component, which corresponds to the amorphous part of the polymer. For strongly irradiated samples the form of the line is similar to the form of the line of unirradiated polyethylene, plotted at a lower temperature. The form of NMR lines plotted at −196°C changes little with irradiation. From the dynamic modulus the authors calculated the number of cross-links in the structure of the polymer. Below we give data characterizing the change in the properties of polyethylene on irradiation with different neutron fluxes (in neutrons/cm^2):

Radiation, neutrons· 10^{-18}/cm^2						
0	0.3	0.6	1.2	2.8	5.5	8.0
No. of cross-links per 100 C atoms						
0	4	10	20	30	—	—
ΔH_2^2, Oe^2:						
at 77°K						
27.2	23.2	25.0	27.4	—	27.5	28.0
at 295°K						
14.5	11.3	13.4	13.3	14.3	15.2	17.0
Degree of crystallinity, %:						
according to NMR						
59	54	—	55	64	67	76
according to x-ray pattern						
55	50	25	—	10	0	0

The authors explain the action of radiation on polyethylene by the superimposition of two processes, namely, disruption of the crystallites and the formation of cross-links. With a flux of 0.3 · 10^{18} neutrons/cm^2 the first process predominates and the second moment of the NMR line of an irradiated sample is somewhat less than that of an unirradiated sample. With a flux of 1.2 · 10^{18} neutrons/cm^2 and above, the second process predominates and an increase in ΔH_2^2 is observed. Thus, the index of the high-frequency rigidity N_{rig}/N changes with the flux.

Analogous results were also obtained by Slichter and Mandell [647], who studied the effect on polyethylene of electron and neutron fluxes. As Fig. 139 shows, after irradiation with comparatively low doses (5 · 10^8 rep) there is a decrease in the width of the NMR line at all temperatures above −173°C. With polyethylene that has been subjected to high neutron fluxes (1.8 · 10^{18} neutrons

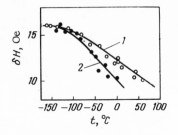

Fig. 139. Temperature dependence of the line width (δH) in the NMR spectrum of polyethylene [647]. 1) Unirradiated polymer; 2) irradiated with an electron beam (dose $5 \cdot 10^8$ rep).

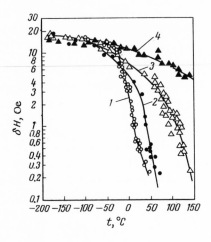

Fig. 140. Temperature dependence of the line width (δH) in the NMR spectrum of polyethylene [647]. 1) Unirradiated polymer; 2,3,4) irradiated with $1.8 \cdot 10^{18}$, $3.8 \cdot 10^{18}$, and $14.9 \cdot 10^{18}$ neutrons/cm^2,

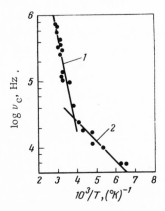

Fig. 141. Relation of the correlation frequency (ν_c) to the reciprocal temperature ($1/T$) for polyethylene irradiated in a reactor with a flux of $1.8 \cdot 10^{18}$ neutrons per cm^2. 1) E = 1.4 kcal/mole; 2) E = 7.4 kcal/mole.

per cm^2 and above), the width of the line at a temperature above −23°C is greater than for unirradiated polyethylene (Fig. 140). To explain the mechanism of the effect of radiation, the authors calculated the correlation frequency ν_c. For polyethylene irradiated with a flux of $1.8 \cdot 10^{18}$ neutrons/cm^2, a graph of log $\nu_c = F(1/T)$ consists of two straight lines (Fig. 141). At low temperatures there is a process with an activation energy E = 1.4 kcal/mole; apparently there are rotational oscillations of low amplitude. At a higher temperature there is the possibility of rotations or oscillations of high amplitude or translational motions of chains and the activation energy equals 7.4 kcal/mole. With a very high flux of $1.49 \cdot 10^{19}$ neutrons/cm^2 there is one straight line for log $\nu_c = F(1/T)$ and E = 1.8 kcal/mole; it is evident that a large number of cross-links strongly hampers the motion of the molecular chains. A detailed study of the effect of neutron radiation on low- and high-density polyethylene by Glick et al. [313] confirmed the conclusions of previous work. The second moment of the NMR line at temperatures below room temperature decreases on irradiation as a result of disruption of the crystallites at low fluxes ($0.25 \cdot 10^{18}$ neutrons/cm^2) and increases with a flux greater than the critical value as a result of the formation of a large number of cross-links (Fig. 142). For high-density polyethylene the critical flux is somewhat higher than for low-density polyethylene.

A decrease in the degree of crystallinity (from NMR) of a sample of high-density polyethylene from 90 to 40% on irradiation with a flux of fast neutrons of $7 \cdot 10^{17}$ neutrons/cm^2 was observed by Kessenikh and his co-workers [61]. As Fig. 143 shows, the broad component of the line disappeared almost completely after irradiation.

Fujiwara [300] observed the effect on polyethylene of an intense beam of deuterons, accelerated in a cyclotron. The irradiated films changed markedly: they lost 2% of hydrogen, the mass increased by 1% as a result of oxidation, and the softening point rose from 105 to 300°C. The change in the second moment of the NMR line with temperature for the initial and irradiated films of polyethylene are shown in Fig. 144.

The second moment of the NMR line of the irradiated polymer is less than for the initial film when plotted at low temperatures, but greater than it is when plotted at high temperatures.

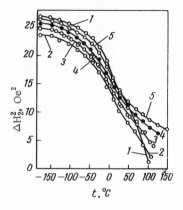

Fig. 142. Temperature dependence of the second moment of the line (ΔH_2^2) in the NMR spectrum of low-density polyethylene irradiated in a nuclear reactor with different neutron fluxes [313]. 1) Unirradiated polymer; 2) irradiated with $0.25 \cdot 10^{18}$; 3) $0.6 \cdot 10^{18}$; 4) $1.3 \cdot 10^{18}$; 5) $2.8 \cdot 10^{18}$ neutrons/cm^2.

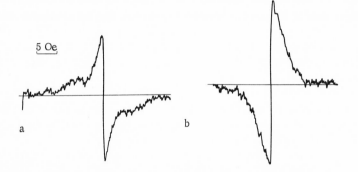

Fig. 143. Change in the form of the NMR line of polyethylene after irradiation with fast neutrons (the spectrum was plotted at room temperature) [61]. a) Unirradiated polymer; b) irradiated with $7 \cdot 10^{17}$ neutrons/cm^2.

This also may be explained by disruption of the crystallites and the formation of cross-links between chains.

Irradiation also has a strong effect on the relaxation times. A change in the form of the curve of $T_1 = F(1/T)$ as a result of irradiation of high-density polyethylene with electrons with an energy of 1.5 MeV [405]. As a result of the formation of cross-links, the high-temperature minimum of T_1, corresponding to motions of the molecules in the crystalline regions, becomes flatter and disappears at high radiation doses. The values of T_1 of irradiated polymers hardly change after annealing; the author believes that the free radicals formed under irradiation conditions do not substantially affect the spin—lattice relaxation process.

The opposite conclusion was reached by Komaki and his co-workers [415], who studied the decrease in T_1 of polyethylene as a result of the effect of paramagnetic centers—radicals of the allyl type, formed by irradiation of the polymer with ^{60}Co γ rays in vacuum with a dose of 10^8 r. When the radicals were destroyed by thermal treatment, T_1 returned to the previous value. The discrepancy between the results of these authors may be explained by different irradiation conditions.

A theoretical calculation of T_1 for an irradiated polymer, whose chain contains local paramagnetic centers, was carried out in the work of Buishvili and Kessenikh [30]. The calculated values of T_1 at liquid helium temperatures agree satisfactorily with experimental data obtained [61] in a study of the effect of paramagnetic centers delocalized along a linear chain of polyethylene, irradiated with fast neutrons, on spin—lattice relaxation.

In a study of irradiated polypropylene, NMR and EPR methods were used for the same samples [332, 335]. The NMR spectra of isotactic polypropylene were plotted over a wide temperature range from −196 to +102°C. There are two falls on the curves of the temperature dependence of the width and second moment of the line. The first fall, which is close to the boiling point of nitrogen, is connected with the beginning of rotation of the methyl groups. Irradiation of polypropylene in a reactor with a flux of up to $1.4 \cdot 10^{19}$ neutrons/cm^2 does not affect this transition in the polymer. In the temperature range from −13 to +17°C there is a second fall on the curves and this is due to motion of segments of the polymer chains. At low radiation doses the second transition in the polymer is shifted to lower temperatures because of the formation of defects in the lattice. High doses of neutrons produce a large number of cross-links, the mobility of the chains is markedly reduced, and the second transition in the polymer is shifted to higher temperatures. At the same time there is a decrease in the width of the EPR line, indicating an increase in the length of the systems of conjugated bonds.

The change in the character of the relations $\delta H = F(T)$ and $\Delta H_2^2 = F(T)$ for polypropylene during its irradiation in a reactor

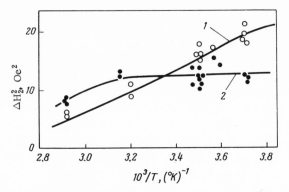

Fig. 144. Relation of the second moment of the line (ΔH_2^2) in the NMR
spectrum of polyethylene to the reciprocal temperature $(1/T)$ [300].
1) Unirradiated polymer; 2) irradiated with a beam of deuterons.

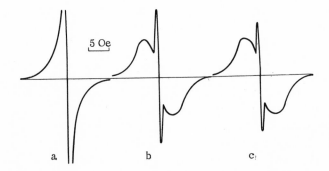

Fig. 145. NMR spectra of SKB rubber after γ irradiation at different
doses in air [57]. a) 50 Mrad (part of the spectrum is given); b) 150
Mrad; c) 300 Mrad.

due to a decrease in the degree of crystallinity and the formation
of cross-links has also been described in [226].

Irradiation of films of polytetrafluoroethylene with a beam of
accelerated deuterons in vacuum leads to broadening of the NMR
line [428]. The author considers that the hindrance to motion in
the polymer is due to the formation of double bonds, rings, and
cross-links. Burget and Sacha [220] also observed a one and a half
fold increase in the width of the NMR line of Teflon after irradiation with neutrons for 5 h in a total flux of $1.8 \cdot 10^{16}$ neutrons/cm^2.

When films of polyvinyl alcohol [298] are irradiated with γ rays with a dose less than the critical value there is a decrease in the second moment of the NMR line and the degree of crystallinity ε_{NMR} of the polymer; doses greater than the critical value produce an increase in ΔH_2^2 and an increase in the fraction of the broad component in the NMR signal. The critical dose equals ≈ 1 Mrad and changes with a change in the irradiation conditions and the dose strength. The sample, which was placed in the channel of a nuclear reactor, was subjected to the simultaneous action of thermal neutrons and γ rays. The ratio of the doses of the two types of radiation depends on the position of the sample in the reactor. By varying the irradiation conditions, it was possible to establish [299] that the change in the molecular weight of polyvinyl alcohol was caused solely by the action of γ rays. Without substantially affecting the degree of polymerization, thermal neutrons considerably reduce the relative intensity of the broad component of the NMR line.

The great effect of the dose strength on the effect of radiation on polymers was also reported by Lösche [445, 447, 448] and Przyborowski [575]. Prolonged (for up to 5 months) irradiation of polyamides from a small source (30 mCi of radium), in which the total absorption equaled 10^{16} quanta, led to a sharp fall in the second moment of the NMR line from 14–16 Oe^2 to 0.07 Oe^2 at room temperature. Apparently hardly any cross-links were formed at the low radiation dose strength, as there is a very low probability of the simultaneous action of two quanta on two neighboring chains in the polymer. The considerable change in the second moment after prolonged action on the polymer of radiation of low strength is of interest and may be used in dosimetry.

Great changes in the form of the NMR line during irradiation were observed by Karpov, Pomerantsev, and Sergeev [57] for synthetic rubbers SKB (polybutadiene containing 60–70% of 1,2-structures) and SKD (polybutadiene containing 90–95% of 1,4-structures). The starting samples gave a narrow line with a width $\delta H \approx 0.2$ Oe. During irradiation with a ^{60}Co γ source there appeared a broad component of the line with $\delta H \approx 9.2$ Oe (Fig. 145). With an increase in the irradiation time the intensity of the broad component increased, while that of the narrow component decreased. It is evident that in an irradiated sample of rubber there are two sorts of proton, namely, "mobile" with a correlation frequency

$\nu_c = 10^5\text{-}10^6$ Hz and "restrained" with $\nu_c = 10\text{-}10^2$ Hz. As a result of radiation cross-linking there is "pumping" of protons from the first state to the second, avoiding the state of protons with an intermediate correlation frequency of $10^2\text{-}10^3$ Hz. The quantum yield of this process equals $(2\text{-}3) \cdot 10^3$ restrained protons per 100 eV. Comparison of this value with data on the kinetics of growth of the vulcanization lattice in rubbers during irradiation [40] shows that the appearance of one cross-link results in the hindrance of 200-300 protons or 50-70 butadiene units. Irradiation hinders the motion of protons in rubber more strongly than vulcanization with sulfur [339]. The authors explain this by the fact that the potential barrier to rotation about C—C bonds is considerably greater than about C—S bonds.

Radiation produces breakdown of the polymer chains in polyisobutene. As Sergeev and Karpov [106] showed, the width of the NMR line (δH) then falls from 0.4 to 0.1 Oe (very high doses of γ radiation up to 240 Mrad were used). However, the decrease in δH was not as great as might have been expected, judging by the sharp fall in the molecular weight. The "microviscosity" apparently falls little: despite the breakdown the entanglement of the chains is preserved and the character of the motion of the segments changes little.

During the irradiation of polydimethylsiloxanes with an electron beam [366] with a dose up to 600 Mrep, the width and second moment of the NMR line at low temperature do not change. Evidently the formation of even a large number of cross-links (up to 1 cross-link to 5 monomer units) does not substantially affect the structure of the polymer. The width of the line at room temperature for irradiated samples is greater than for the starting samples. The activation energy of molecular motion, calculated from the temperature dependence of the width of the NMR line, falls during irradiation. The results obtained are explained by an increase in the local viscosity of the polymer.

The NMR method has been used to study the action of radiation on cellulose [668] and polymethyl methacrylate [577]. Using high-resolution NMR spectra, Pomerantsev and his co-workers [95] observed the formation of CF_3 groups during γ irradiation of perfluorooctadiene and perfluorododecadiene.

Free radicals are formed in irradiated polymers and electron paramagnetic resonance and nuclear magnetic resonance may
be observed at the same time. With saturation of the electronic
transitions the intensity of the NMR signal increases (Overhauser
effect). This phenomenon is used to obtain polarized proton tar-
gets and for studying the properties of irradiated polymers [61, 78,
219, 344].

4. Study of Breakdown and Other
Reactions in Polymers [115]

In the study of the aging of polymeric materials under the action of high temperature, atmospheric oxygen, and other factors,
the NMR method may be used both for identification and quantitative determination of the low-molecular products (see, for example
[150, 275, 288, 294, 656, 702]) and, mainly, for assessing the
changes in the polymer during breakdown.

A series of NMR spectra was plotted of polypropylene which
had been subjected to thermo-oxidative breakdown [90]. By means
of linear anamorphoses the NMR lines were separated into two
components, broad and narrow. It was found that the ratio of the
number of "rigid" protons to the total number of protons falls during breakdown. The width of the narrow component decreases during breakdown, while the width of the broad component remains unchanged (Fig. 146). On this basis it was concluded that oxidation
occurs much more rapidly in the amorphous phase of polypropylene
than in the crystalline phase.

In the study of the thermal breakdown of polycarbonate [65],
which was carried out in vacuum at 500°C, NMR spectra were plotted of the initial polymer and residues after breakdown over a wide
range of temperatures. A comparison of the curves of the temperature dependence of the width of the lines of both samples (Fig. 147)
shows that for the solid residue after breakdown the narrowing of
the line in the region of −150 to −50°C, which is due to rotation of
the CH_3 groups, is much less marked than in the initial polymer.
At all temperatures above −100°C the sample after breakdown
gives a broader NMR line; narrowing of the line does not occur
right up to 210°C. The change in the curve of $\delta H = f(t)$ shows that
there is a sharp decrease in the number of methyl groups in the

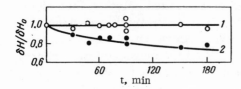

Fig. 146. Relations of the widths of the broad (1) and narrow (2) components of the line (δH) in the NMR spectrum of polypropylene to the duration of oxidative breakdown [90].

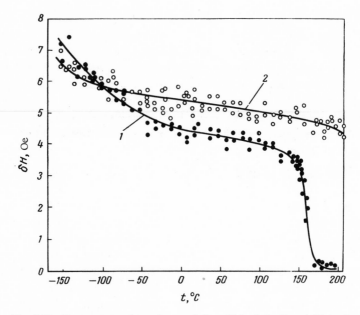

Fig. 147. Temperature dependence of the width of the line (δH) in the NMR spectrum of polycarbonate [65]. 1) Initial polymer; 2) after thermal breakdown for 1 h at 500°C in vacuum.

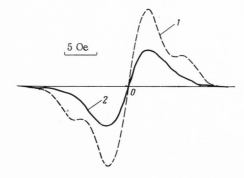

Fig. 148. NMR line of polycarbonate plotted at −196°C [65]. 1) Initial polymer; 2) after thermal breakdown for 1 h at 500°C in vacuum.

polymer and an increase in the rigidity of the structure (evidently due to the formation of polyphenyl structures). The decrease in the number of CH_3 groups in polycarbonate appears even more clearly in the change in the form of the NMR line at −196°C. After breakdown the line with outer extrema, which is characteristic of polymers with CH_3 groups, disappears (Fig. 148). The conclusions based on NMR spectra were confirmed by analysis of the gaseous pyrolysis products and infrared spectroscopic data.

During the thermo-oxidative breakdown of polyamide film [89] there is a decrease in the width and the second moment of the NMR line in the temperature range from −120 to +140°C, indicating loosening of the structure and an increase in the mobility of the molecular chains. At a temperature above 80°C the NMR line of the polyamide film consists of two components, broad and narrow. For a sample subjected to thermo-oxidative breakdown the narrow component appears at a lower temperature and is more intense than for the initial sample (Fig. 149), also confirming the loosening of the structure. Evidently, under the conditions under which breakdown occurred, the effect of the chain scission process predominated, and not the formation of intermolecular and intramolecular bonds, as in the case of the thermal breakdown of polycarbonate examined above.

In the oxidation of rubbers there is a decrease in the amplitude of the NMR signal as a result of structuralization processes. In a series of communications of Zalukaev and his co-workers [31, 47-49, 51, 52] it was shown that the NMR method may be used to study the kinetics of the structuralization process and to estimate the effect of inhibitors. It was pointed out that the NMR method is more sensitive than the volumetric method of studying the oxidation of rubbers.

In the ultraviolet oxidation of natural rubber in air, the amplitude of the signal increases somewhat, while the spin—lattice relaxation time T_1 falls [576]. The author explains the results obtained by the superimposition of two processes, namely, the formation of cross-links, leading to a decrease in the mobility of the methyl groups, and the formation of low-molecular products, which give an intense narrow NMR line.

Maklakov and Pimenov [85] studied the thermal breakdown of polyacrylonitrile in vacuum. It was found that after pyrolysis the second moment of the NMR line decreases in the low-temperature region and increases in the high-temperature region. The change in the second moment of the NMR line is explained satisfactorily by the process

Work to study breakdown includes investigations of the stabilization of polymers. A correlation was found [224, 225, 500] in a series of stabilizers for cellulose esters (substituted 2-hydroxybenzophenones) between the chemical shift of the signal from the protons of the hydroxyl group in the NMR spectrum of the stabilizer and the time for which a stabilized sample retained 75% of its strength under the action of ultraviolet radiation in a weatherometer. The authors consider that the efficiency of the stabilizer, like the chemical shift of the signal of the hydroxyl group, depends on the strength of the intramolecular hydrogen bond

where x and y are $- OH$, $- OCH_3$, $- CH_2C_6H_5$)

and proposed the use of the NMR method for preliminary assessment of the stabilizing power of newly synthesized samples instead of very long aging tests.

The NMR method has been used to elucidate the mechanism and kinetics of reactions of polymers in bulk and in solution. A study was made [125] of the cyclization of natural rubber in a decalin—phenol medium at 180°C with phosphorus pentoxide as the catalyst. Samples with different degrees of residual unsaturation

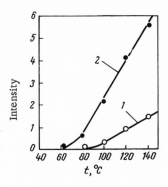

Fig. 149. Temperature dependence of the intensity (in arbitrary units) on the narrow component of the NMR line of a polyamide film [89]. 1) Initial film; 2) after thermo-oxidative breakdown for 2 h at 210°C.

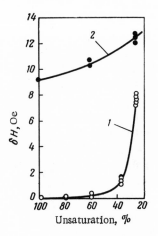

Fig. 150. Relation of the width of the line (δH) in the NMR spectrum of cyclized natural rubber to the residual unsaturation [125]. 1) At room temperature; 2) at −196°C.

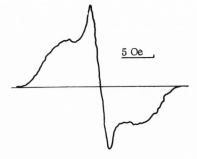

Fig. 151. NMR line of partly cyclized natural rubber with a residual unsaturation of 36% [125].

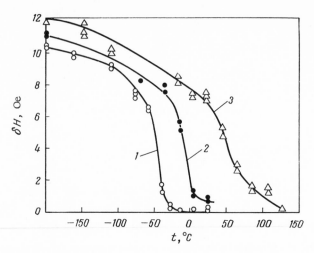

Fig. 152. Temperature dependence of the width of the line (δH) in the
NMR spectrum of cyclized rubber [125]. 1) Rubber with 60% unsaturation; 2,3) low- and high-molecular rubbers with 25% unsaturation.

(with different double bond contents) were withdrawn and the NMR
spectra plotted at temperatures from −196 to +120°C. The initial
natural rubber gave a narrow line (δH ≈ 0.09 Oe) at room temperature. In the initial stage of cyclization the form of the line did not
change and the width increased only insignificantly (Fig. 150), but
with a residual unsaturation less than 50% there was a sharp increase in the line width. In the early stages of the process, mono-
cyclic structures are evidently formed and these do not substantially decrease the mobility of the macromolecules. At a high degree of cyclization condensed polycyclic structures are formed
with a low mobility at room temperature. A sample with a residual unsaturation of 36% gave a two-component NMR line (Fig. 151).
The cyclized rubber remained amorphous to x-rays and the broad
component of the NMR line could be ascribed only to regions of the
polymer where cyclization had occurred to the greatest extent.

Figure 152 shows the change in the width of the lines with
temperature. In the low-temperature region there is a slow decrease in the width of the line due to the motion of methyl groups
and then the line narrows rapidly due to the beginning of motion of
methyl groups and then the line narrows rapidly due to the beginning of motion of segments of the macromolecules. The temperature at which rapid narrowing occurs is −63 to −43°C for partly

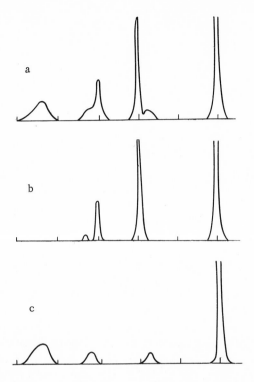

Fig. 153. NMR spectra of solutions in trifluoroacetic acid [33]. a) Poly-
benzylglutamate; b) benzyl ester of trifluoroacetic acid; c) polyglutamic
acid.

cyclized rubber with a degree of unsaturation of 60% and 37 to 77°C
for fully cyclized rubber (degree of unsaturation 25%). For low-
molecular cyclized rubber the curve of $\delta H = f(t)$ occupies an intermediate position. It is evident that in cyclization, as in vulcanization, the possibility of molecular motion of quite large sections of
the chains is retained, but because of the high activation energy the
mobility which can be discerned in the NMR spectrum is observed
only at high temperature. High-resolution NMR spectra [153, 317-
320] (together with infrared spectra) have also been used to elucidate the mechanism of cyclization, cyclohydrochlorination, and hy-
drochlorination of rubbers.

Abramova, Bazhenov, and Shul'gin [6, 8] successfully used
the NMR method to study structural changes in cellulose fibers in
technological processes.

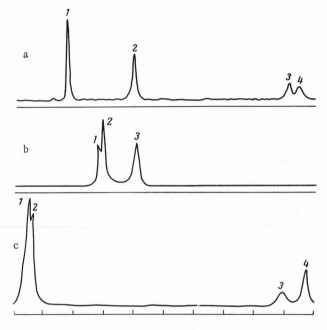

Fig. 154. NMR spectra of the ion-exchange resin sulfonated polystyrene-divinylbenzene [325]. a) Resin in H form, swollen in 5% aqueous solution of tert-butanol (peaks: 1-water in resin; 2-outer water; 3-outer butanol; 4-butanol in resin); b) resin in H form swollen in water (1,2-water in resin; 3-outer water); c) resin in $N(CH_3)_4^+$ form, swollen in an aqueous solution of $N(CH_3)_4Cl$ [1-outer water; 2-water in resin; 3-outer $N(CH_3)_4^+$; 4-$N(CH_3)_4^+$ in resin].

Analysis of high-resolution NMR spectra makes it possible to study chemical reactions in polymer solutions. Interesting results were obtained by Vol'kenshtein and his co-workers [33, 34] in a study of chemical reactions of poly-γ-benzyl-L-glutamate (PBG) with a trifluoroacetic acid in benzene. If trifluoroacetic acid was added to a benzene solution of PBG, at an acid concentration greater than 80% turbidity of the solution was observed. It was found that after the PBG molecules had changed from a rigid spiral form to statistically coiled balls, the PBG is hydrolyzed by traces of water:

$$\{-CO-CH[(CH_2)_2COOCH_2C_6H_5]-NH-\}_n + H_2O \rightarrow$$
$$\rightarrow \{-CO-CH[(CH_2)_2COOH]-NH-\}_n + C_6H_5CH_2OH$$

The poly-L-glutamic acid formed by hydrolysis remains in solution, while the benzyl alcohol and trifluoroacetic acid form an ester, which is suspended in the viscous solution

$$C_6H_5CH_2OH + CF_3COOH \rightarrow C_6H_5CH_2OCOCF_3 + H_2O$$

As Fig. 153 shows, the spectrum of the reaction products (a) is the sum of the separate NMR spectra of the benzyl ester of trifluoroacetic acid (b) and poly-L-glutamic acid (c), demonstrating the accuracy of the reaction mechanism proposed.

Valuable information may be obtained from the NMR spectra of ion-exchange resin—solvent systems [325, 326] (Fig. 154). From the magnitude of the chemical shifts, the width, and the intensity of the signals, it is possible to draw conclusions on the degree of mobility of counterions and water sorbed by the resin, to determine the concentrations of components in the inner and outer phases, to calculate the distribution coefficients between the resin and phases for a mixed solvent, the selectivity factors, etc.

Kotin and Nagasawa [421] estimated the degree of ionization from the NMR spectra of solutions of polystyrenesulfonic acid. It was shown that the "true" degree of ionization is close to unity, while the apparent value, determined by the potentiometric method is less than 0.5 due to the electrostatic interaction between the hydrogen ions and the polyions in solution.

The high-resolution NMR method has also been used to study proton exchange between polyacrylamide and water molecules in solution [203], the denaturation of proteins [450], the formation of peptide bonds [295] and the formation of hydrogen bonds [218] in polymer solutions, the photochlorination of polymethyl methacrylate [543], maleate—fumarate isomerization in unsaturated polyesters [259], etc.

Chapter VI

Use of the NMR Method
for Quantitative Analysis of Polymers

1. Determination of the Composition of Two-Phase Systems

In the analysis of two-phase systems by the NMR method, two cases may be ecnountered: 1) there are different nuclei in the two phases, e.g., ^{1}H in one phase and ^{19}F in the other; 2) the nuclei are the same in the two phases.

The first case is simplest. It is possible to obtain an NMR signal from each phase separately and the problem is reduced to determining the number of resonating nuclei N from the signal intensity. As was shown in Chapter I (see page 51), the integral intensity of an NMR absorption line is proportional to the number of nuclei. The proportionality coefficient could be determined by recording the NMR line of a standard sample under the same conditions. However, the accuracy of the determination would be low because of the effect of saturation, the effect of the properties of the sample on the sensitivity of the spectrometer (in work with an autodyne head), the effect of the form of the sample, and other sources of error. Better results are obtained in practice by calculation from a calibration curve of the relation of the signal amplitude to the concentration, which is constructed by plotting the spectra of a series of standard samples, close in composition to that investigated.

An ingenious method of quantitative analysis, the idea of which is similar to the method of accurate weighing on balances with unequal arms, was developed by Borodin and Skripov [25]. Two RF coils with samples are connected in series into the circuit of the generator of the instrument. One of the samples contains the substance investigated I and the other, a standard S_1. The ratio of the amplitudes of the signals A_I/A_{S_1} is determined. Then the substance investigated is replaced by a second standard S_2 with a known concentration C_{S_2} and the ratio of the amplitudes of the signals A_{S_1}/A_{S_2} is measured. The concentration in the sample investigated is determined from the formula:

$$C_I = C_{S\,2}\left(\frac{A_I}{A_{S_1}}\right)\left(\frac{A_{S_1}}{A_{S_2}}\right) \tag{VI-1}$$

The use of this method eliminates errors due to the nonuniformity of the field and a possible change in the operating conditions of the spectrometer. In the determination of hydrogen-containing impurities in compounds of the type $C_4F_6Cl_2$, the sensitivity achieved was 0.01 wt.% and the accuracy of the determination was about 10% of the value determined.

If the two phases contain identical nuclei, then the analysis may be carried out by means of the high-resolution NMR spectrum (see below), but for samples giving a broad line (polymers in bulk), the analysis is only possible when the line obtained has a complex form, which may be resolved into two components.

To determine the composition of the system, the complex NMR line of the system is divided into two components and the first moment of each of these is calculated and their ratio determined. The following assumptions are then made: 1) it is assumed that the NMR line of the system consists of the sum of the lines of the separate components; 2) it is assumed that the complex line may be divided into its two constituent components unequivocally.

An experimental check was made [111] of additivity during the recording of the line of two-phase systems and the conditions determining the possibility of dividing up the complex line were examined. Figure 155 shows that additivity is observed quite well in spectra plotted on a TsLA NMR spectrometer. The line of the composite sample practically coincides with the sum of the lines plotted separately. Figure 156 gives the typical forms of NMR lines encountered in two-phase systems. A line may be divided up correctly only when it has a form as in Fig. 156b, i.e., there are two maxima on the curve of the derived absorption function (lines of this form have been presented previously, page 163). Otherwise, it is necessary to use special methods (page 169), and the dividing of the line is arbitrary to some extent.

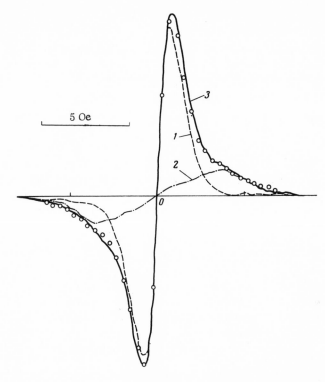

Fig. 155. Check on the additivity of the lines of fluorine nuclei in the NMR spectrum of a fluoro rubber and in the NMR spectrum of calcium fluoride [111]. 1) NMR line of fluoro rubber; 2) NMR line of calcium fluoride; 3) line of a sample of a mixture of the fluoro rubber and calcium fluoride. The points are the sum of curves 1 and 2.

The form of a complex line depends on the ratio of the widths of the lines of each phase $\delta H_{wide}/\delta H_{nar}$ and the ratio of the number of nuclei in the phase in which the mobility is greater (the line is narrower) to the total number of nuclei $N_{nar}/(N_{wide} + N_{nar})$. Curves 1-4 in Fig. 157 show the left-hand limits of the region where the parameters have values at which a derivative with two maxima is obtained for Gaussian and Lorentzian forms of the separate components. This region is also bounded on the right by curves 5 and 6. If the value of $\delta H_{wide}/\delta H_{nar}$ is greater than the value corresponding to line 5, then with a signal-to-noise ratio equal to 50 for the narrow component, the broad component is no longer visible due to the noise of the instrument, and a line is obtained as in Fig. 156d.

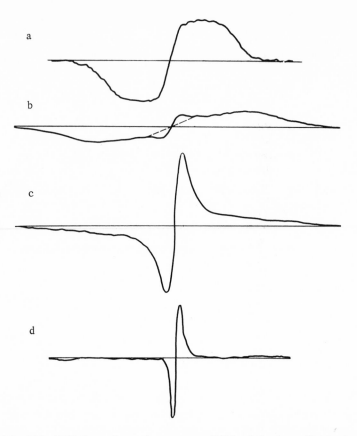

Fig. 156. NMR lines of two-phase systems [111]. a) Partly cyclized rubber at −19°C; b) Kapron with a moisture content of 1.7% at 20°C; c) polypropylene at 30°C; d) mixture of 350 mg polyethylene terephthalate and 12 mg natural rubber at 20°C.

An experiment on model systems confirmed that there is a limited region in which a curve of the derivative with two maxima is obtained (Fig. 158). If the form of the narrow component of the line is close to Gaussian, then the approximate limiting values are

$$(3-5) < \frac{\delta H_{\text{wide}}}{\delta H_{\text{nar}}} < \left(\frac{N_{\text{wide}}}{N_{\text{nar}}} \cdot \frac{[\text{signal}]}{[\text{noise}]} \right)^{1/2} \qquad (\text{VI-2})$$

A complex two-component structure is indicated by the appearance of four maxima on the curve of the second derivative of the absorption function. As was shown in our work [113], the re-

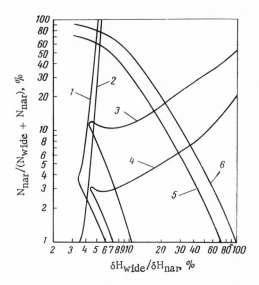

Fig. 157. Calculated limits of regions of parameters in which a complex NMR line with two maxima is obtained [111]. 1) Two Gaussian lines; 2) narrow line, Gaussian; broad line, Lorentzian; 3) narrow line, Lorentzian; broad line, Gaussian; 4) two Lorentzian lines; 5) signal-to-noise ratio, 50; 6) signal-to-noise ratio, 100.

gion characteristic of a two-component system is considerably greater when the second derivative is recorded than when the first derivative is used. For example, if for the Gaussian form of both components $\delta H_{wide}/\delta H_{nar} = 3$, then there will not be two maxima on the curve of the first derivative with any composition of the system, while on the curve of the second derivative there will be four maxima at a content of the mobile phase from 4 to 50%.

The form of the NMR line of the system polymer (solid phase)—low-molecular substance (liquid or gas) depends on the degree of mobility of the molecules of the low-molecular substance. Three possible variants may be selected arbitrarily:

1) a liquid or gas which is poorly bonded to the solid phase and completely retains its mobility;

2) the structure of the low-molecular substance is modified somewhat under the effect of the solid and its mobility is partly limited;

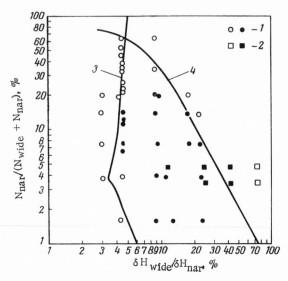

Fig. 158. Relation of the form of the NMR lines of model systems to composition and the ratio of the widths of the lines [111]. 1) NMR line of a mixture of fluoro rubber and calcium fluoride; 2) line from the H spectrum of a mixture of natural rubber and polyethylene terephthalate; 3) calculated limit for Gaussian lines; 4) calculated limit for a signal-to-noise ratio, 50; the black points show where lines with two maxima were obtained.

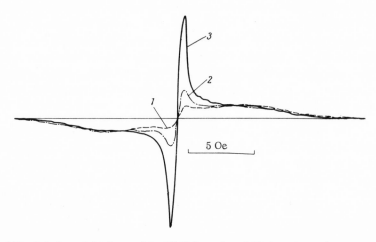

Fig. 159. Relation of the form of the NMR line of cord fiber to the moisture content [119]. 1) Moisture content 1.8%; 2) 2.8%; 3) 4.5%.

3) the liquid or gas is firmly bound to the solid.

All three variants are encountered in real systems and the NMR method makes it possible to obtain data on the composition and interaction of the phases in the system.

If the sample consists of a solid and a liquid weakly bound to it, as a rule the NMR absorption curve will have a form as in Fig. 156c or 156d. The amplitude of the signal is practically dependent only on the amount of the low-molecular-weight substance in the system and by using a calibration curve, it is possible to make a quantitative determination.

The NMR method has been used to determine the moisture content of plant products [32, 73, 278, 617, 672] and is particularly convenient for analyzing granulated materials [614]. The method of determining the moisture content from the NMR absorption curve has a series of advantages [225]: the determination takes no more than a few minutes, the sample is not harmed, and the method is universal. Simplified instruments have been described for determining moisture content by the NMR method [73, 590] and industrial NMR spectrometers are produced for the rapid determination of moisture content [591].

A method has been described [79, 119] for determining the moisture content of plastics, molding powders, and fillers from the amplitude of the NMR signal. As Fig. 159 shows, with an increase in the moisture content of a Kapron sample there is an increase in the narrow component of the line (signal of water protons) while the broad line remains practically unchanged (signal of polymer protons). The moisture content of Kapron in the range of 1-10% may be determined with an accuracy of about 0.5% by means of a calibration curve (Fig. 160).

Similar graphs were constructed for molding powders, wood flour, etc. With a high moisture content the amplitude of the signal is proportional to the water content of the sample. In the region of low moisture contents there is a deviation from linearity, which is explained by the interaction of the water with the solid. This interaction limits the mobility of the water molecules and leads to an increase in the width of the NMR line of water. This has been observed for many systems [161, 362, 387, 616].

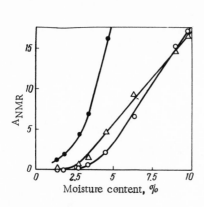

Fig. 160. Calibration curves for determining the moisture content from the NMR signal amplitude (the different curves refer to different recording conditions) [119].

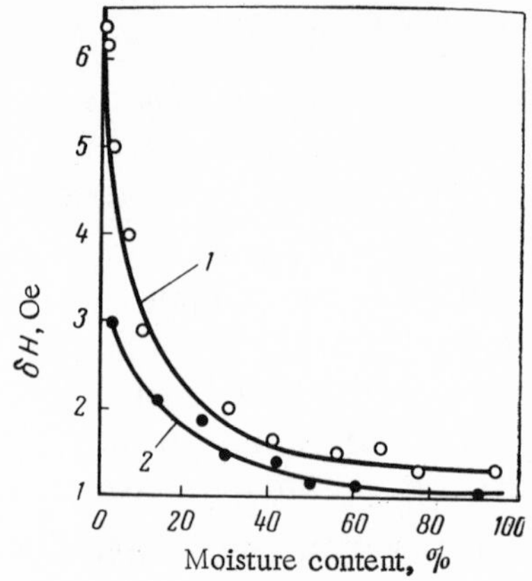

Fig. 161. Relation of the width of the proton line in the NMR spectrum of water adsorbed on cellulose (1) and charcoal (2) to the relative moisture content of the air [680].

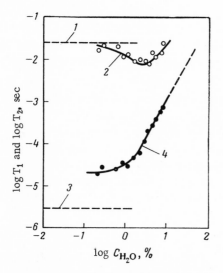

Fig. 162. Relation of the spin–lattice (T_1) and spin–spin (T_2) relaxation times to the composition of the system polyamide–water [315]. 1,2) T_1 for polyamide and for water; 3,4) T_2 for polyamide and for water.

When the molecules of the low-molecular substance are firmly bound to the polymer, for example adsorbed on its surface, and their mobility is markedly reduced, then there is a considerable increase in the width of the line. Figure 161 shows the relation of the width of the line of water adsorbed on charcoal and cellulose to the equilibrium moisture content of the air [680]. It is evident

that even with a relatively large amount of adsorption (40% moisture content the line still remains broad. The author compared
the form of the curve of δH against the moisture content with the
form of the adsorption isotherm and came to the conclusion that
polymolecular layers are formed on cellulose. Measurements of
the width and second moment of the NMR line of viscose fibers with
a change in the moisture content from 4 to 8%, made by Abramova,
Bazhenov, and Shul'gin [6], showed that a stable structure of cellulose with water molecules is formed in this range of water contents.

Additional information on the character of the interaction between a polymer and a low-molecular substance adsorbed on it may
be obtained by studying the relaxation times. The theory of NMR
in a system in which there is an adsorption equilibrium [183, 481,
748] shows that the relaxation time depends on the rate of exchange
between the free and adsorbed phases. With slow exchange there
remain two spin—lattice relaxation times T_{11} and T_{12}, corresponding to the two phases, and the relaxation process is described by
the sum of the two exponents.

With rapid exchange we obtain one relaxation time T_1 and

$$\frac{1}{T_1} = \frac{p_1}{T_{11}} + \frac{p_2}{T_{12}} \qquad \text{(VI-3)}$$

where p_1 and p_2 are the fractions of protons in phases 1 and 2.

Analogous relations hold for the spin—spin relaxation time.

From the measured values of T_1 and T_2 it is possible to estimate the lifetime of molecules in the adsorbed state. In the system cellulose—water, which was studied by the high-resolution
NMR method [519] and the spin-echo method [596], the water molecules were found to have a low mobility. The mean lifetime of a
water molecule adsorbed on cellulose is a few hundredths of a second. Analysis of the relation of T_1 and T_2 to the moisture content
of silk, wool, and nylon fibers [617] shows that water is adsorbed
not as separate molecules, but as groups of molecules at active
centers.

From the relation of T_1 and T_2 to the composition of the system polyamide—water [315] (Fig. 162), it was concluded that with
a water content up to 0.5% the H_2O molecules are firmly bound to
the sorbent. The H—H distance in the sorbed molecules is reduced

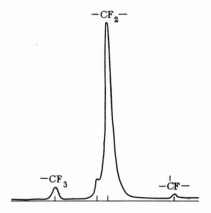

Fig. 163. NMR spectrum of a copolymer of tetrafluoroethylene with hexafluoropropylene (9 ± 1.5 mol.%) at 310°C [716].

to 1.3 Å in comparison with 1.5 Å for free water. The sorption centers are evidently nitrogen atoms. With an increase in the water content from 0.5 to 8% the mobility gradually increases and approaches the mobility of free water molecules.

2. Determination of the
Composition of Copolymers

The use of high-resolution NMR for analytical purposes offers great possibilities. This problem has been discussed in detail in a monograph [96] and reviews [94, 102, 122]; conditions affecting the accuracy of the results have been examined [395, 547, 712]. We will only examine one problem that is typical of polymer chemistry, namely, the determination of the composition of copolymers.

Figure 163 gives the spectrum of a copolymer of tetrafluoroethylene with hexafluoropropylene [716], recorded at 310°C, i.e., somewhat higher than the melting point of the crystallites. The ratio of the areas of the signals from the CF_3 and CF_2 groups corresponds to a content of 9 ± 1.5 mol.% of hexafluoropropylene. In Chapter III we gave the high-resolution NMR spectrum of an acetone solution of a copolymer of vinylidene fluoride with hexafluoropropylene (see page 197). The composition of the copolymer may be determined by two methods:

1) from the ratio of the intensities (I) of the signals from the CF_3 and CF_2 groups:

$$C_{VF} = \frac{3\Sigma I_{CF_2} - 2\Sigma I_{CF_3}}{3\Sigma I_{CF_3}}$$

where C_{VF} is the content of vinylidene fluoride, mol.%.

2) from the ratio of the intensities of the peaks b/c or h/c (where b is the peak corresponding to fluorine nuclei in the groups $C-CF_3^*$, c corresponds to the groups $CH_2CF_2^*CH_2CF_2^*$, and h to the groups $CF_2CF_2^*CF$). The accuracy of the determination is 1%.

Kubota and Takamura [425] developed a method of determining phenyl and methyl groups in methylphenylsiloxane polymers from the high-resolution NMR spectra of solutions of these polymers in dioxane. The solvent is simultaneously an internal standard for determining the intensity. The method gives results close to those obtained by analysis of the infrared spectra.

Methods of determining composition from high-resolution NMR spectra have also been developed for analyzing copolymers of vinyl chloride with vinylidene chloride [238] (see page 197), butadiene with isoprene [228] (see page 199), ethylene with vinyl acetate and with ethyl acrylate [230, 557], copolymers of diazo oxides with cyclic ethers [663], styrene with butadiene [154, 612], and with methyl methacrylate [399], formaldehyde with ethylene oxide [150], α-methylstyrene with styrene [213], isobutene with propylene [166], and vinyl acetate with esters of unsaturated acids [266], and for determining acids and glycols in polyester resins [548].

Bibliography

1. Abdrashitov, R. A., Bazhenov, N. M., Vol'kenshtein, M. V., Kol'tsov, A. I., and Khachaturov, A. S., Vysokomolek. soed., 5:405 (1963).
2. Abdrashitov, R. A., Bazhenov, N. M., Vol'kenshtein, M. V., Kol'tsov, A. I., and Khachaturov, A. S., Vysokomolek. soed., 6:1871 (1964).
3. Abdrashitov, R. A., Bazhenov, N. M., Vol'kenshtein, M. V., Kol'tsov, A. I., and Khachaturov, A. S., Abstracts of Proceedings of the Ninth Conference of the Institute of Macromolecular Compounds of the Academy of Sciences, USSR, Leningrad (1962), p. 34.
4. Abdrashitov, R. A., Bazhenov, N. M., Vol'kenshtein, M. V., Kol'tsov, A. I., and Khachaturov, A. S., Abstracts of Proceedings of the Tenth Scientific Conference of the Institute of Macromolecular Compounds of the Academy of Sciences, USSR, Leningrad (1963), p. 45.
5. Abraham, A., Nuclear Magnetism [Russian translation], Izdatinilit (1963).
6. Abramova, E. A., Bazhenov, N. M., and Shul'gin, E. I., Abstracts of Proceedings of Tenth Scientific Conference of the Institute of Macromolecular Compounds of the Academy of Sciences, USSR, Leningrad (1963), p. 8.
7. Abramova, E. A., Bazhenov, N. M., and Shul'gin, E. I., Khim. volokna, No. 2:33 (1962).
8. Abramova, E. A., Bazhenov, N. M., and Shul'gin, E. I., Khim. volokna, No. 4:51 (1964).
9. Aleksandrov, I. V., Theory of Nuclear Magnetic Resonance, Izd. Nauka (1964).
10. Aleksandrov, N. M., and Skripov, F. I., Usp. fiz. nauk, 75:585 (1961).
11. Bazhenov, N. M., Vol'kenshtein, M. V., Dolgopol'skii, I. M., Kol'tsov, A. I., and Khachaturov, A. S., Abstracts of Proceedings of the Ninth Conference of the Institute of Macromolecular Compounds of the Academy of Sciences, USSR, Leningrad (1962), p. 35.
12. Bazhenov, N. M., Vol'kenshtein, M. V., and Kol'tsov, A. I., Abstracts of Proceedings of the Tenth Scientific Conference of the Institute of Macromolecular Compounds of the Academy of Sciences, USSR, Leningrad (1963), p. 5.
13. Bazhenov, N. M., Vol'kenshtein, M. V., Kol'tsov, A. I., and Khachaturov, A.S., Vysokomolek. soed., 1:1048 (1959).
14. Bazhenov, N. M., Vol'kenshtein, M. V., Kol'tsov, A. I., and Khachaturov, A.S., Vysokomolek. soed., 3:290 (1961).
15. Bazhenov, N. M., Vol'kenshtein, M. V., and Khachaturov, A. S., Vysokomolek. soed., 5:1025 (1963).

16.   Bazhenov, N. M., Vol'kenshtein, M. V., Kol'tsov, A. I., and Khachaturov, A.S.,
       Abstracts of Proceedings of the Ninth Conference of the Institute of Macro-
       molecular Compounds of the Academy of Sciences, USSR, Leningrad (1962),
       p. 33.
17.   Baroni, E. E., Ksenofontov, V. A., Kucheryaev, A. G., Oliferchuk, N. L., and
       Shuander, Yu. A., Zh. strukt. khim., 4:459 (1963).
18.   Bartenev, G. M., and Luk'yanov, I. A., Plastmassy v mashinostr., No. 12:46
       (1963).
19.   Berlin, A. A., Usp. khim., 29:1189 (1960).
20.   Berlin, A. A., Progress in the Chemistry and Technology of Polymers, Collec-
       tion 2, Goskhimizdat (1957), p. 13.
21.   Bovey, F. A., Khim. i tekhnol. polim., No. 7:45 (1965).
22.   Bovey, F. A., and Tiers, G. V. D., Khim. i tekhnol. polim., No. 9:3 (1964).
23.   Borodin, P. M., Investigation of Chemical Shifts and Fine Structure of NMR
       Signals of Fluorine in a Series of Its Compounds, Candidate's Dissertation,
       LGU (1955).
24.   Borodin, P. M., Nikitin, M. K., and Sventitskii, E. I., Zh. strukt. khim.,
       6:188 (1965).
25.   Borodin, P. M., and Skripov, F. I., Zavod. lab., 29:164 (1963).
26.   Borodin, P. M., and Skripov, F. I., Izv. vyssh. uch. zav., Radiofizika, 1(4):69
       (1958).
27.   Borodin, P. M., and Skripov, F. I., Fiz. sbornik L'vovskogo universiteta,
       No. 3:78 (1957).
28.   Bresler, S. E., Kazbekov, E. I., Saminskii, E. M., and Sukhodolova, A. T.,
       Vysokomolek. soed., 4:419 (1962).
29.   Bresler, S. E., Pavlova, S. A., and Finogenov, P. L., Zh. tekhn. fiz., 21:1061
       (1951).
30.   Buishvili, L. L., and Kessenikh, A. V., Fiz. tverd. tela, 6:3016 (1964).
31.   Buryatina, A. S., Zalukaev, L.P. Pivnev, V. I., Shestakova, O. G., Korbanova,
       Z. N., and Reznikov, V. S., Abstracts of Proceedings of Conference on the
       Ageing and Stabilization of Polymers, Moscow (1964), p. 63.
32.   Vasil'ev, B. V., and Onuchin, V. Ya., Trudy Ural'skogo politekhnicheskogo
       instituta, Collection 111, Sverdlovsk (1961), p. 123.
33.   Vol'kenshtein, M. V., Kol'tsov, A. I., and Marshal', Zh., Vysokomolek. soed.,
       4:944 (1962).
34.   Vol'kenshtein, M. V., Kol'tsov, A. I., and Marshal', Zh., Abstracts of Proceed-
       ings of Ninth Scientific Conference of the Institute of Macromolecular Com-
       pounds of the Academy of Sciences, USSR, Leningrad (1962), p. 32.
35.   Vol'kenshtein, M. V., Kol'tsov, A. I., and Khachaturov, A. S., Vysokomolek.
       soed., 7:296 (1965).
36.   Vol'kenshtein, M. V., and Ptitsyn, O. B., in: Polymer Physics, Izdatinlit (1960).
37.   Vol'pin, M. E., Dulova, V. G., and Kursanov, D. N., Izv. Akad. Nauk SSSR,
       Otd. khim. nauk, No. 4:727 (1963).
38.   Glazkov, V. I., Vysokomolek. soed., 5:120 (1963).
39.   Glazkov, V. I., Dokl. Akad. Nauk SSSR, 142:387 (1962).

40. Dogadkin, B. A., Tarasova, Z. N., Kaplukov, M. L., Karpov, V. L., and Klauzen, N. A., Kolloidn. zh., 20 : 258 (1958).

41. D'yachkovskii, F. S., Yarovitskii, P. A., and Bystrov, V. F., Vysokomolek. soed., 6 : 659 (1964).

42. Ermilova, G. A., Urman, Ya. G., and Slonim, I. Ya., Plastmassy v mashinostr., No. 11 : 28 (1964).

43. Zhernovoi, A. I., and Latyshev, G. D., Nuclear Magnetic Resonance in a Flowing Liquid, Atomizdat (1964).

44. Zhidomirov, G. M., Tsvetkov, Yu. D., and Lebedev, Ya. S., Zh. strukt. khim., 2 : 696 (1961).

45. Zhurkov, S. N., and Egorov, E. A., Vysokomolek. soed., 5 : 772 (1963).

46. Zhurkov, S. N., and Egorov, E. A., Dokl. Akad. Nauk SSSR, 152 : 1155 (1963).

47. Zalukaev, L. P., Esina, T. I., Moiseev, V. V., and Pivnev, V. I., Abstracts of Proceedings of Conference on the Ageing and Stabilization of Polymers, Moscow (1964), p. 45.

48. Zalukaev, L. P., and Pivnev, V. I., Vestnik Sovnarkhoza, Voronezh, No. 6 : 40 (1962).

49. Zalukaev, L. P., and Pivnev, V. I., Vysokomolek. soed., 6 : 538 (1964).

50. Zalukaev, L. P., and Pivnev, V. I., Trudy laboratorii khimii vysokomolek. soed., Voronezhskii universitet, No. 1, Voronezh (1962), p. 61.

51. Zalukaev, L. P., and Pivnev, V. I., Abstracts of Proceedings of Conference on the Ageing and Stabilization of Polymers, Moscow (1964), p. 62.

52. Zalukaev, L. P., Pivnev, V. I., Reznikov, V. S. Shestakova, O. G., Korbanova, Z. I., and Buryagina, L. S., Kauchuk i rezina, No. 3 : 19 (1964).

53. Zelenev, Yu. V., and Bartenev, G. M., Vysokomolek. soed., 6 : 915 (1964).

54. Zelenev, Yu. V., and Bartenev, G. M., Vysokomolek. soed., 6 : 1047 (1964).

55. Kazaryan, L. G., and Urman, Ya. G., Zh. strukt. khim., 5 : 534 (1964).

56. Kargin, V. A., and Slonimskii, G. L., Zh. fiz. khim., 23 : 563 (1949).

57. Karpov, V. L., Pomerantsev, N. M., and Sergeev, N. M., Vysokomolek. soed., 5 : 100 (1963).

58. Karpov, V. L., Sergeev, N. M., and Yurkevich, V. G., Dokl. Akad. Nauk SSSR, 152 : 655 (1963).

59. Kennedy, Minkler, Wanless, and Thomas, Khim. i tekhnol. polim., No. 2 : 113 (1964).

60. Kessenikh, A. V., Nuclear Magnetic Resonance, Izd. Znanie (1965).

61. Kessenikh, A. V., Lushchikov, V. I., Manenkov, A. A., and Taran, Yu. V., Fiz. tverd. tela, 5 : 443 (1963).

62. Kitaigorodskii, A. I., Dokl. Akad. Nauk SSSR, 124 : 861 (1959).

63. Kitaigorodskii, A. I., X-Ray Diffraction Analysis of Finely Crystalline and Amorphous Substances, Gostekhteoretizdat (1952), p. 10.

64. Kitaigorodskii, A. I., Progress in the Chemistry and Technology of Polymers, Collection 2, Goskhimizdat (1957), p. 191.

65. Kovarskaya, B. M., Zhigunova, I. E., Slonim, I. Ya, Urman, Ya. G., and Neiman, M. B., Chemical Properties and the Modification of Polymers, Collection, Izd. Nauka (1964), p. 33.

66.   Kol'tsov, A. I., and Vol'kenshtein, M. V., Vysokomolek. soed., 7:250 (1965).
67.   Kopfermann, H., Nuclear Moments, Academic Press (1958).
68.   Korshak, V. V., and Sosin, S. L., Vysokomolek. soed., 7:232 (1965).
69.   Korshak, V. V., Sosin, S. L., and Vasnev, V. A., Dokl. Akad. Nauk SSSR,
       152:872 (1963).
70.   Kostochko, A. V., Prokop'ev, V. P., Shreibert, A. I., and Tishkov, P. G., Tr.
       Kazansk. khim.-tekhnol. inst., No. 33:219 (1964).
71.   Kocharyan, N. M., Pikalov, A. P., Kagramanyan, A. V., and Markosyan, E. A.,
       Dokl. Akad. Nauk Arm.SSR, 40:25 (1965).
72.   Kropacheva, E. N., Ermakova, I. N., Nel'son, K. V., and Dolgoplosk, B. A.,
       Internat. Symp. Macromolec. Chem., Prague, 1965, Abstract A627, Preprint
       P408.
73.   Kuznetsov, S. V., Vestn. c/kh nauk, 6:117 (1961).
74.   Lando, Khim. i tekhnol. polim., No. 2:124 (1964).
75.   Lösche, A., Izv. Akad. Nauk SSSR, Ser. fiz., 21:1064 (1957).
76.   Lösche, A., Nuclear Induction [Russian translation], Izdatinlit (1963).
77.   Lobkov, V. D., Klebanskii, A. L., and Kogan, E. V. Vysokomolek. soed., 7:290
       (1965).
78.   Lushchikov, V. I., Manenkov, A. A., and Taran, Yu. V., Fiz. tverd. tela,
       3:3503 (1961).
79.   Lyubimov, A. N., Varenik, A. F., and Slonim, I. Ya., Zavodsk. lab., 29:991
       (1962).
80.   Lyubimov, A. N., Novikov, A. S., Galil-Ogly, F. A., Gribacheva, A. V., and
       Varenik, A. F., Vysokomolek. soed., 3:1511 (1961).
81.   Lyubimov, A. N., Novikov, A. S., Galil-Ogly, F. A., Gribacheva, A. V., and
       Varenik, A. F., Vysokomolek. soed., 5:687 (1963).
82.   Magnetic Resonance and Its Application, Izd. Ural'skogo Politekhn. Inst.,
       Sverdlovsk (1961).
83.   Maklakov, A. I., Voskresenskii, V. A., Khienkina, B. D., and Egorova, L. Ya.,
       Vysokomolek. soed., 6:923 (1964).
84.   Maklakov, A. I., and Pimenov, G. G., Vysokomolek. soed., 7:536 (1965).
85.   Maklakov, A. I., and Pimenov, G. G., Dokl. Akad. Nauk SSSR, 157:1413 (1964).
86.   Maklakov, A. I., Pimenev, G. G., and Sheplev, V. I., Vysokomolek. soed.,
       7:1894 (1965).
87.   International Symposium on Macromolecular Chemistry, Moscow (1960),
       Section 3.
88.   Milinchuk, V. K., Pshezhetskii, S. Ya., Kotov, A. G., Tupikov, V. I., and
       Tsivenko, V. I., Vysokomolek. soed., 5:71 (1963).
89.   Mochalova, O. A., Diploma work, MITKhT (1962).
90.   Neiman, M. B., Likhtenshtein, G. I., Konstantinov, Yu. S., Karpets, N. P.,
       and Urman, Ya. G., Vysokomolek. soed., 5:1706 (1963).
91.   Nefedov, O. M., Kolesnikov, S. P., Khachaturov, A. S., and Petrov, A. D.,
       Dokl. Akad. Nauk SSSR, 154:1389 (1964).
92.   Novikov, A. S., Galil-Ogly, F. A., and Shashkov, A. S., Internat. Symp.
       Macromolec. Chem., Prague, 1965, Abstract A709.
93.   Paramagnetic Resonance, Izd. Kazanskogo Universiteta, Kazan (1960).

94.   Pomerantsev, N. M., Zavod. lab., 28:167 (1962).

95.   Pomerantsev, N. M., Khramchenkov, V. A., Sumin, L. V., and Zimin, A. V.,
      Dokl. Akad. Nauk SSSR, 137:1153 (1961).

96.   Pople, J., Schneider, W., and Bernstein, H., High-Resolution Nuclear Magnetic
      Resonance, McGraw-Hill, New York (1959).

97.   Prittschek, A., Chim. i tekhnol. polim., No. 3:3 (1959).

98.   Roberts, J., Introduction to the Analysis of Spin–Spin Splitting in High-
      Resolution Nuclear Magnetic Resonance Spectra [Russian translation],
      Izdatinlit (1963).

99.   Roberts, J., Nuclear Magnetic Resonance: Applications to Organic Chemistry,
      McGraw-Hill, New York (1959).

100. Romanov, B. S., Program of the All-Union Conference on the Use of Radio-
      spectroscopy in Chemistry, Moscow (1965), p. 8.

101. Rowland, T. J., NMR in Metals, Pergamon, New York (1962).

102. Samitov, Yu. Yu., Zavod. lab., 26:950 (1960).

103. Sergeev, N. M., Optika i spektroskopiya, 17:784 (1964).

104. Sergeev, N. M., and Karpov, V. L., Vysokomolek. soed., 6:310 (1964).

105. Sergeev, N. M., and Karpov, V. L., Zh. strukt. khim., 5:230 (1964).

106. Sergeev, N. M., and Karpov, V. L., Fiz. tverd. tela, 6:2179 (1964).

107. Skripov, F. I., Course of Lectures on Radiospectroscopy, Izd. LGU (1964).

108. Slichter, C. P., in: Polymer Physics [Russian translation], Izdatinlit (1960),
      p. 171.

109. Slichter, C. P., Khim. i tekhnol. polim., No. 5:50 (1960).

110. Slonim, I. Ya., Vysokomolek. soed., 6:1371 (1964).

111. Slonim, I. Ya., Vysokomolek. soed., 6:1379 (1964).

112. Slonim, I. Ya., Usp. khim., 31:609 (1962).

113. Slonim, I. Ya., Lyubimov, A. N., Urman, Ya. G., Konovalov, A. G., and
      Varenik, A. F., Vysokomolek. soed., 7:245 (1965).

114. Slonim, I. Ya., and Urman, Ya. G., Zh. strukt. khim., 4:216 (1963).

115. Slonim, I. Ya., and Urman, Ya. G., Abstracts of Proceedings of Conference on
      The Ageing and Stabilization of Polymers, Moscow (1964), p. 61.

116. Slonim, I. Ya., Urman, Ya. G., Vonsyatskii, V. A., Liogon'kii, B. I., and
      Berlin, A. A., Dokl. Akad. Nauk SSSR, 154:914 (1964).

117. Slonim, I. Ya., Urman, Ya. G., and Ermolaev, A. D., Zh. strukt. khim.,
      6:531 (1965).

118. Slonim, I. Ya, Urman, Ya. G., Ermolaev, A. D., and Akutin, M. S., Zh.
      strukt. khim., 6:192 (1965).

119. Slonim, I. Ya., Urman, Ya. G., and Konovalov, A. G., Plastich. massy,
      No. 5:58 (1963).

120. Slonim, I. Ya., Urman, Ya. G., Konovalov, A. G., Heterochain High-
      Molecular Compounds, Collection, Izd. Nauka (1964), pp. 209-212.

121. Slonim, I. Ya., and Chmutov, K. V., Zh. fiz. khim., 25:296 (1951).

122. Stepanov, A. P., Trudy Ural'skogo politekhn. inst., Collection 111, Sverdlovsk
      (1961), p. 130.

123. Suprun, A. P., and Soboleva, T. A., Vysokomolek. soed., 6:1128 (1964).

124. Tikhomirova, N. N., and Voevodskii, V. V., Optika i spektroskopiya, 7:829 (1959).

125. Tutorskii, I. A., Slonim, I. Ya., Urman, Ya. G., Kudryavtseva, E. P., and Dogadkin, B. A., Dokl. Akad. Nauk SSSR, 152:674 (1963).

126. Uo, D. S., and Fedin, E. I., Fiz. tverd. tela, 4:2233 (1962).

127. Wall, L. A., and Florin, R. E., in: Analytical Chemistry of Polymers, Vol. 12 (G. Kline, ed.), Wiley, New York (1962).

128. Urazovskii, S. S., and Ezhik, I. I., Vysokomolek. soed., 3:150 (1961).

129. Urazovskii, S. S., and Ezhik, I. I., Ukr. khim. Zh., 29:329 (1962).

130. Urman, Ya. G., Slonim, I. Ya., and Ermolaev, A. D., Vysokomolek. soed., 6:2107 (1964).

131. Urman, Ya. G., Slonim, I. Ya., and Ermolaev, A. D., Vysokomolek. soed., 8:251 (1966).

132. Urman, Ya. G., Slonim, I. Ya., and Konovalov, A. G., Vysokomolek. soed., 6:1651 (1964).

133. Urman, Ya. G., Slonim, I. Ya., and Konovalov, A. G., in: Heterochain High-Molecular Compounds, Izd. Nauka (1964), pp. 227-232.

134. Khazanovich, T. N., Vysokomolek. soed., 5:112 (1963).

135. Khachaturov, A. S., Zavod. lab., 31:948 (1965).

136. Khayasi and Okamura, Khim. i tekhnol. polim., No. 4:89 (1964).

137. Khutsishvilli, G. R., Fiz. tverd. tela, 5:2713 (1963).

138. Tsvankin, D. Ya., Vysokomolek. soed., 5:129 (1963).

139. Tsvankin, D. Ya., and Fedin, E. I., Zh. strukt. khim., 3:101 (1962).

140. Chernitsyn, A. I., Maklakov, A. I., Voskresenskii, V. A., and Orlova, E. M., Vysokomolek. soed., 6:2185 (1964).

141. Chmutov, K. V., and Slonim, I. Ya., Dokl. Akad. Nauk SSSR, 69:223 (1949).

142. Chuvaev, V. F., Ivanova, L. V., and Zubov, P. I., Vysokomolek. soed., 6:1501 (1964).

143. Shepelev, V. I., and Maklakov, A. I., Zh. strukt. khim., 6:298 (1965).

144. Shigorin, D. N., Pomerantsev, N. M., and Sumin, L. V., Vysokomolek. soed., 3:260 (1961).

145. Andrew, E. R., Nuclear Magnetic Resonance, Cambridge Univ. Press (1956).

146. NMR and EPR Spectroscopy, Izd. Mir (1964).

147. Abdrashitov, R., Bazhenov, N., Khatchaturov, A., Kolsov, A., and Vol'kenstein, M., Z. Phys. Chem., 220:413 (1962).

148. Abraham, M., McCausland, M.A. U., and Robinson, F. N. H., Phys. Rev. Letters, 2:449 (1959).

149. Allen, G., Connor, T. M., and Pursey, H., Trans. Faraday Soc., 59:1525 (1963).

150. Allen, G., Warren, R., and Taylor, K.J., Chem.and Ind., No. 15:623 (1964).

151. Alpert, N. L., Phys. Rev., 72:637 (1947).

152. Alpert, N. L., Phys. Rev., 75:398 (1949).

153. Angelo, R. J., Chem. Eng. News, 41:43 (1963).

154. Angelo, R. J., Ikeda, R. M., and Wallach, M. L., Polymer, 6:141 (1965).

155. Aoki, K., Satoh, S., and Chujo, R., Nippon butsuri gakkai. Dai 18 kai nénkai koén ekosyu, No. 4:536 (1963).

156. Ashikari, N., Kinematsu, T., Yamagisawa, K., Nakagawa, K., Okamoto, H., Kobayachi, S., and Nishioka, A., J. Polymer Sci., A2:3009 (1964).
157. Assioma, F., Marschal, J., Schue, F., and Maillard, A., Internat. Symp. Makromolec. Chem., Prague, 1965, Abstract A202.
158. Aufdermarsh, C. A., and Pariser, R., J. Polymer Sci., A2:4727 (1964).
159. Bacskai, R., J. Polymer Sci., A1:2777 (1963).
160. Bacskai, R., and Lapporte, S. J., J. Polymer Sci., A1:2225 (1963).
161. Balazs, E. A., Bothner-By, A. A., and Gergely, J., J. Molec. Biol., 1:147 (1959).
162. Bamford, C. H., Bibby, A., and Eastmond, G. C., Internat. Symp. Makromolec. Chem., Prague, 1965, Abstract A360, Preprint P291.
163. Banas, E. M., Mrowca, B. A., and Guth, E., Phys. Rev., 98:265 (1955).
164. Banas, E. M., Mrowca, B. A., and Guth, E., Phys. Rev., 98:1548 (1955).
165. Bargon, J., Hellwege, K. H., and Johnsen, U., Makromolek. Chem., 85:291 (1965).
166. Barral, E. M., Porter, R. S., and Johnson, J. F., J. Chromatogr., 11:177 (1963).
167. Barrante, J. R., and Rochow, E. G., J. Organometal. Chem., 1:273 (1963).
168. Bartz, K. W., and Chamberlain, N. F., Analyt. Chem., 36:2151 (1964).
169. Bateman, I., Richards, R. E., Farrow, G., and Ward, I., Polymer, 1:63 (1960).
170. Bedford, G. R., and Katritzky, A. R., Nature, 200:652 (1963).
171. Benedek, G. B., Magnetic Resonance at High Pressure, Interscience, New York (1963).
172.. Beredjik, N., J. Appl. Polymer Sci., 9:439 (1965).
173. Berendsen, H. J. C., J. Chem. Phys., 36:3297 (1962).
174. Bhacca, N. S., Johnson, L. F., and Shoolery, J. M., High Resolution NMR Spectra Catalog, Varian Assoc., Palo Alto (1962).
175. Bhacca, N. S., Hollas, D. T., Johnson, L. F., and Dear, E. A., NRM Spectra Catalog, Vol. 2, National Press, Palo Alto (1963).
176. Bhacca, N. S., and Williams, D. H., Applications of NMR Spectroscopy in Organic Chemistry: Illustrations from Steroid Field, Holden Day, San Francisco (1964).
177. Bible, R. H., Interpretation of NMR Spectra: An Empirical Approach, Plenum Press, New York (1965).
178. Bloembergen, N., Nuclear Magnetic Relaxation, Benjamin, New York (1961).
179. Bloembergen, N., Purcell, E. M., and Pound, R. V., Phys. Rev., 73:679 (1948).
180. Böckman, O. Ch., J. Polymer Sci., A3:3399 (1965).
181. Bodor, G., Grell, M., and Radics, L., Internat. Symp. Macromolec. Chem., Prague, 1965, Abstract A611.
182. Bonera, G., Chierico, A., and Rigamonti, A., J. Polymer Sci., A2:2963 (1964).
183. Bonera, G., Chiodi, L., Lanzi, G., and Rigamonti, A., Nuovo Cimento, 19:234 (1961).
184. Bonera, G., and Guilotto, J., Nuovo Cimento, 14:435 (1959).
185. Bonera, G., De Stefano, P., Mascheretti, P., and Rigamonti, A., Magnetic and Electric Resonance and Relaxation, Amsterdam (1963), p. 220.
186. Bonera, G., De Stefano, P., Mascheretti, P., and Rigamonti, A., Proc. Colloq. AMPERE, 11:220 (1963).

187.   Bonera, G., De Stefano, P., and Rigamonti, A., Arch. Sci., 14, fasc. spec., 375 (1961).

188.   Bonera, G., De Stefano, P., and Rigamonti, A., J. Chem. Phys., 37:1226 (1962).

189.   Bonera, G., De Stefano, P., and Rigamonti, A., Nuovo Cimento, 22:847 (1961).

190.   Borghini, M., and Abraham, A., Compt. Rend., 248:1803 (1959).

191.   Borsa, F., and Lanzi, G., J. Polymer Sci., A2:2623 (1964).

192.   Bovey, F. A., J. Polymer Sci., 46:59 (1960).

193.   Bovey, F. A., J. Polymer Sci., 47:480 (1960).

194.   Bovey, F. A., J. Polymer Sci., 62:197 (1962).

195.   Bovey, F. A., J. Polymer Sci., A1:843 (1963).

196.   Bovey, F. A., Nature, 192:324 (1961).

197.   Bovey, F. A., Anderson, E. W., Douglas, D. C., and Manson, J. A., J. Chem. Phys., 39:1199 (1963).

198.   Bovey, F. A., Hood, F. P., Anderson, E. W., and Snyder, L. C., J. Chem. Phys., 42:3900 (1965).

199.   Bovey, F. A., and Tiers, G. V. D., Chem. and Ind., No. 42:1826 (1962).

200.   Bovey, F. A., and Tiers, G. V. D., Fortschr. Hochpolymer. Forsch., 3:139 (1963).

201.   Bovey, F. A., and Tiers, G. V. D., J. Am. Chem. Soc., 81:2870 (1959).

202.   Bovey, F. A., and Tiers, G. V. D., J. Polymer Sci., 44:173 (1960).

203.   Bovey, F. A., and Tiers, G. V. D., J. Polymer Sci., A1:849 (1963).

204.   Bovey, F. A., Tiers, G. V. D., and Filipovich, G., J. Polymer Sci., 38:73 (1959).

205.   Boye, C. A., and Goodlett, V. W., J. Appl. Phys., 34:59 (1963).

206.   Brady, G. W., and Salovey, R., J. Am. Chem. Soc., 86:3499 (1964).

207.   Brame, E. G., Sudol, R. S., and Vogl, O., J. Polymer Sci., A2:5337 (1964).

208.   Brandrup, J., and Goodman, M., Am. Chem. Soc. Polymer Preprints, 5:1119 (1964).

209.   Brandrup, J., and Goodman, M., J. Polymer Sci., B2:123 (1964).

210.   Braun, D., Herner, M., Johnsen, U., and Kern, W., Makromolek. Chem., 51:15 (1962).

211.   Braun, D., and Heufer, G., Makromolek. Chem., 80:98 (1964).

212.   Braun, D., Heufer, G., Johnsen, U., and Kolbe, K., Ber. Bunsenges. phys. Chem., 68:959 (1964).

213.   Braun, D., Heufer, G., Johnsen, U., and Kolbe, K., Kolloid. Z., 195:134 (1964).

214.   Brownstein, S., Bywater, S., and Worsfold, D.J., J. Phys. Chem., 66:2067 (1962).

215.   Brownstein, S., Bywater, S., and Worsfold, D.J., Makromolek. Chem., 48:127 (1961).

216.   Brownstein, S., and Willes, D. M., J. Polymer Sci., A2:1901 (1964).

217.   Bruce, J. M., and Farren, D. W., Polymer, 6:509 (1965).

218.   Buc, H., Martin, G., Mavel, G., and Neel, J., J. chim. phys. phys.-chim. biol., 59:284 (1962).

219.   Burget, J., Odehnal, M., Petricek, V., and Šacha, J., Arch. Sci., 14, fasc. spec., 487 (1961).

220.   Burget, J., and Šacha, J., Českosl. časop. fys., 9:749 (1959).

221. Carazzolo, G., Leghissa, S., and Mammi, M., Makromolek. Chem., 60:171 (1963).
222. Carrelli, A., and Ragozzino, R., Nuovo Cimento, 35: 731 (1965).
223. Chan, K. S., Ranby, G., Brumberger, H., and Odajima, A., J. Polymer Sci., 61(S29) (1962).
224. Chandet, J. H., Newland, G. C., Patton, H. W., and Tamblyn, J. W., SPE Trans., 1:26 (1961).
225. Chandet, J. H., and Tamblyn, J. W., SPE Trans., 1:57 (1961).
226. Chappel, S. E., Sauer, J. A., and Woodward, A. E., J. Polymer Sci., A1:2805 (1963).
227. Chem. Eng. News., 41:36 (1963).
228. Chen Hung Yu, Analyt. Chem., 34:1134 (1962).
229. Chen Hung Yu, Analyt. Chem., 34:1793 (1962).
230. Chen Hung Yu, and Lewis, M. E., Analyt. Chem., 36:1394 (1964).
231. Chien, J. C. W., and Walker, J. F., J. Polymer Sci., 45:239 (1960).
232. Chudesima, M., Kakigzahi, M., Nippon butsuri gakkai. Dai 18 kai nénkai ekosyu, No. 4: 543 (1963).
233. Chujo, R., J. Phys. Soc. Japan, 18:124 (1963).
234. Chujo, R., Busséiron kénkyu, 5:520 (1959).
235. Chujo, R., Nippon gomu kekatsi, 31:430 (1958); RZhKhim., No. 1:3442 (1960).
236. Chujo, R., Aoki, K., Satoh, S., and Nagai, E., J. Polymer Sci., B1:501 (1963).
237. Chujo, R., Satoh, S., and Nagai, E., J. Polymer Sci., A2:895 (1964).
238. Chujo, R., Satoh, S., Ozeki, T., and Nagai, E., J. Polymer Sci., 61:S12 (1962).
239. Chujo, R., and Sudzuki, T., Busséiron kénkyu, 10:159 (1961).
240. Clark, H. G., Makromolek. Chem., 63:69 (1963).
241. Clark, H. G., Makromolek. Chem., 86:107 (1965).
242. Cohen-Hardia, A., and Gabillar, R., Compt. Rend., 234:1877 (1952).
243. Coleman, B. D., and Fox, T. G., J. Polymer Sci., A1:3183 (1963).
244. Coleman, B. D., and Fox, T. G., J. Polymer Sci., C(4):I, 345 (1963).
245. Collins, R. L., Bull. Am. Phys. Soc., 1:216 (1956).
246. Collins, R. L., J. Polymer Sci., 27:67 (1958).
247. Collins, R. L., J. Polymer Sci., 27:75 (1958).
248. Collison, R., and McDonald, M. P., Nature, 186:548 (1960).
249. Connor, T. M., Polymer, 5:265 (1964).
250. Connor, T. M., Trans. Faraday Soc., 60:1574 (1964).
251. Connor, T. M., and Blears, D. J., Polymer, 6:385 (1965).
252. Connor, T. M., Blears, D. J., and Allen, G., Trans. Faraday Soc., 61:1097 (1965).
253. Connor, T. M., and McLaucklan, K. A., J. Phys. Chem., 69:1888 (1965).
254. Connor, T. M., Read, B. E., and Williams, G., J. Appl. Chem., 14:74 (1964).
255. Conway, T. F., Cohee, R. F., and Smith, R. J., Manuf. Confectioner, 37(5):27 (1957).
256. Cottam, B. J., Wiles, D. M., and Bywater, S., Canad. J. Chem., 41:1905 (1963).
257. Curry, J. W., J. Am. Chem. Soc., 78:1686 (1956).
258. Curry, J. W., J. Org. Chem., 26:1308 (1961).

259. Curtis, L. G., Edwards, D. L., Simons, R. M., Trent, P. J., and Von-Bramer, P. T., Ind. Eng. Chem., Proc. Res. Develop., 3 : 218 (1964).

260. Danjard, J. C., Rev. gen. caoutch., 35 : 51 (1958).

261. Danno, A., and Hayakawa, N., Bull. Chem. Soc. Japan, 35 : 1748 (1962). A261. Tanaka, N., Enka biniru to porima, 6 : 8 (1966).

262. Daskiewich, O. K., Henne, J. W., Lubas, B., and Szczepkowski, T. W., Nature, 200 : 1006 (1963).

263. Daubeny, R. P., Bunn, C. W., and Brown, C. J., Proc. Roy. Soc., A226 : 531 (1954).

264. David, C., Van der Parren, J., Provoost, F., and Ligotti, A., Polymer, 4 : 341 (1963).

265. Demarquay, J., Pham Quand Tho, M. Guyot de la Handrouyere, A. Guyot, Pretre M., Compt. Rend., 259 : 3509 (1964).

266. Dietrich, M. W., and Keller, R. E., Analyt. Chem., 36 : 2174 (1964).

267. Dinius, R. H., Emerson, M. T., and Choppin, G. R., J. Phys. Chem., 67 : 1178 (1963).

268. Dismore, P. F., and Statton, W. O., J. Polymer Sci., B2 : 1113 (1964).

269. Doskočilova, D., Faserforsch. Textiltechn., 15 : 610 (1964).

270. Doskočilova, D., J. Polymer Sci., B2 : 421 (1964).

271. Doskočilova, D., and Schneider, B., Coll. Czech. Chem. Comm., 29 : 2290 (1964).

272. Doskočilova, D., Štokr, J., Schneider, B., Pivkova, M., Kolinsky, M., Petranek, J., and Lim, D., Internat. Symp. Macromolec. Chem., Prague, 1965, Abstract A14.

273. Doskočilova, D., Štockr, J., Votavovas, Schneider, B., and Lim, D., Internat. Symp. Makromolec. Chem., Prague, 1965, Abstract A13.

274. Dubertaki, A. J., and Milles, C. M., Analyt. Chem., 37 : 1231 (1965).

275. Dyer, E., and Hammond, R. J., J. Polymer Sci., A2 : 1 (1964).

276. Eby, R. K., and Sinnot, K. M., J. Appl. Phys., 32 : 1765 (1961).

277. Edwards, W. R., and Chamberlain, N. F., J. Polymer Sci., 1A : 2299 (1963).

278. Elsken, R. H., and Kunsman, C. H., J. Assoc. Offic. Agr. Chem., 39 : 434 (1956).

279. Farrow, G., McIntosh, J., and Ward, I. M., Makromolek. Chem., 38 : 147 (1960).

280. Farrow, G., McIntosh, J., and Ward, I. M., Reol. Abs., 2 : 26 (1959).

281. Farrow, G., McIntosh, J., and Ward, I. M., Symposium über Makromolekule in Wiesbaden, 1959, Vortrag I-A-6.

282. Farrow, G., and Ward, I. M., Brit. J. Appl. Phys., 11 : 543 (1960).

283. Ferguson, R. C., J. Am. Chem. Soc., 82 : 2416 (1960).

284. Ferguson, R. C., J. Polymer Sci., A2 : 4735 (1964).

285. Ferroni, E., Atti Accad. naz. hincei, Rend. Cl. Sci. fis. mat. e. natur., 26 : 774 (1959).

286. Ferroni, E., and Gabrielli, G., J. Polymer Sci., B2 : 51 (1964).

287. Filipovich, G., J. Polymer Sci., 60 : S37 (1962).

288. Filipovich, G., J. Polymer Sci., A1 : 2279 (1963).

289. Fischer, E. W., Faserforsch. Textiltechn., 15 : 565 (1964).

290. Fischer, E. W., and Peterlin, A., Makromolek. Chem., 74 : 1 (1964).

291. Fluck, E., Die kernmagnetische Resonanz und ihre Anwendung in der anorganischen Chemie, Springer, Berlin (1963).
292. Formula Index to NMR Literature Data, Vol. 1 (J. S. Webb, G. Holwell, and A. Kende, eds.), Plenum Press, New York (1964).
293. Fox, T. G., and Schneko, H. W., Polymer, 3 : 575 (1962).
294. Francis, S. A., and Archer, E. D., Analyt. Chem., 35 : 1363 (1963).
295. Franconi, C., Z. Elektrochem., 65 : 645 (1961).
296. Frigge, K., Z. phys. Chem. (DDR), 224 : 430 (1963).
297. Fujii, K., Fujiwara, Y., and Fujiwara, S., Makromolek. Chem., 89 : 278 (1965).
298. Fujiwara, S., J. Polymer Sci., 44 : 93 (1960).
299. Fujiwara, S., J. Polymer Sci., 51 : S15 (1961).
300. Fujiwara, S., Amamiia, A., and Shinohara, K., J. Chem. Phys., 26 : 1343 (1957).
301. Fujiwara, S., Hayashi, S., and Hattory, G., Kôguô Kagaku, Zasshi, 59 : 803 (1956); Chem. Abs., 52 : 6835 (1958).
302. Fujiwara, S., and Nakajima, M., Bull. Chem. Soc. Japan, 33 : 1615 (1960).
303. Fujiwara, S., and Narasaki, H., J. Polymer Sci., B1 : 139 (1963).
304. Furukawa, J., Iseda, Y., Saegusa, T., and Fujii, H., Makromolek. Chem., 89 : 263 (1965).
305. Fuschillo, N., Rhian, E., and Sauer, J. A., J. Polymer Sci., 25 : 381 (1957).
306. Fuschillo, N., and Sauer, J. A., Bull. Am. Phys. Soc., 2 : 125 (1957).
307. Fuschillo, N., and Sauer, J. A., Bull. Am. Phys. Soc., 4 : 187 (1959).
308. Fuschillo, N., and Sauer, J. A., J. Appl. Phys., 28 : 1073 (1957).
309. Fuschillo, N., and Sauer, J. A., J. Chem. Phys., 26 : 1348 (1957).
310. Gagnaire, D., and Vincendon, M., Bull. Soc. chim. France, No. 2 : 472 (1965).
311. Giulotto, L., J. chim. phys. phys.-chim. biol., 61 : 177 (1964).
312. Glick, R. E., Gupta, R. P., Sauer, J. A., and Woodward, A. E., J. Polymer Sci., 42 : 271 (1960).
313. Glick, R. E., Gupta, R. P., Sauer, J. A., and Woodward, A. E., Polymer, 1 : 340 (1960).
314. Glick, R. E., and Phillips, R. C., J. Polymer Sci., A3 : 1885 (1965).
315. Golling, E., Z. angew. Phys., 14 : 717 (1962).
316. Golub, M. A., Fuqua, S. A., and Bhacca, N. S., J. Am. Chem. Soc., 84 : 4981 (1962).
317. Golub, M. A., and Heller, J., Canad. J. Chem., 41 : 937 (1963).
318. Golub, M. A., and Heller, J., J. Polymer Sci., B2 : 523 (1964).
319. Golub, M. A., and Heller, J., J. Polymer Sci., B2 : 723 (1964).
320. Golub, M. A., and Heller, J., Tetrahedron Letters, No. 30 : 2137 (1963).
321. Goodman, M., and Brondrup, J., J. Polymer Sci., A3 : 327 (1965).
322. Goodman, M., and Masuda, Y., Biopolymers, 2 : 107 (1964).
323. Goodman, M., and You-Ling, Fan, J. Am. Chem. Soc., 86 : 4922 (1964).
324. Goodrich, J. E., and Porter, R. S., J. Polymer Sci., B2 : 353 (1964).
325. Gordon, J. E., Chem. and Ind., No. 6 : 267 (1962).
326. Gordon, J. E., J. Phys. Chem., 66 : 1150 (1962).
327. Graham, R. K., Dunkelberger, D. L., and Panchak, J. R., J. Polymer Sci., 59 : S43 (1962).

328.   Griesbaum, K., Oswald, A. A., and Naegele, W., J. Org. Chem., 29:1887 (1964).

329.   Gruber, U., and Elias, H.-G., Makromolek. Chem., 86:168 (1965).

330.   Gupta, R. P., J. Phys. Chem., 65:1128 (1961).

331.   Gupta, R. P., J. Phys. Chem., 66:1 (1962).

332.   Gupta, R. P., J. Phys. Chem., 66:849 (1962).

333.   Gupta, R. P., J. Polymer Sci., 54:S20 (1961).

334.   Gupta, R. P., Kolloid-Z., 174:73 (1961).

335.   Gupta, R. P., Kolloid-Z., 174:74 (1961).

336.   Gupta, R. P., Makromolek. Chem., 42:248 (1961).

337.   Gupta, V. D., and Beevers, R. B., Chem. Rev., 62:665 (1962).

338.   Gutowsky, H. S., and Meyer, L. H., J. Chem. Phys., 21:2122 (1953).

339.   Gutowsky, H. S., Saika, A., Takeda, M., and Woessner, D. S., J. Chem. Phys., 27:534 (1957).

340.   Haddon, W. F., Jr., Porter, R. S., and Johnson, J. R., J. Appl. Polymer Sci., 8:1371 (1964).

341.   Haeberlen, U., Hausser, R., and Noack, F., Z. Naturforsch., 18a:689 (1963).

342.   Haeberlen, U., Hausser, R., and Noack, F., Z. Naturforsch., 18a:1026 (1963).

343.   Haeberlen, U., Hausser, R., and Pechold, W., Z. Naturforsch., 18a:1345 (1963).

344.   Hardeman, G. E., Philips Res. Rept., 15:587 (1960).

345.   Harwood, H. J., and Ritchey, W. M., J. Polymer Sci., B3:419 (1965).

346.   Havlik, A. J., and Hildebrandt, A. T., J. Polymer Sci., 41:533 (1959).

347.   Heffelfinger, C. J., and Burton, R. L., J. Polymer Sci., 47:289 (1960).

348.   Heller, J., Tieszen, D. Q., and Parkinson, D. B., J. Polymer Sci., A1:125 (1963).

349.   Herring, M. J., and Smith, J. A. S., J. Chem. Soc., 1960:273.

350.   Higdon, W. T., and Robinson, J. D., J. Chem. Phys., 37:1161 (1962).

351.   Hikichi, K., J. Phys. Soc., Japan, 19:2169 (1964).

352.   Hikichi, K., and Furuichi, J., J. Phys. Soc. Japan, 18:742 (1963).

353.   Hikichi, K., and Furuichi, J., J. Polymer Sci., A3:3003 (1965).

354.   Hirai, A., and Kawai, T., Mem. Coll. Sci. Univ. Kyoto, A29:345 (1961).

355.   Hirst, R. C., Grant, D. M., Hoff, R. E., and Burke, W., J. Polymer Sci., A3:2091 (1965).

356.   Hochfrequenzspektroskopie, Akad. Verlag, Berlin (1961).

357.   Holroyd, L. V., Codrington, R. S., Mrowca, B. A., and Guth, E., J. Appl. Phys., 22:696 (1951).

358.   Holroyd, L. V., Codrington, R. S., Mrowca, B. A., and Guth, E., Rubb. Chem. and Technol., 25:767 (1952).

359.   Holroyd, L. V., Mrowca, B. A., and Guth, E., Phys. Rev., 79:1026 (1950).

360.   Honnold, V. R., McCaffrey, F., and Mrowca, B. A., J. Appl. Phys., 25:1219 (1954).

361.   Honnold, V. R., McCaffrey, T., and Mrowca, B. A., Phys. Rev., 94:1414 (1954).

362.   Huggert, A., and Odeblad, E., Acta radiol., 51:385 (1959).

363.   Huggins, C. M., and Carpenter, D. R., General Electr. Res. Lab. Rept., No. 61-RL-2710 (1961).

364.   Huggins, C. M., St. Pierre, L. E., and Bueche, A. M., General Electr. Res. Lab. Rept., No. 60-RL-2412 (1960).

365. Huggins, C. M., St. Pierre, L. E., and Bueche, A. M., J. Phys. Chem., 64:1304 (1960).
366. Huggins, C. M., St. Pierre, L. E., and Bueche, A. M., J. Polymer Sci., A1:2731 (1963).
367. Hunt, B. I., Powles, J. G., and Woodward, A. E., Polymer, 5:323 (1964).
368. Hwa, J. C. H., J. Polymer Sci., 60:S12 (1962).
369. Hwang, C., and Sanders, T. M., Proc. 7th Intern. Conf. on Low Temp. Phys., Toronto, 1960, publ. (1961), p. 148.
370. Hyndman, D., and Origlio, G. F., J. Appl. Phys., 31:1849 (1960).
371. Hyndman, D., and Origlio, G. F., J. Polymer Sci., 39:556 (1959).
372. Hyndman, D., and Origlio, G. F., J. Polymer Sci., 46:259 (1960).
373. Illers, K. H., and Kosfeld, R., Makromolek. Chem., 42:44 (1960).
374. Ilzuka, S., Hatada, M., and Hirota, K., Bull. Chem. Soc. Japan, 36:817 (1963).
375. Ingham, J. D., Lawson, D. D., Manatt, S. L., Rapp, N. S., and Hardi, J. P., Internat. Symp. Macromolec. Chem., Prague, 1965, Abstract A156.
376. Ishibachi, N., Seiyawa, T., Sakai, W., and Ishii, Y., Denki Kagaku, 31:752 (1963).
377. Ito, K., and Yamashita, Y., J. Polymer Sci., B3:625 (1965).
378. Ito, K., and Yamashita, Y., J. Polymer Sci., B3:631 (1965).
379. Ito, K., and Yamashita, Y., J. Polymer Sci., B3:637 (1965).
380. Iwatsuki, S., Yamashita, Y., and Ishii, Y., J. Chem. Soc. Japan, Ind. Chem. Sect., 66:1162 (1963).
381. Iwayanagi, S., Nippon butsuri gakkai. Dai 18 kai nénkai koén ekosyu, No. 4:588 (1963).
382. Iwayanagi, S., and Miura, I., Japan J. Appl. Phys., 4:94 (1965); Chem. Abs., 62:7271 (1965).
383. Iwayanagi, S., and Miura, I., Nippon butsuri gakkai. Dai 18 kai nénkai koén ekosyu, No. 4:544 (1963).
384. Jackman, L. M., Applications of Nuclear Magnetic Resonance Spectroscopy in Organic Chemistry, Pergamon Press, London (1959).
385. Jacobson, B., Anderson, W., and Arnold, J. T., Nature, 173:772 (1954).
386. Jain, P. L., J. Polymer Sci., 36:443 (1959).
387. Jardetzky, C. D., and Jardetzky, O., Biochem. Biophys. Acta, 26:668 (1957).
388. Jardetzky, O., and Jardetzky, C. D., J. Am. Chem. Soc., 79:5322 (1957).
389. Jenner, G., Brini, M., Deluzarche, A., Schue, F., and Maillard, A., Bull. Soc. chim. France, No. 9:2211 (1964).
390. Johnsen, U., J. Polymer Sci., 54:S6 (1961).
391. Johnsen, U., Kolloid-Z., 178:161 (1961).
392. Johnsen, U., and Tessmar, K., Kolloid-Z., 168:160 (1960).
393. Jones, D. W., J. Polymer Sci., 59:271 (1962).
394. Jones, D. W., Polymer, 2:203 (1961).
395. Jungnickel, J. L., and Forbes, J. W., Analyt. Chem., 35:938 (1963).
396. Kail, J. A. E., Sauer, J. A., and Woodward, A. E., J. Phys. Chem., 66:1292 (1962).
397. Kail, J. A. E., Sauer, J. A., and Woodward, A. E., Polymer, 4:413 (1963).
398. Kamashima, K., Japan J. Appl. Phys., 4:259 (1965).

399. Kato, Y., Ashikari, N., and Nishioka, A., Bull. Chem. Soc. Japan, 37:1630
      (1964).

400. Kato, Y., and Nishioka, A., Bull. Chem. Soc. Japan, 37:1614 (1964).

401. Kato, Y., and Nishioka, A., Bull. Chem. Soc. Japan, 37:1622 (1964).

402. Kato, Y., and Nishioka, A., J. Polymer Sci., B3:739 (1965).

403. Kato, Y., Watanabe, H., and Nishioka, A., Bull. Chem. Soc. Japan, 37:1762
      (1964).

404. Kawai, T., J. Phys. Soc. Japan, 16:1220 (1961).

405. Kawai, T., Mem. Coll. Sci. Univ. Kyoto, A30:29 (1962).

406. Kawai, T., Sasaki, M., Hirai, A., Hashi, T., and Odajima, A., J. Phys. Soc.
      Japan, 15:1700 (1960).

407. Kawai, T. Yoshino, Y., and Hirai, A., J. Phys. Soc. Japan, 16:2356 (1961).

408. Kennedy, J. P., and Hinlicky, J. A., Polymer, 6:133 (1965).

409. Kennedy, J. P., Naegele, W., and Elliot, J. J., J. Polymer Sci., B3:729 (1965).

410. Kern, R. J., Am. Chem. Soc. Div. Polymer Chem. Preprints, 4:324 (1963).

411. Kern, R. J., Hawkins, J. J., and Galfee, J. D., Makromolek. Chem., 66:126
      (1963).

412. Kern, R. J., and Pustinger, J. V., Nature, 185:236 (1960).

413. Ketley, A. D., J. Polymer Sci., B2:827 (1964).

414. Komaki, A., and Matsumoto, T., J. Polymer Sci., B1:671 (1963).

415. Komaki, A., Yano, S., Yoshida, H., and Okamura, S., J. Phys. Soc. Japan,
      17:581 (1962).

416. Kontos, E. G., and Slichter, W. P., J. Polymer Sci., 61:61 (1962).

417. Kosfeld, R., Kolloid-Z., 172:182 (1960).

418. Kosfeld, R., and Jenckel, E., Kolloid-Z., 165:136 (1959).

419. Kosfeld, R., and Jenckel, E., Symposium über Makromoleküle in Wiesbaden,
      1959, Vortrag I-A-13.

420. Kosfeld, R., and Vosskötter, G., Z. Elektrochem., 65:642 (1961).

421. Kotin, L., and Nagasawa, M., J. Am. Chem. Soc., 83:1026 (1961).

422. Kratky, O., Kolloid-Z., 64:213 (1933).

423. Kriegsmann, H., Seifert, G., Frigge, K., and Dube, G., Internat. Symp. Macro-
      molec. Chem., Prague, 1965, Abstract A221.

424. Kubo, R., and Tomita, K., J. Phys. Soc. Japan, 9:888 (1954).

425. Kubota, T., and Takamura, T., Bull. Chem. Soc. Japan, 33:70 (1960).

426. Kummer, D., and Rochow, E. G., U.S. Govt. Res. Rept., 39:25 (1964); Chem.
      Abs., 62:5347 (1965).

427. Kusumoto, H., J. Phys. Soc. Japan, 11:1015 (1956).

428. Kusumoto, H., J. Phys. Soc. Japan, 12:826 (1957).

429. Kusumoto, H., and Gutowsky, H. S., J. Polymer Sci., A1:2905 (1963).

430. Kusumoto, H., Hukuda, K., Kawano, I., and Takayanagi, M., J. Chem. Soc.
      Japan, Ind. Chem. Soc., 68:825 (1965).

431. Kusumoto, H., Lawrenson, I. J., and Gutowsky, H. S., J. Chem. Phys., 32:724
      (1960).

432. Ladacki, M., J. Appl. Polymer Sci., 9:1561 (1965).

433. Land, R., Richards, R. E., and Ward, I. M., Trans. Faraday Soc., 55:225 (1959).

434. Lanzavecchia, G., Materie plastiche, 23:561 (1957).

435. Lee, C. L., and Haberland, G. G., J. Polymer Sci., B3 : 883 (1965).
436. Lipscomb, N. T., and Weber, E. C., J. Polymer Sci., A3 : 55 (1965).
437. Liquori, A. M., and Quadrifoglio, F., Polymer, 4 : 448 (1963).
438. Liu, Kong-Jen, and Burlant, W., Am. Chem. Soc. Polymer Preprints, 6 : 326 (1965).
439. Liu, K. J., Szuty, J. S., and Ullman, R., Am. Chem. Soc. Polymer Preprints, 5 : 761 (1964).
440. Liu, Kong-Jen, and Ullman, R., Polymer, 6 : 100 (1965).
441. Loebl, E. M., and O'Neill, J. J., J. Polymer Sci., B1 : 27 (1963).
442. Lombardi, E., and Segre, A., Atti Accad. naz. Lincei Rend Cl. Sci. fis. mat. natur., 34 : 547 (1963).
443. Lombardi, E., Segre, A., Zambelli, A., Marinangeli, A., and Natta, G., Internat. Symp. Macromolec. Chem., Prague, 1965, Abstract A657.
444. Lösche, A., Arch. Sci. (Geneva), 10 : 197 (1957).
445. Lösche, A., Arch. Sci. (Geneva), 12 : 205 (1959).
446. Lösche, A., Exp. Techn. Phys., A4 : 168 (1956).
447. Lösche, A., Kolloid-Z., 165 : 116 (1959).
448. Lösche, A., Symposium über Makromoleküle in Wiesbaden, 1959, Vortrag I-A-11.
449. Lowe, I. J., Brown, L. O., and Norberg, R. E., Phys. Rev., 100 : 1243 (1955).
450. Lumry, R., Matsumiya, H., Bovey, F. A., and Kowalsky, A., J. Phys. Chem., 65 : 837 (1961).
451. Luszczynski, K., Arch. Sci. (Geneva), 12 : 127 (1959).
452. Luszczynski, K., and Powles, J. G., Proc. Phys. Soc., 74 : 408 (1959).
453. Mabuchi, K., Saegusa, T., and Furukawa, J., Makromolek. Chem., 81 : 112 (1965).
454. Marconi, W., Mazzei, A., Yugli, G., and Bruzzone, M., Internat. Symp. Macromolec. Chem., Prague, 1965, Abstract A125.
455. Matsumoto, M., and Fujii, K., Kôgyô Kagaku Zasshi, J. Chem. Soc. Japan, Ind. Chem. Soc., 68 : 843 (1965).
456. Matsushita, A., J. Inst. Electr. Eng. Japan, 81 : 77 (1961).
457. Matsuzaki, K., Uryu, T., Ishida, A., and Ohki, T., J. Polymer Sci., B2 : 1139 (1964).
458. Matsuzaki, K., Uryu, T., Ishida, A., and Takeuchi, M., Internat. Symp. Macromolec. Chem., Prague, 1965, Abstract A197.
459. Matsuzaki, K., Uryu, T., and Takeuchi, M., J. Polymer Sci., B3 : 835 (1965).
460. McCall, D. W., Appl. Phys. Letters, 7 : 153 (1965).
461. McCall, D. W., Anderson, E. W., and Huggins, C. M., J. Chem. Phys., 34 : 804 (1961).
462. McCall, D. W., and Anderson, E. W., J. Polymer Sci., A1 : 1175 (1963).
463. McCall, D. W., and Anderson, E. W., Polymer, 4 : 93 (1963).
464. McCall, D., and Bovey, F. A., J. Polymer Sci., 45 : 530 (1960).
465. McCall, D. W., and Douglass, D. C., Appl. Phys. Letters, 7 : 12 (1965).
466. McCall, D. W., and Douglass, D. C., Polymer, 4 : 433 (1963).
467. McCall, D. W., Douglass, D. C., and Anderson, E. W., J. Chem. Phys., 30 : 771 (1959).

468. McCall, D. W., Douglass, D. C., and Anderson, E. W., J. Chem. Phys., 30:1272 (1959).

469. McCall, D. W., Douglass, D. C., and Anderson, E. W., J. Polymer Sci., 59:301 (1962).

470. McCall, D. W., Douglass, D. C., and Anderson, E. W., J. Polymer Sci., A1:1709 (1963).

471. McCall, D. W., and Hamming, R. W., Acta Crystallogr., 12:81 (1959).

472. McCall, D. W., and Slichter, W. P., J. Polymer Sci., 26:171 (1957).

473. McCall, D. W., and Slichter, W. P., Newer Methods of Polymer Characterisation (Bacon, K., ed.), Interscience, New York (1964), p. 321.

474. McDonald, M. P., and Ward, I. M., Proc. Phys. Soc. (London), 80:1249 (1962).

475. Merker, R. L., Barnes, G. H., Burce, J. G., David, M. P., Parker, G. A., Piccoli, W. A., Scott, M. J., and Daughenbaugh, N. E., Mellon Inst. Res. 52 Annual Rept., S.1:17 (1965).

476. Michel, R. H., J. Polymer Sci., A2:2533 (1964).

477. Miller, R. L., J. Polymer Sci., 56:375 (1962).

478. Miller, R. L., Polymer, 1:135 (1960).

479. Miller, R. L., SPE Trans., 3:123 (1963).

480. Miller, W. L., Brey, W. S,, and Butler, G. B., J. Polymer Sci., 54:329 (1961).

481. Miyake, A., J. Chem. Phys., 27:1425 (1957).

482. Miyake, A., J. Phys. Soc. Japan, 15:1057 (1960).

483. Miyake, A., J. Polymer Sci., 28:476 (1958).

484. Miyake, A., Busséiron Kénkyu, 5:334 (1959).

485. Miyake, A., Kobunshi Kagaku, 16:667 (1959).

486. Miyake, A., and Chujo, R., J. Phys. Soc. Japan, 15:198 (1960).

487. Miyama, H., and Kamachi, N., J. Polymer Sci., B3:241 (1965).

488. Moraglio, G., and Danusso, F., Polymer, 4:445 (1963).

489. Morawetz, A., and Rubin, I., J. Polymer Sci., 57:687 (1962).

490. Mrowca, B. A., and Guth, E., Proc. 3rd Rubber Technol. Conf., London (1954), p. 370.

491. Mrowca, B. A., and Holroyd, L. V., Phys. Rev., 81:303 (1951).

492. Mrowca, B. A., Holroyd, L. V., and Guth, E., Phys. Rev., 79:1026 (1950).

493. Murakami, I., Kawai, A., and Yamamura, H., J. Sci. Hiroshima Univ., Ser. A-II, 27:141 (1964).

494. Murano, M., Kaneichi, Y., and Yamadera, R., J. Polymer Sci., A3:2698 (1965).

495. Naegele, W., and Kennedy, J. P., Am. Chem. Soc. Polymer Preprints, 6:260 (1965).

496. Natta, G., Lombardy, E., Segre, A., Zambelli, A., and Marinangeli, A., Chim. ind. (Milan), 47:378 (1965).

497. Natta, G., Zambelli, A., Lanzi, G., Pasquon, I., Mognaschi, S. R., Segre, A.L., and Centola, P., Makromolek. Chem., 81:161 (1965).

498. Naylor, R. E., and Lasoski, S. W., J. Polymer Sci., 44:1 (1960).

499. Neiman, M. B., Slonim, I. Ya., and Urman, Ya. G., Nature, 202:693 (1964).

500. Newland, G. C., and Tamblyn, J. W., J. Appl. Polymer Sci., 8:1949 (1964).

501. Newman, R., J. Chem. Phys., 18:1303 (1950).

502. Nishioka, A., J. Polymer Sci., 37:163 (1959).

503. Nishioka, A., Ōyō Butsuri, 27:269 (1958); Chem. Abs., 1958:15198.

504. Nishioka, A., Suppl. Prog. Theoret. Phys., No. 10:137 (1959).

505. Nishioka, A., Kato, Y., and Ashikari, N., J. Polymer Sci., 62:S10 (1962).

506. Nishioka, A., Kato, Y., and Mitsuoka, H., J. Polymer Sci., 62:S9 (1962).

507. Nishioka, A., Kato, Y., Uetake, T., and Watanabe, H., J. Polymer Sci., 61:S32 (1962).

508. Nishioka, A., Koike, J., Owaka, M. Naraba, T., and Kato, Y., J. Phys. Soc. Japan, 15:416 (1960).

509. Nishioka, A., Komatsu, H., and Kakiuchi, Y., J. Phys. Soc. Japan, 12:283 (1957).

510. Nishioka, A., Watanabe, H., Abe, K., and Sono, Y., J. Polymer Sci., 48:241 (1960).

511. Nishioka, A., Watanabe, H., Yamaguchi, J., and Shimizu, H., J. Polymer Sci., 45:232 (1960).

512. NMR and EPR Spectroscopy, Pergamon Press, Oxford (1960).

513. Nohara, S., Kobunshi Kagaku, 13:531 (1956); Chem. Abs., 51:18686 (1957).

514. Nohara, S., Kobunshi Kagaku, 14:318 (1957); Chem. Abs., 52:4313 (1958).

515. Nohara, S., Kobunshi Kagaku, 15:105 (1958); Chem. Abs., 53:12837 (1959).

516. Nolle, A. W., Bull. Am. Phys. Soc., 1:109 (1956).

517. Nolle, A. W., Phys. Rev., 98:1560 (1955).

518. Nolle, A. W., and Billings, J. I., J. Chem. Phys., 30:84 (1959).

519. Odajima, A., J. Phys. Soc. Japan, 14:308 (1959).

520. Odajima, A., J. Phys. Soc. Japan, 14:777 (1959).

521. Odajama, A., Suppl. Prog. Theoret. Phys., No. 10:142 (1959).

522. Odajima, A., Inoue, S., and Hayache, J., Hokkaido Daigaky Oyo Denkyi Kenkyus Ihŏ, 10:95 (1958); Chem. Abs., 53:11998 (1959).

523. Odajima, A., Koike, M., and Nagai, M., Bull. Res. Inst. Appl. Elec. (Japan), 6:137 (1954); Chem. Abs., 50:5401 (1956).

524. Odajima, A., and Nagai, M., Kobunshi Kagaku, 14:512 (1957); Chem. Abs., 52:6831 (1958).

525. Odajima, A., and Nagai, M., Oyo Denki Kenkyu Ihŏ, 9:113 (1957); Chem. Abs., 52:1434 (1958).

526. Odajima, A., and Nagai, M., Oyo Denki Kenkyu Ihŏ, 9:195 (1957); Chem. Abs., 52:14335 (1958).

527. Odajima, A., Sauer, J. A., and Woodward, A. E., J. Phys. Chem., 66:718 (1962).

528. Odajima, A., Sauer, J. A., and Woodward, A. E., J. Polymer Sci., 57:107(1962).

529. Odajima, A., Sohma, J., and Koike, M., J. Chem. Phys., 23:1959 (1955).

530. Odajima, A., Sohma, J., and Koike, M., J. Phys. Soc. Japan, 12:272 (1957).

531. Odajima, A., Woodward, A. E., and Sauer, J. A., J. Polymer Sci., 55:181(1961).

532. Odajima, A., Woodward, A. E., and Sauer, J. A., Papers presented at the New York City Meeting, Am. Chem. Soc., 1:63 (1960).

533. Odeblad, E., Nature, 188:579 (1960).

534. Ohnishi, S., and Nukada, K., J. Polymer Sci., B3:179 (1965).

535. Okamoto, S., Hama, Y., and Tamura, N., Bull. Sci. Eng. Res. Lab. Waseda Univ., No. 26:91 (1964).

536. Okuda, K., J. Polymer Sci., A2:1749 (1964).

537. Olf, H. G., and Peterlin, A., J. Appl. Phys., 35:3108 (1964).
538. O'Reilly, D. E., Poole, C. P., Belt, R. F., and Scott, H., J. Polymer Sci., A2:3257 (1964).
539. Oshima, K., and Kusumoto, H., J. Chem. Phys., 24:913 (1956).
540. Oshima, K., and Kusumoto, H., Kogyo Kagaku Zasshi, 59:806 (1956); Chem. Abs., 52:7753 (1958).
541. Oth, J. F. M., Ind. chim. belge, 26:470 (1961).
542. Overberger, C. G., and Yamamoto, N., J. Polymer Sci., B3:569 (1965).
543. Owens, F. H., and Zimmerman, F. E., J. Polymer Sci., A1:2711 (1963).
544. Page, T. F., and Bresler, W. E., Analyt. Chem., 36:1981 (1964).
545. Pake, G. E., Disc. Faraday Soc., 19:252 (1955).
546. Pake, G. E., and Purcell, E. M., Phys. Rev., 74:1184 (1948).
547. Paulsen, P. J., and Cooke, W. D., Analyt. Chem., 36:1713 (1964).
548. Percival, D. F., and Stevens, M. P., Analyt. Chem., 36:1574 (1964).
549. Peterlin, A., Krasovec, F., Pirkmajer, E., and Levstek, I., Makromolek. Chem., 37:231 (1960).
550. Peterlin, A., Krasovec, F., Pirkmajer, E., and Levstek, I., Symposium über Makromoleküle in Wiesbaden, 1959, Vortrag I-A-12.
551. Peterlin, A., and Olf, H. G., J. Polymer Sci., B2:409 (1964).
552. Peterlin, A., and Olf, H. G., J. Polymer Sci., B2:769 (1964).
553. Peterlin, A., and Pirkmajer, E., J. Polymer Sci., 46:185 (1960).
554. Peterlin, A., and Roecl, E., J. Appl. Phys., 34:102 (1963).
555. Pickett, L. W., J. Am. Chem. Soc., 58:2299 (1936).
556. Pino, P., Pucci, S., Benedetti, E., and Bucci, P., J. Am. Chem. Soc., 87:3263 (1965).
557. Porter, R. S., Nicksic, S. W., and Johnson, J. F., Analyt. Chem., 35:1948 (1963).
558. Powles, J. G., Arch. Sci. (Geneva), 9:182 (1956).
559. Powles, J. G., Arch. Sci. (Geneva), 10, fasc. spec., 253 (1957).
560. Powles, J. G., J. Polymer Sci., 22:79 (1956).
561. Powles, J. G., Polymer, 1:219 (1960).
562. Powles, J. G., Proc. Phys. Soc. (London), B69:281 (1956).
563. Powles, J. G., Symposium über Makromoleküle in Wiesbaden, 1959, Vortrag I-A-10.
564. Powles, J. G., and Gutowsky, H. S., J. Chem. Phys., 21:1695 (1953).
565. Powles, J. G., and Hartland, A., Nature, 186:26 (1960).
566. Powles, J. G., Hartland, A., and Kail, J. A. E., J. Polymer Sci., 55:361 (1961).
567. Powles, J. G., and Hunt, B. I., Physics Letters, 14:202 (1965).
568. Powles, J. G., Hunt, B. I., and Sandiford, D. J. H., Polymer, 5:505 (1964).
569. Powles, J. G., and Kail, J. A. E., J. Polymer Sci., 31:183 (1958).
570. Powles, J. G., and Kail, J. A. E., Trans. Faraday Soc., 55:1996 (1959).
571. Powles, J. G., and Luszczynski, K., Physica, 25:455 (1959).
572. Powles, J. G., and Mansfield, P., Polymer, 3:336 (1962).
573. Powles, J. G., and Mansfield, P., Polymer, 3:339 (1962).
574. Powles, J. G., Strange, J. H., and Sandiford, D. J. H., Polymer, 4:401 (1963).
575. Przyborowski, S., Hochfrequenzspektroskopie, Akademic Verlag, Berlin (1961), p. 75.

576. Racoš, M., Coll. Czech. Chem. Comm., 28:1914 (1963).
577. Racoš, M., Českosl. časop. fys., A12:205 (1962).
578. Ramey, K. C., and Field, N. D., J. Polymer Sci., B2:461 (1964).
579. Ramey, K. C., and Field, N. D., J. Polymer Sci., B3:63 (1965).
580. Ramey, K. C., and Field, N. D., J. Polymer Sci., B3:69 (1965).
581. Ramey, K. C., Field, N. D., and Borchert, A. E., J. Polymer Sci., A3:2885 (1965).
582. Ramey, K. C., Field, N. D., and Hasegawa, I., J. Polymer Sci., B2:865 (1964).
583. Ramey, K. C., and Statton, C. L., Makromolek. Chem., 85:287 (1965).
584. Reichel, B., Marbel, C. S., and Greenley, R. Z., J. Polymer Sci., A1:2935 (1963).
585. Reichenberg, D., and Lawrenson, I. J., Trans. Faraday Soc., 59:141 (1963).
586. Reiss, C., and Benoit, H., Compt. Rend., 253:268 (1961).
587. Reiss, C., and Benoit, H., Internat. Symp. Macromolec. Chem. Prague, 1965, Preprint P540.
588. Rempel, R. C., Weaver, H. E., Sands, R. H., and Miller, R. L., J. Appl. Phys., 28:1082 (1957).
589. Rochow, E. G., and LeClair, H. G., J. Inorg. Nucl. Chem., 1:92 (1955).
590. Rollwitz, W. L., Proc. Natl. Electronics Conf., 12:113 (1956).
591. Rubin, H., ISA Journal, 5:64 (1958).
592. Ryan, C. F., and Fleischer, D. C., J. Phys. Chem., 69:3384 (1965).
593. Saha, A., and Das, T. P., Theory and Application of Nuclear Induction. Calcutta (1957).
594. Sakurada, Y., Matsumoto, M., Imai, K., Nishioka, A., and Kato, Y., J. Polymer Sci., B1:633 (1963).
595. Samec, M., Blinc, R., and Brenman, M., J. Polymer Sci., 56:S21 (1962).
596. Sasaki, M., Kawai, T., Hirai, A., Hashi, T., and Odajima, A., J. Phys. Soc. Japan, 15:1652 (1960).
597. Satoh, S., J. Polymer Sci., A2:5221 (1964).
598. Satoh, S. Chujo, R., Ozeki, T., and Nagai, E., J. Polymer Sci., 62:S101 (1962).
599. Sauer, J. A., and Fuschillo, N., Bull. Am. Phys. Soc., 2:318 (1957).
600. Sauer, J. A., Wall, R. A., Fuschillo, N., and Woodward, A. E., J. Appl. Phys., 29:1385 (1958).
601. Sauer, J. A., and Woodward, A. E., Rev. Mod. Phys., 32:88 (1960).
602. Sauer, J. A., Woodward, A. E., and Fuschillo, N., Bull. Am. Phys. Soc., 4:187 (1959).
603. Sauer, J. A., Woodward, A. E., and Fuschillo, N., J. Appl. Phys., 30:1488 (1959).
604. Saunders, M., and Wishnia, A., Ann. N. Y. Acad. Sci., 70:870 (1958).
605. Saunders, M., Wishnia, A., and Kirkwood, J. G., J. Am. Chem. Soc., 79:3289 (1957).
606. Schue, F., Bull. Soc. chim. France, No. 4:980 (1965).
607. Schue, F., Rev. gen. caoutch., 41:261 (1964).
608. Schue, F., and Dole-Robbe, J. P., Bull. Soc. chim. France, No. 5:975 (1963).
609. Schuerch, C., Fowells, W., Yamada, A., Bovey, F. A., Hood, F. P., and Anderson, E. W., Am. Chem. Soc. Polymer Preprints, 5:1145 (1964).

610.   Schuerch, C., Fowells, W., Yamada, A., Bovey, F. A., Hood, E. P., and
       Anderson, E. W., J. Am. Chem. Soc., 86:4481 (1964).
611.   Schulz, R. C., and Schwab, Y., Makromolek. Chem., 87:90 (1965).
612.   Senn, W. L., Jr., Analyt. chim. acta, 29:505 (1963).
613.   Shaw, D. J., and Dunnell, B. A., Canad. J. Chem., 39:1154 (1961).
614.   Shaw, T. M., and Elsken, R. H., J. Appl. Phys., 26:313 (1955).
615.   Shaw, T. M., and Elsken, R. H., J. Chem. Phys., 18:1113 (1960).
616.   Shaw, T. M., and Elsken, R. H., J. Chem. Phys., 21:565 (1953).
617.   Shaw, T. M., Elsken, R. H., and Lundin, R. E., J. Textile Inst. Trans., 51:T562
       (1960).
618.   Shaw, T. M., and Palmer, K. J., Phys. Rev., 83:213 (1951).
619.   Shibata, T., Kimura, I., and Suita, K., J. Phys. Soc. Japan, 13:1546 (1958).
620.   Shimonouchi, T., Tasumi, M., and Abe, Y., Makromolek. Chem., 86:43 (1965).
621.   Shioji, Y., Ohnishi, S. I., and Nitta, I., J. Polymer Sci., A1:3373 (1963).
622.   Sibilia, J. P., and Paterson, A. R., J. Polymer Sci., C(8):41 (1965).
623.   Sinnot, K. M., J. Polymer Sci., 42:3 (1960).
624.   Skell, P., and Suhr, H., Chem. Ber., 94:3317 (1961).
625.   Skorko, M., Polymery tworzywa wielkoczast., 9:449 (1964).
626.   Skorko, M., Polymery tworzywa wielkoczast, 9:497 (1964).
627.   Slichter, W. P., Ann. N. Y. Acad. Sci., 83:60 (1959).
628.   Slichter, W. P., ASTM spec. techn. publ., No. 247, Philadelphia (1959), p. 257.
629.   Slichter, W. P., Fortschr. Hochpolymer. Forsch., 1:35 (1958).
630.   Slichter, W. P., in: Growth and Perfection of Crystals (R. H. Doremus et al.,
       eds), Wiley, New York (1958), p. 558.
631.   Slichter, W. P., J. Appl. Phys., 26:1099 (1955).
632.   Slichter, W. P., J. Appl. Phys., 31:1865 (1960).
633.   Slichter, W. P., J. Appl. Phys., 32:2339 (1961).
634.   Slichter, W. P., J. Polymer Sci., 24:173 (1957).
635.   Slichter, W. P., J. Polymer Sci., 35:77 (1959).
636.   Slichter, W. P., Makromolek. Chem., 34:67 (1959).
637.   Slichter, W. P., Principles of Magnetic Resonance with Examples from Solid
       State Physics, Harper and Row, New York (1963).
638.   Slichter, W. P., Rubb. Chem. Technol, 34:1574 (1961).
639.   Slichter, W. P., SPE Journal, 15:303 (1959).
640.   Slichter, W. P., and Davis, D. D., J. Appl. Phys., 34:98 (1963).
641.   Slichter, W. P., and Davis, D. D., J. Appl. Phys., 35:10 (1964).
642.   Slichter, W. P., and Davis, D. D., J. Appl. Phys., 35:3103 (1964).
643.   Slichter, W. P., and Mandell, E. R., Bull. Am. Phys. Soc., 4:187 (1959).
644.   Slichter, W. P., and Mandell, E. R., J. Appl. Phys., 29:1438 (1958).
645.   Slichter, W. P., and Mandell, E. R., J. Appl. Phys., 30:1473 (1959).
646.   Slichter, W. P., and Mandell, E. R., J. Chem. Phys., 29:232 (1958).
647.   Slichter, W. P., and Mandell, E. R., J. Phys. Chem., 62:334 (1958).
648.   Slichter, W. P., and Mays, I. M., Phys. Rev., 98:1559 (1955).
649.   Slichter, W. P., and McCall, D. W., Bull. Am. Phys. Soc., 2:125 (1957).
650.   Slichter, W. P., and McCall, D. W., J. Polymer Sci., 25:230 (1957).

651. Slonim, I. Ya., Lyubimov, A. N., and Kovarskaya, B. M., Chem. Prum., 13/38:606 (1963).

652. Smith, J. A. S., Chem. Soc. (London), Spec. Publ., No. 12:199 (1958).

653. Smith, J. A. S., Disc. Faraday Soc., N19:207 (1955).

654. Smith, D. C., Ind. Eng. Chem., 48:1161 (1956).

655. Smith, G. W., J. Appl. Polymer Sci., 9:1553 (1965).

656. Smith, P., Muller, N., and Tosch, W. C., J. Polymer Sci., 57:823 (1962).

657. Sobue, H., Matsuzaki, K., and Nakano, S., J. Polymer Sci., A2:3339 (1964).

658. Sobue, H., Uryu, T., and Matsuzaki, K., J. Polymer Sci., B1:409 (1963).

659. Sobue, H., Uryu, T., Matsuzaki, K., and Tabata, Y., J. Polymer Sci., A2:3333 (1964).

660. Statton, W. O., Am. Dyestuff Rept., 54:26 (1965).

661. Statton, W. O., J. Polymer Sci., C(3):3 (1963).

662. Stehling, F. C., J. Polymer Sci., A2:1815 (1964).

663. Stille, J. K., and Cassidy, P., J. Polymer Sci., B1:563 (1963).

664. Stok, J., Doskŏcilova, D., Schneider, B., Kolinsky, M., Ryska, M., Sykora, S., and Lim, D., Internat. Symp. Macromolec. Chem., Prague, 1965, Abstract A98.

665. Story, V. M., and Canty, G., J. Res. Nat. Bur. Stand., A68:165 (1964).

666. Strehlow, H., Magnetische Kernresonanz und chemische Struktur, Steinkopff, Darmstadt (1962).

667. Sugai, S., Kamashima, K., and Hikichi, K., Japan J. Appl. Phys., 2:588 (1963).

668. Sugimatsu, A., Senda, S., and Harada, Y., J. Chem. Soc. Japan, Ind. Chem. Sec., 62:576 (1959).

669. Sumi, M., Chokki, Y., Nakai, Y., Nakabayashi, M., and Kanzava, T., Makromolec. Chem., 78:146 (1964).

670. Suzuishi, M., Oki Rev., 32:21 (1965).

671. Swanenburg, T. J. B., Poulis, N. J., and Drewes, G. W. J., Physica, 29:713 (1963).

672. Swanson, T., Stejskal, E. O., and Tarkow, H., Tappi, 45:929 (1962).

673. Szymanski, H. S., and Bluemle, A., J. Polymer Sci., A3:63 (1965).

674. Tadokoro, H., Yasumoto, T., Marahashi, S., and Nitta, I., J. Polymer Sci., 44:266 (1960).

675. Takeda, M., Kondo, S., and Toykku, Ja, Nippon butsuri gakkai. Dai 18 kai nénkai koén ekosyu, No. 4:542 (1963).

676. Takeda, M., Tanaka, K., and Nagao, R., J. Polymer Sci., 57:517 (1962).

677. Takeuchi, T., and Yamazaki, M., Kogyo Kagaku Zasshi, J. Chem. Soc. Japan, Industr. Chem. Sect., 68:931 (1965).

678. Tamura, N., J. Phys. Soc. Japan, 16:2338 (1961).

679. Tanaka, K., Bull. Chem. Soc. Japan, 33:1702 (1960).

680. Tanaka, K., and Yamagata, K., Bull. Chem. Soc. Japan, 28:90 (1955).

681. Tanaka, K., Yamagata,K., and Kittaka, S., Bull. Chem. Soc. Japan, 29:843 (1956).

682. Tanaka, K., Yamagata, K., Toshido, M., and Odajima, M., Bull. Chem. Soc. Japan, 30:428 (1957).

683. Terada, M., Lussier, B., Bensasson, R., and Brot, C., J. chim. phys. phys.-chim. biol., 60: 701 (1963).

684. Thompson, E. V., J. Polymer Sci., B3: 675 (1965).

685. Thurn, H., Ergebn., Exakt. Naturwiss., 31: 222 (1959).

686. Thurn, H., Kolloid-Z., 165: 140 (1959).

687. Thurn, H., Kolloid-Z., 174: 73 (1961).

688. Thurn, H., Kolloid-Z., 179: 11 (1961).

689. Tiers, G. V. D., and Bovey, F. A., J. Polymer Sci., 47: 479 (1960).

690. Tiers, G. V. D., and Bovey, F. A., J. Polymer Sci., A1: 833 (1963).

691. Tincher, W. C., J. Polymer Sci., 62: S148 (1962).

692. Tincher, W. C., Makromolek. Chem., 85: 20 (1965).

693. Tincher, W. C., Makromolek. Chem., 85: 34 (1965).

694. Tincher, W. C., Makromolek. Chem., 85: 46 (1965).

695. Trappeniers, N. J., Gerritsma, C. J., and Oosting, P. H., Physica, 30: 997 (1964).

696. Ullman, R., Am. Chem. Soc. Polymer Preprints, 6: 331 (1965).

697. Varoqui, R., and Benoit, H., Compt. Rend., 254: 3541 (1962).

698. Villiers, J. P., and Parrish, J. R., J. Polymer Sci., A2: 1331 (1964).

699. Vogl, O., J. Polymer Sci., A2: 4591 (1964).

700. Wada, Y., J. Phys. Soc. Japan, 16: 1226 (1961).

701. Walling, C., and Tanner, D. D., J. Polymer Sci., A1: 2271 (1963).

702. Wanless, G. G., J. Polymer Sci., 62: 263 (1962).

703. Wanless, G. G., and Kennedy, J. P., Polymer, 6: 111 (1965).

704. Ward, I. M., J. Chem. Phys., 31: 858 (1959).

705. Ward, I. M., Lab. Pract., 13: 957 (1964).

706. Ward, I. M., Trans. Faraday Soc., 56: 648 (1960).

707. Watanabe, H., Kato, Y, and Nischioka, A., Kogyo Kagaku zassi, 65: 270 (1962).

708. Weill, G., and Reiss, C., Compt. Rend., 256: 2816 (1963).

709. Wiberg, K. B., and Nist, B. J., The Interpretation of NMR Spectra, Benjamin, New York (1962).

710. Wiles, D. M., and Brownstein, S., J. Polymer Sci., B3: 951 (1965).

711. Wiles, D. M., and Bywater, S., Polymer, 3: 175 (1962).

712. Williams, R. W., Ann. N. Y. Acad. Sci., 70: 890 (1958).

713. Williams, G., Connor, T. M., and Read, B. E., Polymer, 5: 384 (1964).

714. Williams, M. L., Landel, R. F., and Ferry, J. D., J. Am. Chem. Soc., 77: 3701 (1955).

715. Wilson, C. W., Am. Chem. Soc., 148th Meeting, Div. Analyt. Chem., Abstracts of Papers, 1964, 29B61.

716. Wilson, C. W., J. Polymer Sci., 56: S16 (1962).

717. Wilson, C. W., J. Polymer Sci., 61: 403 (1962).

718. Wilson, C. W., J. Polymer Sci., A1: 1305 (1963).

719. Wilson, C. W., and Pake, G. E., J. Chem. Phys., 27: 115 (1957).

720. Wilson, C. W., and Pake, G. E., J. Polymer Sci., 10: 503 (1953).

721. Wilson, C. W., and Pake, G. E., Phys. Rev., 89: 896 (1953).

722. Wilson, C. W., and Santee, E. R., J. Polymer Sci., C(8): 97 (1965).

723. Wishnia, A., J. Chem. Phys., 32: 871 (1960).

724.   Woessner, D. E., J. Phys. Chem., 67:1365 (1963).
725.   Woodbrey, J. C., J. Polymer Sci., B2:315 (1964).
726.   Woodbrey, J. C., Higginbottom, H. P., and Culbertson, H. M., J. Polymer Sci., A3:1079 (1965).
727.   Woodbrey, J. C., and Trementozzi, Q. A., Am. Chem. Soc., 148th Meeting, Div. Analyt. Chem., Abstracts of Papers, 1964, 29B62.
728.   Woodbrey, J. C., and Trementozzi, Q. A., J. Polymer Sci., C(8):113 (1965).
729.   Woodward, A. E., SPE Trans., 2:86 (1962).
730.   Woodward, A. E., Trans. N. Y. Acad. Sci., 24:250 (1962).
731.   Woodward, A. E., Glick, R. E., Sauer, J. A., and Gupta, R. P., J. Polymer Sci., 45:367 (1960).
732.   Woodward, A. E., Gupta, R. P., Odajima, A. J., and Sauer, J. A., Papers presented at the New York City Meeting, Am. Chem. Soc., 1:69 (1960).
733.   Woodward, A. E., Odajima, A. J., and Sauer, J. A., J. Phys. Chem., 65:1384 (1961).
734.   Woodward, A. E., and Sauer, J. A., J. Polymer Sci., C(8):137 (1965).
735.   Yamadera, R., and Murrano, M., J. Polymer Sci., B3:821 (1965).
736.   Yamagata, K., J. Appl. Phys. Japan, 30:940 (1961).
737.   Yamagata, K., J. Appl. Phys. Japan, 31:395 (1962).
738.   Yamagata, K., and Hirota, S., J. Appl. Phys. Japan, 29:866 (1960).
739.   Yamagata, K., and Hirota, S., J. Appl. Phys. Japan, 30:261 (1961).
740.   Yen, S. P. S., Berry, G. C., and Fox, T. G., Mellon Inst. Res. 52 Annual Rept., S1:15 (1965).
741.   Yokota, K., and Ishii, Y., J. Polymer Sci., B3:771 (1965).
742.   Yokota, K., Sakai, Y., and Ishii, Y., J. Polymer Sci., B3:839 (1965).
743.   Yoshida, M., Odajima, M., and Tanaka, K., Bull. Chem. Soc. Japan, 30:197 (1957).
744.   Yoshino, T., and Komiyama, J., J. Polymer Sci., B3:311 (1965).
745.   Yoshino, T., Komiyama, J., and Shinomiya, M., J. Am. Chem. Soc., 86:4482 (1964).
746.   Yoshino, T., Kyogoku, H., Komiyama, J., and Manabe, Y., J. Chem. Phys., 38:1026 (1963).
747.   Yoshino, T., Shinomiya, M., and Komiyama, T., J. Am. Chem. Soc., 87:387 (1965).
748.   Zimmerman, J. R., and Britton, W. E., J. Phys. Chem., 61:1328 (1957).

Additional Bibliography

A1.   Alekseeva, V. P., Sosin, S. L., and Korshak, V. V., Vysokomolek. soed., 8:1920 (1966).

A2.   Borodin, P. M., Nikitin, M. K., and Sventitskii, E. I., in: Nuclear Magnetic Resonance, No. 1, Izd. LGU, Leningrad (1965), p. 83.

A3.   Gabuda, S. P., Kovrov, B. G., and Lundin, A. G., in: Radiospectroscopy of Solids, Atomizdat, Moscow (1967), p. 139.

A4.   Grad, N. M., and Al'shits, I. M., Vysokomolek. soed., A9:832 (1967).

A5.   Gul', V. E., and Lyubeshkina, E. G., Dokl. Akad. Nauk SSSR, 165:110 (1965).

A6.   Egorov, E. A., and Zhizhenkov, V. V., Fiz. tverd. tela, 8:3583 (1966).

A7.   Zhuravleva, I. P., Zgadzai, É. A., Maklakov, A. I., and Pimenov, G. G., in: Final Sci. Conference of Kazan Univ. of 1962, Kazan (1963), p. 39.

A8.   Zhurkov, S. N., Egorov, E. A., and Madeeva, É. N., in: Radiospectroscopy of Solids, Atomizdat, Moscow (1967), p. 130.

A9.   Zalukaev, L. P., and Pivnev, V. I., Nuclear Magnetic Resonance in Elastomers Izd. Voronizh Univ., Voronizh (1965).

A10.   Kazaryan, L. G., and Urman, Ya. G., Zh. strukt. khim., 7:54 (1966).

A11.   Kol'tsov, A. I., Vysokomolek. soed., B9:97 (1967).

A12.   Kol'tsov, A. I., Kamalov, S., and Vol'kenshtein, M. V., Vysokomolek. soed., A9:131 (1967).

A13.   Koton, M. M., Andreeva, I. V., Getmanchuk, Yu. P., Modorskaya, L. Ya., Pokrovskii, E. I., Kol'tsov, A. I., and Filatova, V. A., Vysokomolek. soed., 7:2039 (1965).

A14.   Koton, M. M., Andreeva, I. V., Getmanchuk, Yu. P., Madorskaya, L. Ya., and Pokrovskii, E. I., Vysokomolek. soed., 8:1389 (1966).

A15.   Kocharyan, N. M., Pikalov, A. P., Yan, S. A., Kagramanyan, A. V., and Markosyan, É. A., Vysokomolek. soed., 8:635 (1966).

A16.   Kocharyan, N. M., Pikalov, A. P., Kagramanyan, A. V., Markosyan, É. A., and Yan, S. A., Vysokomolek. soed., 8:640 (1966).

A17.   Lezhnev, N. N., Yampol'skii, B. Ya., Lyalina, N. M., Drevina, V. P., and Kogotkova, L. I., Dokl. Akad. Nauk SSSR, 160:861 (1965).

A18.   Lyubimov, A. N., Belitskii, I. Z., Slonim, I. Ya., Varenik, A. F., and Fedorov, V. I., Zavod. lab., 32:1163 (1966).

A19.   Maklakov, A. I., Maklakov, L. I., Nikitina, V. I., Balakireva, R. S., Shepelev, V. I., and Kurzhunova, Z. Z., Izv. vyssh. uchebn. zavedenii, Khimiya i khim. tekhnologiya, 10:90 (1967).

A20. Maklakov, A. I., and Nagumnova, É. I., Vysokomolek. soed., 7 : 2102 (1965).

A21. Maklakov, A. I., Pimenov, G. G., and Shepelev, V. I., in: Radiospectroscopy of Solids, Atomizdat, Moscow (1967), p. 136.

A22. Maklakov, A. I., and Chenborisova, L. Ya., Dokl. Akad. Nauk SSSR, 165 : 868 (1965).

A23. Maksimov, V. L., Votinov, M. P., and Dokukina, A. F., Vysokomolek. soed., 8 :1117 (1966).

A24. Maksimov, V. L., Dolgopol'skii, I. M., Votinov, M. P., and Rabinovich, R.L., Vysokomolek. soed., 8 : 620 (1966).

A25. Mikhailov, G. P., and Shevelev, V. A., Vysokomolek. soed., 8 : 763 (1966).

A26. Mikhailov, G. P., and Shevelev, V. A., Vysokomolek. soed., 8 : 1542 (1966).

A27. Morgunova, M. M., Zhuzhgov, É. L., Zaev, E. E., Zhinkin, D. Ya., and Bubnov, N. N., Khimiya geterotsiklich. soed., No. 6 : 943 (1965).

A28. Prokop'ev, V. P., Kostochko, A. V., Shestakova, A. D., and Tishkov, P. G., Tr. Kazansk. khim.-tekhnol. inst., No. 34 : 354 (1965).

A29. Prokop'ev, V. P., Tishkov, P. G., Shreibert, A. I., and Khardin, A. P., Vysokomolek. soed., 8 : 787 (1966).

A30. Sagitov, R. Ya., and Maklakov, A. I., Vysokomolek. soed., 8 : 1003 (1966).

A31. Sergeev, N. M., and Grinberg, A. I., Zh. strukt. khim., 7 : 356 (1966).

A32. Slonim, I. Ya., in: Ageing and Stabilization of Polymers, Izd. Khimiya, Moscow (1966), p. 196.

A33. Slonim, I. Ya., in: Radiospectroscopy of Solids, Atomizdat, Moscow (1967), p. 123.

A34. Sorokin, A. Ya., Andreeva, I. A., Volkova, L. A., Kolotsov, A. I., Rudakov, A. P., Pyrkov, L. M., and Frenkel', S. Ya., Khim. volokna, No. 6 : 22 (1965).

A35. Spitsyn, Vikt. I., Zubov, P. I., Kabanov, V. Ya., and Grozinskaya, Z. P., Vysokomolek. soed., 8 : 604 (1966).

A36. Urman, Ya. G., and Slonim, I. Ya., in: Radiospectroscopy of Solids, Atomizdat, Moscow (1967), p. 133.

A37. Fedotov, V. D., and Ionkin, V. S., Teor. i éksperim. khim., 3 : 134 (1967).

A38. Khachaturov, A. S., Bazhenov, N. M., Vol'kenshtein, M. V., Dolgopol'skii, I. M., and Kol'tsov, A. I., Kauchuk i resina, No. 12 : 6 (1965).

A39. Chenborisova, L. Ya., Ionkin, V. S., Maklakov, A. I., and Voskresenskii, V. A., Vysokomolek. soed., 8 : 1810 (1966).

A40. Chenborisova, L. Ya., Maklakov, A. I., Teplov, B. F., Ovchinkov, Yu. V., and Ionkin, V. S., Vysokomolek. soed., B9 : 368 (1967).

A41. Shashkov, A. S., Galil-Ogly, F. A., and Novikov, A. S., Vysokomolek. soed., 8 : 267 (1966).

A42. Shepelev, V. I., Maklakov, A. I., Nasirpov, F. M., and Davydov, B. É., Elektrokhimiya, 2 :1468 (1966).

A43. Shumilovskii, N. I., Skripko, A. L., Korol', V. S., and Kovalev, G. V., Methods of Nuclear Magnetic Resonance, Energiya, Moscow (1966).

A44. Yukel'son, I. I., Gladkovskii, V. S., and Pivnev, V. I., Izv. vyssh. uchebn. zav. Khimiya i khim. tekhnologiya, 8 : 1006 (1965).

A45. Yan, Nyan'-tsy, Chzhan Tsyun', and Khuan Yun-zhén; Khuaduk shida syuébao, Tsyzhan' késyué, Huadong Shida, khievao, No. 1 : 57 (1965).

A46. Nuclear Magnetic Resonance (P. M. Borodin, ed.), No. 1, Izd. LGU, Leningrad (1965).

A47. Abe, H., Imai, K., and Matsumoto, M., J. Polymer Sci., B3:1053 (1965).

A48. Abe, Y, Tasumi, M., Shimonouchi, T., Satoh, S., and Chujo, R., J. Polymer Sci., Part A-1, 4:1413 (1966).

A49. Advances in Magnetic Resonance, Vol. 1 (J. S. Waugh, ed.),Academic Press, New York-London (1965).

A50. Anderson, J. E., and Slichter, W. P., J. Phys. Chem., 69:3099 (1965).

A51. Aso, C., Kunitake, T., Ito, K., and Ishimoto, Y., J. Polymer Sci., B4:701 (1966).

A52. Assioma, F., Marchal, J., and Schue, F., Compt. Rend., 261:1315 (1965).

A53. Banas, E. M., and Juveland, O. O., J. Polymer Sci., Part A-1, 5:397 (1967).

A54. Baney, R. H., and Haberland, G. G., J. Organometal Chem., 5:320 (1966).

A55. Bargon, J., Hellwege, K.-H., and Johnson, U., Makromolek. Chem., 95:187 (1966).

A56. Bargon, J., Hellwege, K.-H., and Johnson, U., Kolloid-Z und Z. Polymere, 213:51 (1966).

A57. Bauer, R. C., Harwood, H. J., and Ritchey, W. M., Am. Chem. Soc., Polymer Preprints, 7:973 (1966).

A58. Bergmann, K., and Nawotki, K., in: Molecular Relaxation Process, Academic Press, London-New York (1966), p. 135.

A59. Binder, J. L., J. Polymer Sci., B4:19 (1966).

A60. Bovey, F. A., Pure and Appl. Chem., 12:525 (1966).

A61. Bovey, F. A., Hood, F. P., Anderson, E. W., and Kornegay, R. J., J. Phys. Chem., 71:312 (1967).

A62. Brame, E. G., Jr., J. Makromolek. Chem., A1:277 (1967).

A63. Brame, E. G., Jr., Ferguson, R. C., and Thomas, G. J., Jr., Analyt. Chem., 39:517 (1967).

A64. Brame, E. G., Jr., and Vogl, O., Am. Chem. Soc. Polymer Preprints, 7:227 (1966).

A65. Brownstein, S., and Wiles, D. M., Canad. J. Chem., 44:153 (1966).

A66. Burget, J., Petricek, V., and Saha, J., Phys. and techn. low temperatures, Czechosl. Acad. Sci., Prague (1964), p. 246.

A67. Caputa, K., Daszkiewitcz, O. K., Hennel, J. W., Lubas, B., and Szezepkowski, Proc. Colloc. AMPERE, 13:396 (1964).

A68. Caraculacu, A. A., J. Polymer Sci., Part A-1, 4:1829 (1966).

A69. Cefelin, P., Doskočilova, D., and Sebenda, J., Chem. Prumysl., 17:73 (1967).

A70. Cernicki, B., Tehnika, 21(9) (1966).

A71. Chen, Hung Yu., J. Polymer Sci., B4:891 (1966).

A72. Chen, Hung Yu., J. Polymer Sci., B4:1007 (1966).

A73. Chen, L. W., and Kumanotani, J., J. Appl. Polymer Sci., 9:3519 (1965).

A74. Chiba, A., Hasegawa, A., Hihichi, K., and Furuichi, J., J. Phys. Soc. Japan, 21:1777 (1966).

A75. Chierico, A., Del-Nero, G., Lanzi, G., and Mognaschi, E. R., European Polymer J., 2:339 (1966).

A76.   Chierico, A., Lanzi, G., and Mognaschi, E. R., Rend. Ist. Lombardo Sci.
        Lettre, A99:236 (1965).
A77.   Chujo, R., J. Chem. Soc. Japan, Ind. Chem. Sect., 68:1343 (1965).
A78.   Chujo, R., J. Phys. Soc. Japan, 21:2669 (1966).
A79.   Cleron, V., Coston, C. J., and Drickamer, H. G., Rev. Scient. Instrum.,
        37:68 (1966).
A80.   Connor, T. M., Polymer, 7:426 (1966).
A81.   Connor, T. M., and McLauchlan, K. A., in: Nuclear Magnetic Resonance in
        Chemistry (B. Pesce, ed.), Academic Press, New York (1965), p. 319.
A82.   Davidson, E. B., J. Polymer Sci., B4:175 (1966).
A83.   Davies, A. G., and Wassermann, A., J. Polymer Sci., Part A-1, 4:1887 (1966).
A84.   Dietschi, H. G., Diss. Doct. techn. Wiss., Eidgenöss. techn. Hochschule.
        Zurich, Juris verl (1966).
A85.   Dole Roble, J.-P., Peintures, pigmentes, vernis, 42:783 (1966).
A86.   Doskočilova, D., Stokr., J., Schneider, B., Pivcova, H., Kolinsky, M.,
        Petranek, J., and Lim, D., J. Polymer Sci., C(16):215 (1966).
A87.   Dudek, T. J., and Bueche, F., J. Polymer Sci., A2:811 (1964).
A88.   Dumitru, E. T., and Wahr, J. C., Am. Chem. Soc., Div. Org. Coatings,
        Plastics Chem., Preprints, 25:147 (1965).
A89.   Ebert, I., and Seifert, G., Kernresonanz im Festkörper, Leipzig, Akad.
        Verlagges (1966).
A90.   Emsley, J. W., Feeney, J., and Sutcliffe, L. H., High Resolution Nuclear
        Magnetic Resonance Spectroscopy, Vols. 1 and 2, Pergamon Press, Oxford
        (1965).
A91.   Enomoto, S., Asahina, M., and Satoh, S., J. Polymer Sci., Part A-1, 4:1373
        (1966).
A92.   Enomoto, S., Chem. Ind. (Japan), 17:633 (1966).
A93.   Epstein, G., and Weinberg, I., SPI Reinf. Plast. Div., 22 Ann. Techn. Conf.,
        January 1967, 4-E, p. 1.
A94.   Ferguson, R. C., Kautschuk und Gummi, 18:723 (1965).
A95.   Filipovich, G., Knutsen, C. D., and Spitzer, D. M., Jr., J. Polymer Sci.,
        B3:1065 (1965).
A96.   Fischer, T., Kinsinger, J. B., and Wilson, C. W., J. Polymer Sci., B4:379
        (1966).
A97.   Flory, P. J., J. Am. Chem. Soc., 88:2873 (1966).
A98.   Flory, P. J., Mark, J. E., and Abe, A., J. Polymer Sci., B3:973 (1965).
A99.   Formula Index to NMR Literature Data, Vol. 2 (M. G. Howell, A. S. Kende, and
        S. S. Webb, eds.), Plenum Press, New York (1966).
A100.   Frata, M., Vidotto, C., and Talamini, G., Chim. ind. (Ital.), 48:42 (1966)
A101.   Friedmann, G., Bull. Soc. chim. France, No. 2:698 (1967).
A102.   Friedmann, G., Schue, F., Brini, M., Deluzarche, A., and Maillard, A., Bull.
        Soc. chim. France, No. 5:1343 (1965).
A103.   Frisch, H. L., Mallows, C. L., and Bovey, F. A., J. Chem. Phys., 45:1565
        (1966).
A104.   Fujii, K., Shibatani, K., Fujiwara, Y., Ohyanagi, Y., Ukida, J., and Matsu-
        moto, M., J. Polymer Sci., B4:787 (1966).

A105. Fujimoto, K., Yoshimura, N., and Inomata, I., J. Soc. Materials Sci., Japan, 14:338 (1965).

A106. Fujiwara, S., Fujiwara, Y., Fujii, K., and Fakuroi, T., J. Molec. Spectroscopy, 19:294 (1966).

A107. Fujiwara, Y., J. Chem. Soc. Japan, Ind. Chem. Sect., 68:1352 (1965).

A108. Fujiwara, Y., Fujiwara, S., and Fujii, K., J. Polymer Sci., Part A-1, 4:2577 (1966).

A109. Gagnaire, D., and Vincendon, M., Bull. Soc. chim. France, No. 1:204 (1966).

A110. Gechele, G. B., Stea, G., and Manescalchi, F., European Polymer J., 2:1 (1966).

A111. Genser, E. E., J. Molec. Spectroscopy, 16:56 (1965).

A112. Geuskens, G., Lubikulu, J. C., and David, C., Polymer, 7:63 (1966).

A113. Gilson, D. F. R., Polymer Seminar, McGill Univ., High Polymer Div. Chem. Depart., 1:1 (Nov. 1965).

A114. Glass, C. A., Canad. J. Chem., 43:2652 (1965).

A115. Golub, M. A., J. Polymer Sci., B4:227 (1966).

A116. Grassie, N., McNeill, I. C., and McLaren, I. F., J. Polymer Sci., B3:897 (1965).

A117. Grassie, N., Torrance, B. J. D., Fortune, J. D., and Gemmel, J. D., Polymer, 6:653 (1965).

A118. Gupta, R. P., J. Appl. Polymer Sci., 10:1535 (1966).

A119. Gupta, R. P., and Laible, R. C., J. Polymer Sci., A3:3951 (1965).

A120. Haberland, G. G., and Carmichael, J. B., Am. Chem. Soc., Polymer Preprints, 6:637 (1965).

A121. Haberland, G. G., and Carmichael, J. B., J. Polymer Sci., C(14):291 (1966).

A122. Haney, C. P., Johnson, F. A., and Baldwin, M. G., J. Polymer Sci., Part A-1, 4:1791 (1966).

A123. Harwood, H. J., Angew. Chem., 13:1124 (1965).

A124. Harada, T., and Ueda, N., Chem. High Polymers (Japan), 22:685 (1965).

A125. Hatada, K., Kobunshi, High Polymers, Japan, 16:284 (1967).

A126. Hatada, K., Ota, K., and Yuki, H., J. Polymer Sci., B5:225 (1967).

A127. Haubenstock, H., and Naegele, W., Makromolek. Chem., 97:248 (1966).

A128. Heinze, D., Makromolek. Chem., 101:166 (1967).

A129. Hellwege, K.-H., Johnsen, U., and Kolbe, K., Kolloid-Z. Z. Polymere, 214:45 (1966).

A130. Hendus, H., Illers, K.-H., and Ropte, E., Kolloid-Z. Z. Polymere, 216-217:110 (1967).

A131. Hill, D. A., Hasher, B. A., and Hwang, C. F., Phys. Lett., 23:63 (1966).

A132. Hoffmann, W., and Kimmer, W., Plaste Kautschuk, 13:519 (1966).

A133. Holahan, F. S., Stivala, S. S., and Levi, D. W., J. Polymer Sci., A3:3993 (1965).

A134. Hudson, B. E., Jr., Makromolek. Chem., 94:172 (1966).

A135. Ihashi, Y., Sawa, K., and Morita, S., J. Chem. Soc. Japan, Ind. Chem. Sect., 68:1427 (1965).

A136. Illman, J. C., J. Appl. Polymer Sci., 10:1519 (1966).

A137. Ingham, J. D., Lawson, D. D., Manatt, S. L., Rapp, N. S., and Hardy, J. P.,
       J. Macromol. Chem., I: 75 (1966).
A138. Ino, M., and Tokura, N., Bull. Chem. Soc. Japan, 38:1094 (1965).
A139. Ishigure, K., Tabata, Y., and Oshima, K., J. Polymer Sci., B4: 669 (1966).
A140. Ito, K., and Yamashita, Y., J. Chem. Soc. Japan, Ind. Chem. Sect., 68:1469
       (1965).
A141. Ito, K., and Yamashita, Y., J. Polymer Sci., Part A-1, 4: 631 (1966).
A142. Iwayanagi, S, and Sakurai, I., J. Polymer Sci., C(15): 29 (1966).
A143. Jacques, R., Double Liaison, No. 129: 653 (1966)(Chem. Abstr., 65: 7293, 1966).
A144. Jenner, G., Bull. Soc. chim. France, 2851 (1965).
A145. Johnsen, K. E., Lovinger, J. A., Parker, C. O., and Baldwin, M. G.,
       J. Polymer Sci., B4: 977 (1966).
A146. Johnsen, U., Kolloid-Z. Z. Polymere, 210:1 (1966).
A147. Johnsen, U., Ber. Bunsenges. phys. Chem., 70: 320 (1966).
A148. Johnsen, U., and Kolbe, K., Kolloid- Z. Z. Polymere, 216-217: 97 (1967).
A149. Jones, D. W., and Pearson, J. E., in: Nuclear Magnetic Resonance in
       Chemistry (B. Pesce, ed.), Academic Press, New York (1965), p. 331.
A150. Kamide, K., and Sanada, M., Chem. High Polymers (Japan), 23: 481 (1966).
A151. Kanai, H., Makimoto, T., and Tsurata, T., J. Chem. Soc. Japan, Ind. Chem.
       Sect., 68:1947 (1965).
A152. Kato, Y., and Nishioka, A., J. Chem. Soc. Japan, Ind. Chem. Sect., 68:1461
       (1965).
A153. Kato, Y., and Nishioka, A., Bull. Chem. Soc. Japan, 39:1426 (1966).
A154. Kawai, T., Makromolek. Chem., 90:288 (1966).
A155. Kennedy, J. P., and Isaacson, R. B., Am. Chem. Soc., Polymer Preprints,
       7: 419 (1966).
A156. Kennedy, J. P., and Isaacson, R. B., J. Macromol. Chem., 1: 541 (1966).
A157. Kistler, J.-P., Bull. Soc. chim. France, 764 (1965).
A158. Knutson, C.D., and Spitzer, D. M., J. Chem. Phys., 45: 407 (1966).
A159. Kobayashi, S., Kato, Y., Watanabe, H., and Nishioka, A., J. Polymer Sci.,
       Part A-1, 4: 245 (1966).
A160. Kondo, S., Ishii, T., Chokki, Y., Tanaka, K., and Takeda, M., Bull. Chem.
       Soc., Japan, 39:1866 (1966).
A161. Konishi, K., and Kano, Y., Bunseki Kagaku. Japan Analyst, 15:1110 (1966).
A162. Kotake, Y., Yoshihara, T., Sato, H. Yamada, N., and Yoh, Y., J. Polymer
       Sci., B5 :163 (1967).
A163. Kuntz, I., J. Polymer Sci., B4: 427 (1966).
A164. Kusumoto, N., Shirano, K., And Takayanagi, M., J. Chem. Soc. Japan, Ind.
       Chem. Sect., 68:1553 (1965).
A165. Lando, J. B., Olf, H. G., and Peterlin, A., Am. Chem. Soc. Polymer Preprints,
       6: 910 (1965).
A166. Lando, J. B., Olf, H. G., and Peterlin, A., J. Polymer Sci., Part A-1, 4:914
       (1966).
A167. Lanzi, G., Nucl. Magnetic Resonance Chem. Proc. Symp., Cagliari, Italy
       (1964), p. 325.
A168. Lehmann, J., Kolloid- Z. Z. Polymere, 212: 167 (1966).

A169.   Lim, D., Obereigner, B., and Doskočilova, D., J. Polymer Sci., B3 : 893 (1965).

A170.   Manatt, S. L., Lawson, D. D., Ingham, J. D., Rapp, N. S., and Hardy, J. P.,
        Analyt. Chem., 38 : 1063 (1966).

A171.   Mark, J. E., Wessling, R. A., and Hughes, R. E., J. Phys. Chem., 70 : 1895
        (1966).

A172.   Marlborough, D. I., Orrell, K. G., and Ryden, H. N., Chem. Commun.,
        21 : 518 (1965).

A173.   Marx, R., and Tchiboakdjion, D., J. Chim. phys. phys.-chim. biol., 62 : 1102
        (1965).

A174.   Mathias, A., and Mellor, N., Analyt. Chem., 38 : 472 (1966).

A175.   Matsuzaki, K., and Fujinami, K., J. Chem. Soc. Japan, Ind. Chem. Sect.,
        68 : 1456 (1965).

A176.   Matsuzaki, K., Uryu, T., Tameda, K., and Takeuchi, M., J. Chem. Soc.
        Japan, Ind. Chem. Sect., 68 : 1466 (1965).

A177.   Matsuzaki, K., Uryu, T., Okada, M., Ishigure, K., Ohki, T., and
        Takeuchi, M., J. Polymer Sci., B4 : 487 (1966).

A178.   Matsuzaki, K., Uryu, T., Ishigure, K., and Takeuchi, M., J. Polymer Sci.,
        B4 : 93 (1966).

A179.   Matsuzaki, K., and Uryu, T., J. Polymer Sci., B4 : 255 (1966).

A180.   Matsuzaki, K., Yoshimura, M., and Sobue, H., Kogyo Kagaku Zasshi,
        67 : 944 (1965).

A181.   Mattes, R., and Rochow, E. G., J. Polymer Sci., Part A-2, 4 : 375 (1966).

A182.   Mavel, G., Theories Moleculares de la resonance Magnetique Nucleare:
        application a la chimie structuralle, Dunod, Paris (1966).

A183.   McCall, D. W., Douglass, D. C., and Falcone, D. R., J. Phys. Chem., 71 : 998
        (1967).

A184.   McClanahan, J. L., and Previtera, S. A., J. Polymer Sci., A3 : 3919 (1965).

A185.   McDonald, C. C., Phillips, W. D., and Penswik, J., Biopolymers, 3 : 609 (1965).

A186.   McMahon, P. E., J. Polymer Sci., Part A-2, 4 : 501 (1966).

A187.   McMahon, P. E., J. Polymer Sci., B4 : 75 (1966).

A188.   McMahon, P. E., J. Polymer Sci., Part A-2, 4 : 639 (1966).

A189.   McMahon, P. E., J. Polymer Sci., B4 : 43 (1966).

A190.   McMahon, P. E., and Tincher, W. C., J. Molec. Spectroscopy, 15 : 180 (1965).

A191.   Merril, L. J., Sauer, J. A., and Woodward, A. E., J. Polymer Sci., A3 : 4243
        (1965).

A192.   Mijs, W. J., Chem. Weekb., 63 : 225 (1967).

A193.   Misono, A., Uchida, Y., and Yamada, K., Bull. Chem. Soc. Japan, 39 : 2458
        (1966).

A194.   Miyamoto, Y., and Takemura, T., Technol. Repts. Kyushu Univ., 39 : 9
        (1966).

A195.   Model, F. S., Redl., G., and Rochow, E. G., J. Polymer Sci., Part A-1, 4 : 639
        (1966).

A196.   Modena, M., Carraro, G., and Cossi, G., J. Polymer Sci., B4 : 613 (1966).

A197.   Moerdritzer, K., and Van Wazer, R., J. Org. Chem., 30 : 3920 (1965).

A198.   Murahashi, S., Nozakura, S., and Kotake, Y., Chem. High Polymers (Japan),
        22 : 652 (1965).

A199.   Murahashi, S., Nozakura, S., Sumi, M., Yuki, H., and Hatada, K., Chem.
        High Polymers (Japan), 23 : 605 (1966).
A200.   Murahashi, S., Nozakura, S., Sumi, M., Yuki, H., and Hatada, K., J. Poly-
        mer Sci., B4 : 65 (1966).
A201.   Murahashi, S., Nozakura, S., Okamoto, T., and Kotake, Y., Chem. High
        Polymers (Japan), 23 : 354 (1966).
A202.   Murano, M., and Yamadera, R., Chem. High Polymers (Japan), 23 : 497 (1966).
A203.   Nakata, T., and Otsu, T., J. Macromol. Chem., 1 : 563 (1966).
A204.   Nishioka, A., Kobunshi High Polymers (Japan), 15 : 309 (1966).
A205.   Nishioka, A., and Kato, Y., J. Chem. Soc. Japan, Ind. Chem. Sect., 68 : 1348
        (1965).
A206.   Nishioka, A., and Kato, Y., J. Chem. Soc. Japan, Ind. Chem. Sect., 68 : 1452
        (1965).
A207.   Nuclear Magnetic Resonance in Chemistry (B. Pesce, ed.), Academic Press,
        New York (1965).
A208.   Nuclear Magnetic Resonance and Relaxation in Solids (Van Gerven, ed.),
        North-Holland Publisch., Amsterdam (1965).
A209.   O'Donnell, J. H., Proc. Roy. Australian Chem. Inst., 33 : 122 (1966).
A210.   Ohnishi, S., and Nukada, K., J. Polymer Sci., B3 : 1001 (1965).
A211.   Ohsumi, Y., Higashimura, T., and Okamura, S., J. Polymer Sci., A3 : 3729
        (1965).
A212.   Ohsumi, Y., and Okamura, S., J. Polymer Sci., Part A-1, 4 : 923 (1966).
A213.   Okada, M., Yamashita, Y., and Ishii, Y., Makromolek. Chem., 94 : 181 (1966).
A214.   Okuto, K., Makromolek. Chem., 98 : 148 (1966).
A215.   Olf, H. G., and Peterlin, A., Kolloid- Z. Z. Polymere, 212 : 12 (1966).
A216.   Olf, H. G., and Peterlin, A., Kolloid- Z. Z. Polymere, 215 : 97 (1967).
A217.   Overberger, C. G., and Yamamoto, N., J. Polymer Sci., Part A-1, 4 : 3101
        (1966).
A218.   Peffley, W. M., J. Polymer Sci., Part A-1, 4 : 977 (1966).
A219.   Peterlin, A., and Olf, H. G., J. Polymer Sci., Part A-2, 4 : 587 (1966).
A220.   Peterson, J., and Ranby, B., Makromolek. Chem., 102 : 83 (1967).
A221.   Pivcova, H., and Schneider, B., Coll. Czech. Chem. Comm., 30 : 2045 (1965).
A222.   Pivcova, H., and Schneider, B., Coll. Czech. Chem. Comm., 31 : 3154 (1966).
A223.   Porter, R. S., J. Polymer Sci., Part A-1, 4 : 189 (1966).
A224.   Porter, R. S., Cantow, J. R., and Johnson, J. F., Makromolek. Chem., 94 : 143
        (1966).
A225.   Porter, R. S., and Johnson, J. F., Chem. Rev., 66 : 1 (1966).
A226.   Pritchard, J. G., Vollmer, R. L., Lawrence, W. C., and Black, W. B, J. Polymer
        Sci., Part A-1, 4 : 707 (1966).
A227.   Rakos, M., Czech. J. Phys., 16 : 864 (1966).
A228.   Ramey, K. C., J. Phys. Chem., 70 : 2525 (1966).
A229.   Ramey, K. C., and Messick, J., J. Polymer Sci., Part A-2, 4 : 155 (1966).
A230.   Ravve, A., and Fitkoc, W., J. Polymer Sci., Part A-1, 4 : 2533 (1966).
A231.   Reddisch, W., Powles, J. G., and Hunt, B. I., J. Polymer Sci., B3 : 671
        (1965).

A232. Reinmöller, M., and Fox, T. G., Am. Chem. Soc. Div. Polymer Chem.,
       Polymer Preprints, 7: 987 (1966).
A233. Reinmöller, M., and Fox, T. G., Am. Chem. Soc. Div. Polymer Chem.,
       Polymer Preprints, 7: 994 (1966).
A234. Reinmöller, M., and Fox, T. G., Am. Chem. Soc. Div. Polymer Chem.,
       Polymer Preprints, 7: 1006 (1966).
A235. Reiss, C., J. chim. phys. phys.-chim. biol., 63: 1307 (1966).
A236. Rhodes, J. H., and Chiang, R., J. Polymer Sci., B4: 393 (1966).
A237. Richards, D. W., Salter, D. A., and Williams, R. L., Chem. Commun.,
       No. 2: 38 (1966).
A238. Ritchey, W. M., and Ball, L. E., J. Polymer Sci., B4: 557 (1966).
A239. Ritchey, W. M., and Knoll, F. J., J. Polymer Sci., B4: 853 (1966).
A240. Rochow, E. G., Mh. Chem., 95: 750 (1964).
A241. Rosten, F., SPI Reinf. Plast. Div., 21st ann. techn. conf. Proc., Sect. 12-D, 1
       (1966).
A242. Rupprecht, A., Acta chem. Scand., 20: 582 (1966).
A243. Schaefer, J., J. Phys. Chem., 70: 1975 (1966).
A244. Segre, A. L., Ciampelli, F., and Dall'Asta, G., J. Polymer Sci., B4: 633 (1966).
A245. Sheldon, R. A., Fueno, T., Tsunetsugu, T., and Furukawa, J., J. Polymer Sci.,
       B3: 23 (1965).
A246. Shiba, T., and Shih, C. C., J. Chem. Soc. Japan, Ind. Chem. Sect., 69: 1003
       (1966).
A247. Shimonouchi, T., Pure Appl. Chem., 12: 289 (1966).
A248. Shimonouchi, T., Kobunshi High Polymers (Japan), 15: 776 (1966).
A249. Sigal, P., Masciantonio, P., and Fugassi, P., J. Polymer Sci., Part A-1, 4: 761
       (1966).
A250. Sillescu, H., Kernmagnetische Resonanz. Einführung in die Theoretischen
       Grundlagen, Springer-Verlag, Berlin (1966).
A251. Sinnott, K. M., J. Appl. Phys., 37: 3385 (1966).
A252. Slichter, C., Principles of Magnetic Resonance, 2nd ed., Harper and Row,
       New York (1966).
A253. Slichter, W. P., Am. Chem. Soc. Polymer Preprints, 6: 632 (1965).
A254. Slichter, W. P., J. Polymer Sci., C(14): 33 (1966).
A255. Stehling, F. C., and Bartz, K. W., Analyt. Chem., 38: 1467 (1966).
A256. Stubbs, W. H., Gore, C. R., and Marvel, C. S., J. Polymer Sci., Part A-1,
       4: 1898 (1966).
A257. Suhr, H., Anwendung der Kernmagnetischen Resonanz in der Organischen
       Chemie, Springer, Berlin-Heidelberg-New York (1965).
A258. Sýkora, S., Collection (in press).
A259. Takeuchi, T., and Yamazaki, M., J. Chem. Soc. Japan, Ind. Chem. Sect.,
       68: 1478 (1965).
A260. Takeuchi, T., Yamazaki, M., and Mori, S., J. Polymer Sci., B4: 695 (1966).
A261. Tanaka, N., Enka biniru to porima, 6: 8 (1966).
A262. Tanaka, N., and Chiba, T., J. Chem. Soc. Japan, Ind. Chem. Sect., 68: 1495
       (1965).

A263.   Taylor, J. D., Pluhar, M., and Rubin, L. E., J. Polymer Sci., B5: 77 (1967).

A264.   Thompson, E. V., J. Polymer Sci., Part A-2, 4:199 (1966).

A265.   Trappeniers, N. J., Physics Noncrystalline Solids, Amsterdam (1965), p. 197.

A266.   Trautvetter, W., Kunststoffe Plastics, 13: 54 (1966).

A267.   Tsutia, M., and Yamamoto, K., Oyo Butsuri, 34: 424 (1965).

A268.   Ullman, R., J. Chem. Phys., 43: 3161 (1965).

A269.   Ullman, R., J. Chem. Phys., 44: 1558 (1966).

A270.   Vincendon, M., Peintures, pigments, vernis, 43:164 (1967).

A271.   Vosskötter, G., Kosteld, R., Kolloid-Z. Z. Polymere, 216-217: 85 (1967).

A272.   Weil, G., and Reeb, R., Compt. Rend., C263 :21 (1966).

A273.   Weiner, M., and Rotaru, M., Rev. roumaine phys., 10: 921 (1965).

A274.   Yamada, N., Ortio, Z., and Minami, S., J. Polymer Sci., A3 :4173 (1965).

A275.   Yamashita, Y., and Ito, K., Chemistry (Japan), 21: 804 (1966).

A276.   Yamashita, Y., and Ito, K., Chemistry (Japan), 21: 947 (1966).

A277.   Yamashita, Y., and Ito, K., Kagaku (Kyoto), No. 10: 994 (1966).

A278.   Yokota, K., and Ishii, Y., J. Chem. Soc. Japan, Ind. Chem. Sect., 68: 1473
        (1965).

A279.   Yokota, K., and Ishii, Y., J. Chem. Soc. Japan, Ind. Chem. Sect., 69: 1053
        (1966).

A280.   Yokota, K., and Ishii, Y., J. Chem. Soc. Japan, Ind. Chem. Sect., 69: 1083
        (1966).

A281.   Yoshimoto, T., Fujimori, H., Tanaka, K., and Imamura, T., Nippon Gomu
        Kyokaishi, 39: 403 (1966).

A282.   Yoshimoto, T., Tanaka, K., Fujimori, H., and Imamura, T., Nippon Gomu
        Kyokaishi, 39: 397 (1966).

A283.   Yoshino, T., Kikuchi, Y., and Komiyama, J., J. Phys. Chem., 70: 1059
        (1966).

A284.   Yoshino, T., and Komiyama, J., J. Am. Chem. Soc., 88: 176 (1966).

A285.   Yoshino, T., and Kuno, F., J. Am. Chem. Soc., 87: 4404 (1965).

Subject Index to the Bibliography

Theoretical Problems of the
NMR Method

1-7,* 184.

Books on NMR

5, 9, 43, 60, 67, 76, 82, 93, 96, 98, 99, 101, 107, 145, 146, 171, 174, 175, 176, 177, 178, 291, 292, 356, 384, 512, 593, 637, 666, 709, A9, A43, A46, A49, A89, A99, A182, A207, A208, A250, A252, A257.

Reviews on NMR in Polymers

18, 21, 22, 108, 109, 112, 127, 135, 200, 260, 434, 447, 473, 504, 541, 561, 601, 607, 625, 626, 627, 628, 630, 636, 638, 639, 685, 700, 705, 715, 729, 730, A32, A33, A60, A70, A77, A94, A107, A113, A123, A125, A143, A146, A147, A192, A204, A205, A209, A247, A248, A254, A275, A276, A277.

Articles on the Investigation of
Specific Polymers by the NMR Method

Butadiene polymers and copolymers

31, 47, 49, 50, 51, 52, 53, 54, 57, 72, 104, 105, 110, 123, 151, 152, 154, 157, 163, 228, 274, 331, 357, 359, 360, 361, 442, 454, 490, 491, 492, 513, 606, 608, 612, 642, 676, A11, A17, A18, A37, A52, A59, A101, A102, A128, A130, A142, A256, A281, A282.

---

*Nos. 1-7 refer to the literature cited on page 144 at the end of Part I.

266, 293, 311, 327, 329, 345, 357, 368, 377, 378, 379, 391,
392, 399, 400, 402, 403, 404, 406, 419, 420, 436, 438, 439,
441, 457, 458, 459, 477, 480, 487, 505, 506, 507, 510, 511,
523, 525, 526, 529, 530, 531, 532, 542, 543, 558, 560, 567, 568,
572, 577, 592, 609, 610, 619, 621, 623, 643, 645, 657, 658,
659, 683, 684, 686, 689, 701, 707, 710, 711, 735, 740, 741,
742, 745, 747, A12, A21, A28, A29, A47, A56, A57, A58, A65,
A95, A103, A117, A122, A126, A133, A139, A140, A141, A151,
A152, A158, A162, A169, A171, A173, A176, A177, A178,,
A179, A191, A202, A217, A227, A228, A229, A232, A233,
A234, A236, A238, A244, A253, A259, A264, A278, A279,
A280, A283, A284, A285.

Polyaldehydes

109, 110, 118, 130, 131, 132, 150, 207, 208, 209, 232, 249, 250,
321, 353, 499, 537, 552, 624, 695, 699, 734, A36, A62, A64,
A74, A135, A196, A197, A265, A275.

Polyamides

45, 46, 84, 89, 119, 144, 268, 287, 288, 306, 312, 314, 315,
330, 373, 393, 394, 445, 448, 463, 514, 570, 575, 613, 617,
631, 635, 648, 660, 661, 731, A6, A8, A19, A30, A69, A118,
A149, A186, A187, A188, A189, A221, A230.

Polybutene and other polyolefins

59, 223, 277, 324, 340, 367, 409, 413, 430, 495, 602, 628, 641,
702, 703, 733, A134, A164, A253, A255.

Polyesters

46, 55, 65, 79, 84, 86, 110, 120, 133, 142, 149, 155, 159, 169,
205, 217, 219, 231, 236, 250, 251, 252, 253, 254, 259, 279, 280,
281, 282, 304, 346, 352, 353, 372, 375, 433, 456, 493, 494,
514, 515, 535, 544, 548, 578, 611, 663, 679, 704, 706, 713, A4,
A6, A10, A25, A30, A50, A63, A80, A81, A124, A127, A137,
A161, A163, A170, A174, A181, A213, A271.

Polyethylene

61, 78, 100, 110, 137, 151, 219, 230, 233, 234, 239, 245, 246,
247, 289, 290, 300, 301, 303, 305, 306, 307, 308, 309, 313, 336,
338, 341, 342, 343, 349, 360, 369, 374, 381, 382, 397, 405, 415,
416, 449, 462, 465, 466, 467, 468, 469, 472, 482, 484, 501, 509,

A148, A161, A174, A184, A193, A204, A213, A217, A223, A238, A243, A255, A259, A260, A270, A277.

Investigation of Cross-Linking
of Chains

A4, A93, A136, A241, A281, A282.

Investigation of Crystallinity

A5, A6, A30, A74, A142, A154, A168, A187, A189, A215, A216, A262, A251.

Investigation of the Effects
of Radiation

A35, A66, A131, A218, A267, A273.

Investigation of Polymerization

A28, A29, A47, A51, A55, A61, A78, A82, A100, A122, A126, A139, A144, A145, A157, A163, A167, A173, A180, A201, A202, A203, A211, A217, A232, A233, A234, A244, A245, A246, A270, A276, A284, A285.

Molecular Motion and Nuclear
Magnetic Relaxation Time in Polymers

A22, A25, A26, A28, A37, A39, A50, A80, A104, A160, A167, A183, A225, A231, A253, A262, A268, A269.

Molecular Motion in Swollen
Polymers and Solutions

A40, A152, A272, A281, 282.

Molecular Motion and Width of
NMR Lines in Polymers

A6, A15, A16, A19, A20, A21, A36, A38, A42, A54, A58, A73, A74, A75, A76, A77, A79, A95, A111, A120, A121, A142, A158, A164, A173, A181, A188, A186, A191, A194, A195, A215, A216, A251, A254, A261, A271, A274.

Index